International Workshop on Embodied Intelligence

IOP Conference Series: Materials Science and Engineering
Volume 1261

Online
24 September 2021

ISBN: 978-1-7138-6157-7
ISSN: 1757-8981

Printed from e-media with permission by:

Curran Associates, Inc.
57 Morehouse Lane
Red Hook, NY 12571

Some format issues inherent in the e-media version may also appear in this print version.

Institute of Physics
Dirac House, Temple Back
Bristol BS1 6BE UK

Phone: 44 1 17 929 7481
Fax: 44 1 17 920 0979

techtracking@iop.org

Additional copies of this publication are available from:

Curran Associates, Inc.
57 Morehouse Lane
Red Hook, NY 12571 USA
Phone: 845-758-0400
Fax: 845-758-2633
Email: curran@proceedings.com
Web: www.proceedings.com

TABLE OF CONTENTS

IOP Conf. Series: Materials Science and Engineering 1261 (2022) 011001 doi:10.1088/1757-899X/1261/1/011001

Preface

Whilst providing a universal, clear and meaningful definition of 'Embodied Intelligence' is challenging, there is however a clear direction and motivation of the research community to further our understanding of systems that exhibit embodied intelligence and to understand how it can be designed and leveraged. This is a challenging topic, and requires an interdisciplinary and cooperative approach from researchers from engineering, social sciences, computer sciences, biological, materials science and more.

Furthering our understanding or Embodied Intelligence requires advancement in the philosophical and theoretical domains to provide frameworks intelligence, consciousness and learning. However the inherently physicality of embodied intelligence relies on the creation of physical technologies and real world interactions, requiring increasingly smart and novel technologies. The intersection of these domains, in a world where AI and machine intelligence are playing an ever growing role in our daily lives makes this research field timely, fascinating and significant.

In this book we provide a representative collection of papers written by the leading researchers in the field who attended and participated in the Embodied Intelligence Workshop, held virtually in March 2021. This includes short reviews, perspectives and original research articles. These aim to provide thought provoking, interdisciplinary and novel views and perspectives on this field, its future direction and some of the key concepts needed to deliver breakthroughs in our understanding or practical implementation.

The book first touches on perspectives and outlooks for Embodied Intelligence before including examples of soft and bio-inspired robotics. This is followed by chapters discussing the principles of Embodied Intelligence in robotics and also examples of embodiment for real-world applications. Finally, the last two chapters focus on cognition and learning and also the philosophical and conceptual issues surrounding Embodied Intelligence.

We would like to thank all of the participants of the seminar, the authors, and the reviewers for their excellent contributions to this volume.

Josie Hughes,
Arsen Abdulali,
Ryman Hashem,
Fumiya Iida.

IOP Conf. Series: Materials Science and Engineering 1261 (2022) 011002 doi:10.1088/1757-899X/1261/1/011002

Peer Review Statement

All papers published in this volume have been reviewed through processes administered by the Editors. Reviews were conducted by expert referees to the professional and scientific standards expected of a proceedings journal published by IOP Publishing Publishing.

- **Type of peer review:** Single Anonymous
- **Conference submission management system:** Morressier
- **Number of submissions received:** 27
- **Number of submissions sent for review:** 24
- **Number of submissions accepted:** 27
- **Acceptance Rate (Submissions Accepted / Submissions Received × 100):** 100
- **Average number of reviews per paper:** 2.2
- **Total number of reviewers involved:** 33
- **Contact person for queries:**
  Name: Josie Hughes
  Email: josie.hughes@epfl.ch
  Affiliation: EPFL - IGM

IOP Conf. Series: Materials Science and Engineering 1261 (2022) 012001 doi:10.1088/1757-899X/1261/1/012001

Embodied Artificial Intelligence: Enabling the Next Intelligence Revolution

Josie Hughes[1], Arsen Abdulali[2], Ryman Hashem[2] & Fumiya Iida[2]

[1] CREATE Lab, EPFL, Lausanne, Switzerland.

[2] Bio-inspired Robotics Lab (BIRL), University of Cambridge, UK.

E-mail: josie.hughes@epfl.ch, rh805@cam.ac.uk, aa2335@cam.ac.uk, fi224@cam.ac.uk

Abstract. There has been a long-standing debate about the extent to which the human mind controls the human body and vice-versa. Today this debate still powers a profound scientific desire to deepen our understanding of the nature of both humans' and animals' intelligent, adaptive behavior. In order to gain further insights into intelligence and explore how our brain and bodies develop through physical interactions with the world, Embodied Intelligence places the physical entity of the human body at the center of this subject. In the age of AI and Machine Learning, Embodied Intelligence research remains highly important as it can deliver valuable input which enhances the impact of conventional AI technology. Embodied Intelligence provides potential solutions for current AI technologies that are dependent on large amounts of data and reliable output whilst that output is reliant on uncertain, unstructured tasks and situations aligned closely to human users both as individuals and communities. In the following Chapter we discuss current trends and directions within this interdisciplinary community, and also the future challenges and opportunities.

1. Introduction

The concept and notion of computation and intelligence has historically seen significant evolution which continues to this day [17, 19]. It is shaped by our social awareness, needs and challenges, and also developments in both our understanding of intelligence, its origins, and technologies that show intelligent behaviors or interactions. At a time period in which we need solutions to global health, climate and societal challenges, there is a pressing need to advance the capabilities of intelligent embodied systems and to understanding how embodied systems can be used in understandable, fair and ethical ways [26].

The development of Artificial Intelligence (AI) technologies has been highly successful in developing general-purpose computing machines that can execute tasks with speed and scalability. AI is also increasingly being leveraged for widespread applications in the physical world, where it has significant potential. However, when compared to biological systems, AI based machines still struggle to complete unknown or previously unseen tasks, or to exploit their interactions with the environment. For example, AI based machines are still unable to autonomously recover from failures whereas animals can quickly and robustly find alternative solutions by exploiting their physicality. Most of the challenges of developing or fully utilizing the strengths of AI are related to the need to consider the algorithmic capabilities and also the physical 'embodiment' of the system. This has led to the emergence of the concept of "Embodied Intelligence" (EI) where the physical system, materials and design offers some

IOP Conf. Series: Materials Science and Engineering 1261 (2022) 012001 doi:10.1088/1757-899X/1261/1/012001

physical intelligence of robustness [11, 5]. Pioneers in this field have investigated how embodied interactions are used in nature to develop 'physical controllers', cognitive processes and how machines can perform computation through 'raw materials'[16].

The emergence of embodied intelligence has led to the development of smart materials or soft materials where passivity can be exploited to achieve highly complex interactions, or walking motion can be programmed physically removing the need for complex controllers. New fabrication methods such as multi-material 3D printing or micro fabrication have enabled significant acceleration of EI research through the development of unconventional structures, materials and systems[29]. These approaches also introduce new challenges as it becomes increasingly complex to model and understand interactions due to the scales, complexity or soft nature of the machines. Although there are many exciting capabilities offered by EI, to truly push the boundaries of conventional AI, we must understand how to design the EI of a body in conjunction with the algorithmic intelligence of the 'brain' [10] and how to understand and leverage the capabilities to better understand or interact with the surrounding environment.

The landscape of the EI community is hugely interdisciplinary and also ever growing. This leads to significant diversity in the technologies developed to exploit EI, and also the frameworks, perspectives and methodologies created to explain and develop and explain EI. In this Chapter we review some of the current trends in EI research, before commenting on future challenges and opportunities.

2. Current Trends

2.1. Bio-inspiration & Soft Robotics

Nature and biological systems show highly complex and nuanced interactions with the environment through imaginative and intelligent ways is made possible through their physical structure and embodied intelligence [4]. Analyzing biology allows us to develop artificial agents that have similar capabilities [20] and also to understand how EI arises in these beings through 'robo-physics'. There is increasing awareness and acceptance that plants offer and demonstrate many aspect of EI which can provide many insights into EI [21]. Furthermore, their 'brain-less' structure still enables computation and intelligence, challenging views and perspective of standard concepts and ideas of intelligence.

Soft robotics is closely tied to bio-inspiration, with many animals exploiting soft body interactions[13]. However, soft robotics also allows for the development of technologies which are not purely bio-inspired. Exciting developments have been shown in manipulation, locomotion and also application orientated capabilities such as manipulation of food items. However, there are still unresolved changes in the scale-ability of sensing, actuation and control in comparison to biological systems [8].

2.2. Artificial Life & Bio-hybrid Systems

Developing artificial life, or 'living-creatures' opposed to human-made materials, structures and fabrication would be a truly embodied approach to the design and creation of intelligence systems. This could enable a move towards systems that are Self-X, with capabilities in self-replication, self-healing or self-sensing for example[28]. Key developments with works such as the Xenobots [3], have shown how 'robots' can be formed from living cells, with some ability for these 'bots' to contribute to the creation of others. This is accompanied by an increase in the development of bio-hybrid systems, intersecting the key advances of biological and artificial systems. Although facing many technical hurdles, the integration of living muscle cells into robotic systems has been shown gao2021recent. In term of EI, this allows exploration of the balance and distribution of 'brain' and 'body' but also between different types of 'brain' and 'body', combining analog and digital computation.

2.3. Neuroscience & Cognition
The complexity of the behaviors and functions of human brains is incredible with billions of neurons, trillions of synapses operating with the human cerebral cortex, the outer layer of the brain that is associated with our highest mental capabilities [22]. To be able to design and develop artificial system with comparable capabilities we must better understand the behaviors and function of the brain. Analysis and exploration of human developmental behavior and cognition provides a means of understanding the importance and contribution of embodiment, emergence and continuous autonomous development [14]. Exploring the concept of embodiment through the lens of neuroscience leads provides the key concept that embodied intelligence provides the body with an means of dealing with open-ended and even unexpected interactions, as EI does not depend on specific inputs/outputs as in supervised learning which is the core of modern deep learning. Furthermore, EI has been show to have a role in 'shaping the way we think' [25], with congition and learning inherently tied to the body. This is a complex and multi-faceted research area, but is central to furthering our understanding of EI and human intelligence.

2.4. Understanding Intelligence & Philosophy
The term "higher level" intelligence is used to designate behavior that are purely sensory-motor, such as problem solving and reasoning, or even natural language, emotion, and consciousness. However, there is an intersection between the physical experience, its translation to memory and the long term implication in terms of this 'higher level' of intelligent behavior [18]. Links between physical interaction, sensory-motor control and memory are advancing through the creation of models that explain these responsive capabilities [9]. Likewise, the definition or requirements for consciousness is being explored within a number of disciplines, leading to diverse views and approaches.

2.5. Application & Deployment
As technologies that exhibit or exploit EI become more widespread and accessible, the ability to develop more application oriented systems becomes increasingly possible. One area that has seen growth is in manipulation. Soft Robotics Inc. demonstrated the ability to exploit jamming and variable stiffness to provide a universal gripper solution which relies on environmental interactions [2]. This has been followed by many soft or compliant solutions which exploit EI for robust and intelligence interactions.

Developments in haptic devices has provided technologies for embodied interactions between machines and humans, and also between humans[7]. In addition to the application enabled haptic devices they offer a platform to understand human embodied interactions, sensory-motor co-ordination, learning and more. This use of soft robotics as a tool to understand intelligence is also a growing field. This is seen in the use of soft wearable devices for wearers undergoing MRI studies[12]. Thus, the development of technologies that demonstrate and enabled embodied interactions has the potential to be used as tools to further our understanding of our response to embodied interactions.

3. Future Challenges & Opportunities
3.1. The Quest for a Unifying Concept
As engineers and scientists we seek a definitive, quantitative approach to representing and defining Embodied Intelligence in a tidy methodical manner. However, what does it mean to create a unifying theory or concept of EI? A number of approaches has been proposed including physical embodied intelligence [27], physical modeling of intrinsic and extrinsic interactions [15], energy levels, constraint driven actions [1] or behavioral lensing [24]. Whilst many of these show significant similarities, it highlights the range of approaches in terms of philosophical, level

IOP Conf. Series: Materials Science and Engineering 1261 (2022) 012001 doi:10.1088/1757-899X/1261/1/012001

of abstraction and also level of fidelity. In 2010, Friston presented his work entitled 'The free-energy principle: a unified brain theory?'[6] which presents perhaps the most universal approach, that accounts for action, perception and learning. This work identifies how optimization is a common theme throughout, where systems and perpetually seeking to optimize or minimize value (expected reward, expected utility) or its complement, surprise (prediction error, expected cost). This presents an exciting unified direction to approach the concept of embodied intelligence.

3.2. Design Creativity & Diversity
Biology and nature shows a vast range of creative and diverse solutions and forms. One observable limitation in robotics is the lack of diversity and innovation in structure and form in comparison to nature. However, there is not a clear solution for how, as a research field, we can develop a wider variety of forms. As already discussed, bio inspired and bio-hybrid robots is one approach; having a greater focus on plants could be beneficial as they show a hugely diverse form factor and have been less closely imitated in robotic form. There is also the need to have a more diverse set of human engineers to form these robots who have different backgrounds and experiences which provides them with different stimuli and experiences on which to form designs. Finally, there is the potential to explore learning based or large scale data-driven approaches to identify new design forms and structures.

3.3. Connection with Conventional AI
Conventional AI and machine learning has huge potential to be leveraged in conjunction with EI. To best interface these systems we must have an understanding on how to design or distribute intelligence, and how to incorporate or understand the role of EI. In addition, it is important to identify why tasks are a challenge for standalone AI, and would clearly benefit from EI. This will aid in demonstrating the role of EI alongside AI, and also for improve the standing of EI within the scientific and industrial community. Another consideration is the alignment of AI with human society. If we want AI systems to be more inline, accepted and understood by human society, we naturally must consider the intersection and interplay between embodiment and EI more carefully.

3.4. Researcher Education & Awareness
Embodied intelligence requires an interdisciplinary viewpoint with broadened views on the origins of intelligence. The requires researchers that are equipped with theoretical understanding and practical skills that straddle a number of disciplines ranging from materials science, biology to computer science [23]. To address this, we must develop methods for training the future generation of researchers to build and sustaining, active and growing EI discipline. There have been a number of mechanisms exploring this, for example summer schools and consortium for the training of early stage researchers. We must also consider the more fundamental aspects, for example what common language or theories tie together EI researchers to allow for interdisciplinary conversations, research and progress.

4. Discussion
In the last ten years there has been significant technological developments that have shown bio-hybrid robots, artificial bots and increasingly capable manipulators, swimming robots, growing robots and much more. Technology developments have also driven our capabilities to understand and characterize EI through development robotics, haptic devices and human-robot interactions. Going forwards we must further unite this research community to increase diversity, viewpoints and areas of expertise to address the key research questions in this area. We require both developments in application oriented solution but also further exploration of the fundamental

understanding and definitions of EI and physical cognition. The creation of a grand challenge for EI which drives both fundamental and applied EI research could provide one means driving this field forwards, whilst also finding and illustrating the role of EI alongside more conventional AI.

References

[1] Jessica Abele et al. "Environmental Constraints Exploitation". In: (2019).

[2] John Amend et al. "Soft robotics commercialization: Jamming grippers from research to product". In: *Soft robotics* 3.4 (2016), pp. 213–222.

[3] Douglas Blackiston et al. "A cellular platform for the development of synthetic living machines". In: *Science Robotics* 6.52 (2021), eabf1571.

[4] Matteo Cianchetti et al. "Design and development of a soft robotic octopus arm exploiting embodied intelligence". In: *2012 IEEE International Conference on Robotics and Automation*. IEEE. 2012, pp. 5271–5276.

[5] Dario Floreano et al. "Evolution of embodied intelligence". In: *Embodied artificial intelligence*. Springer, 2004, pp. 293–311.

[6] Karl Friston. "The free-energy principle: a unified brain theory?" In: *Nature reviews neuroscience* 11.2 (2010), pp. 127–138.

[7] Gowri Shankar Giri, Yaser Maddahi, and Kourosh Zareinia. "An application-based review of haptics technology". In: *Robotics* 10.1 (2021), p. 29.

[8] Elliot W Hawkes, Carmel Majidi, and Michael T Tolley. "Hard questions for soft robotics". In: *Science robotics* 6.53 (2021), eabg6049.

[9] James B Heald, Máté Lengyel, and Daniel M Wolpert. "Contextual inference underlies the learning of sensorimotor repertoires". In: *Nature* 600.7889 (2021), pp. 489–493.

[10] Toby Howison et al. "Reality-assisted evolution of soft robots through large-scale physical experimentation: a review". In: *Artificial Life* 26.4 (2020), pp. 484–506.

[11] Dongdong Jin and Li Zhang. "Embodied intelligence weaves a better future". In: *Nature Machine Intelligence* 2.11 (2020), pp. 663–664.

[12] Sanne Kikkert et al. "Hand and face somatotopy shown using MRI-safe vibrotactile stimulation with a novel Soft Pneumatic Actuator (SPA)-Skin interface". In: *bioRxiv* (2021).

[13] Sangbae Kim, Cecilia Laschi, and Barry Trimmer. "Soft robotics: a bioinspired evolution in robotics". In: *Trends in biotechnology* 31.5 (2013), pp. 287–294.

[14] Yasuo Kuniyoshi. "Fusing autonomy and sociality via embodied emergence and development of behaviour and cognition from fetal period". In: *Philosophical Transactions of the Royal Society B* 374.1771 (2019), p. 20180031.

[15] Cecilia Laschi. "The Physics of Interactions in Soft Robotics". In: *Bulletin of the American Physical Society* (2022).

[16] Cecilia Laschi and Matteo Cianchetti. "Soft robotics: new perspectives for robot bodyware and control". In: *Frontiers in bioengineering and biotechnology* 2 (2014), p. 3.

[17] Max Lungarella et al. "AI in the 21 st Century–With Historical Reflections". In: *50 Years of Artificial Intelligence*. Springer, 2007, pp. 1–8.

[18] Riccardo Manzotti. "Embodied AI beyond embodied cognition and enactivism". In: *Philosophies* 4.3 (2019), p. 39.

[19] Fernando Martınez-Plumed et al. "The facets of artificial intelligence: A framework to track the evolution of AI". In: *International Joint Conferences on Artificial Intelligence*. 2018, pp. 5180–5187.

[20] Barbara Mazzolai and Cecilia Laschi. "A vision for future bioinspired and biohybrid robots". In: *Science robotics* 5.38 (2020), eaba6893.

[21] Barbara Mazzolai et al. "Roadmap on soft robotics: multifunctionality, adaptability and growth without borders". In: *Multifunctional Materials* (2022).

[22] Lotte Meteyard et al. "Coming of age: A review of embodiment and the neuroscience of semantics". In: *Cortex* 48.7 (2012), pp. 788–804.

[23] Aslan Miriyev and Mirko Kovač. "Skills for physical artificial intelligence". In: *Nature Machine Intelligence* 2.11 (2020), pp. 658–660.

[24] Thrishantha Nanayakkara. "Section 10–Realtime mechanical impedance control in soft robots for information gain and efficient control". In: *Roadmap on Soft Robotics: multifunctionality, adaptability and growth without borders* ().

[25] Rolf Pfeifer and Josh Bongard. *How the body shapes the way we think: a new view of intelligence.* MIT press, 2006.

[26] Jahanzaib Shabbir and Tarique Anwer. "Artificial intelligence and its role in near future". In: *arXiv preprint arXiv:1804.01396* (2018).

[27] Metin Sitti. "Physical intelligence as a new paradigm". In: *Extreme Mechanics Letters* 46 (2021), p. 101340.

[28] Seppe Terryn et al. "A review on self-healing polymers for soft robotics". In: *Materials Today* 47 (2021), pp. 187–205.

[29] Michael Wehner et al. "An integrated design and fabrication strategy for entirely soft, autonomous robots". In: *nature* 536.7617 (2016), pp. 451–455.

IOP Conf. Series: Materials Science and Engineering 1261 (2022) 012002 doi:10.1088/1757-899X/1261/1/012002

Embodied Intelligence in soft robotics: joys and sorrows

Cecilia Laschi

National University of Singapore
Department of Mechanical Engineering

9, Engineering Drive 1, 117575 Singapore

mpeclc@nus.edu.sg

Abstract. Embodied intelligence and soft robotics are strictly interconnected, as soft bodyware is key to obtain emergent behaviours from external interactions. The overall advantages of embodied intelligence in robotics are balanced by additional challenges in robot design and control. A main contribution of embodied intelligence to robotics consists of simplifying principles for robot control, in terms of control parameters. At the same time, robot design becomes more complex, involving the complex dynamic interaction of the robot with the environment and the task. These two form the ecological niche where embodied intelligence is effective, but they also limit the robot task space. To what extent can we enlarge the ecological niche without missing the benefits of embodied intelligence? Emergent behaviours are what we expect to simplify control, but what level of controllability can we still achieve? And how far are we from a mathematical description of embodied intelligence, to the benefit of soft robot design and control? Few trade-offs between contrasting needs in the use of embodied intelligence in soft robotics are discussed here, in view of directions of future progress.

1. Introduction

Embodied intelligence is one of the grand lessons that we can take from nature, for improving robot behaviour. What we observe in nature are complex living beings behaving in complex environments, smoothly, fast, effectively and efficiently. In robotics, we tend to take the opposite approach, i.e. keeping the complexity of the system at an affordable and controllable level, in terms of number of sensors and degrees of freedom (DOFs), and simplifying the environment to ideally structured scenarios. The simplifying principles that nature puts in place stand as the main lessons we can take. In nervous systems, we find many elegant solutions that shorten the path of the nervous signal in responding to external stimuli so that sensory-motor behaviour is faster. That is not necessarily a simplification of the neural circuitry, which is in fact more complex than required to be, e.g., redundant. The set of principles that simplifies sensory-motor response is named simplexity [1]. But, in fact, it is not about simplification. It is about simplifying principles that make the behaviour of a complex system in a complex environment possible. Embodied intelligence stands among such simplifying principles in nature. Sensory-motor behaviour emerges from interaction with the environment and part of control is outsourced to the body [2]. What is simplified is the control of sensory-motor behaviour, in terms of processing of sensory data and in terms of control parameters to compute.

So where is embodied intelligence in a robot system? Fig.1 shows the typical representation of a robot as having its mechanical system (robot mechanisms and actuators), its sensory system (sensors and sensory processing software), some form of high-level controller or planner using information from

IOP Conf. Series: Materials Science and Engineering 1261 (2022) 012002 doi:10.1088/1757-899X/1261/1/012002

a world model, and of course an environment where the robot operates. The robot sensory system senses the environment and its mechanical system acts on the environment. We generally consider a sensory-motor behaviour control loop going up from the sensory system, through the planner that generates motor commands based on the sensory input, and down to the mechanical system. A low-level controller typically guarantees that the motor commands are executed correctly by closing a low-level control loop on the robot position error, using proprioceptive sensors. The environment may also act on the robot mechanical system, but the low-level control system cancels such possible external forces, considered as disturbances. Embodied intelligence, instead, makes good use of the external interaction forces. Interaction control techniques can do it [3], but taking full advantage of embodied intelligence means closing a very short control loop at the mechanical level, as shown by the dotted line in Fig.1, with no involvement of the control system.

At this point, it becomes clear how our view of the robot bodyware has to be reversed. While robots need to be rigid and to reject external disturbances, for accurate motion control, with embodied intelligence body compliance becomes instrumental to take advantage of interaction forces. In this perspective, soft robots are particularly suitable for the implementation of embodied intelligence. Soft robotics technologies can help implement embodied intelligence, by providing deformability under external forces. According to the soft robotics definition[1] from the RoboSoft community [4], such deformability does not need to rely only on the use of soft materials, but also do so on structures that are highly deformable (braids, for example) despite their materials having high Young's modulus.

A number of contrasting needs emerge, however, in the balance between the advantages of embodied intelligence and the complexity of soft robot design and control. The next sections try to outline four of such 'trade-off' challenges in embodied intelligence and soft robotics, analysing the underlying problems, discussing cases in the current state of the art and suggesting directions to explore.

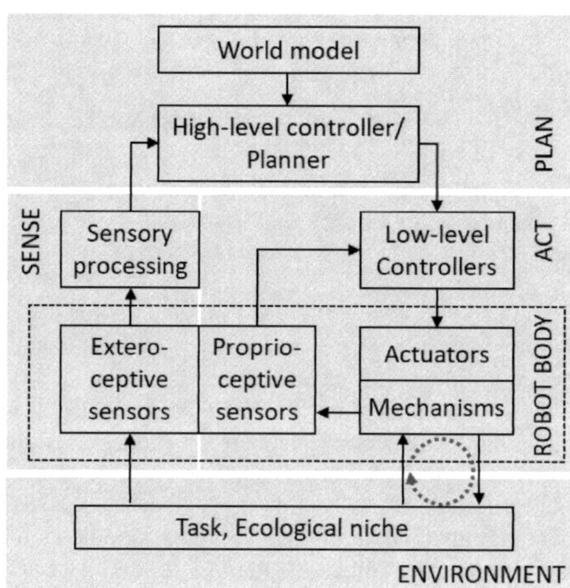

Figure 1. A typical representation of a robot system: The information flows from the environment to the sensory system, which sends processed sensory inputs to the high-level controller or planner, which generates the motor commands for the mechanical system, which acts back on the environment. Embodied intelligence is represented here as the very short control loop at the mechanical level, shown by the dotted line.

[1] "Soft robot/devices that can actively interact with the environment and can undergo 'large' deformations relying on inherent or structural compliance."

IOP Conf. Series: Materials Science and Engineering 1261 (2022) 012002 doi:10.1088/1757-899X/1261/1/012002

2. Ecological niche vs. rich environment and multiple tasks

Extreme examples of embodied intelligence like the passive walker [5] show that the ecological niches where embodied intelligence can be taken full advantage of can be very narrow. The two legs of the passive walker can passively produce sequential steps only on a slope, where gravity can perturb the system balance. To what extent can the ecological niche be tweaked so as to use embodied intelligence and, at the same time, preserve a suitably rich environment and space of tasks for a robot?

A few works show that the passive walker's ecological niche can be extended to flat terrains, by adding a limited number of active movements [6] and making it simple to control, without losing the advantages of embodied intelligence. Gravity is still used and control is simplified, as part of the leg movement derives from gravity forces. Energy consumption is greatly reduced as well, thanks to this use of embodied intelligence.

A good balance between benefitting from an ecological niche and the need for a robot to operate in multiple environments and perform multiple tasks can be found by considering the complete agent scheme of Fig.1, building robot controllers on top of the embodied intelligence mechanical loop.

3. Emergent behaviour vs. controllability

Emergent behaviour is what we expect from embodied intelligence, in order to simplify robot control. However, we also wish to control, to some extent, our robot movements. What is the balance here? Still referring to Fig.1, controlling a soft robot with embodied intelligence means enlarging the low-level control loop to include the embodied intelligence loop.

Two typical approaches can be taken for soft robot control, i.e. based on models or model-free, based on learning [7]. In the first case, models have to include the robot deformations not only as produced by its actuators, but also as generated by external forces. In the second case, training has to similarly include internally- and externally-induced deformations. Examples in literature suggest that learning-based methods capture emerging behaviour and, in dynamic control, let self-stabilizing trajectories emerge [8].

Both directions of evolution can bring progress in the use of embodied intelligence in soft robots, to the extent that external interactions are properly taken into account. Computational modelling techniques can help stress the modelling of soft robots and their dynamic interactions, and control theory can address the inclusion of interaction forces and body deformations. A faster progress can beneficially come from the combination of the two approaches, by integrating learning-based components into model-based control systems.

4. Simple behaviour vs. complex design

We stated that embodied intelligence implies simplifying control, but design may become more complex. Making use of embodied intelligence in robotics involves a three-constituent design [9], where the environment, the task and the robot itself have to be accounted for, at the same time. The robot body is designed according to the expected interaction with the environment, in view of the task to perform. For instance, gravity may be used to perform the task (think of locomotion) and the robot body has to be designed accordingly. The same holds for contact forces, which may help perform the desired movements (think of grasping). Thus, design involves a more complex consideration of the environment, the tasks, and the robot body with its required compliance. If control becomes simpler on one side, design becomes more complex, dealing with a larger space of design parameters, including interaction forces and material properties. Referring to Fig.1, designing a robot with embodied intelligence means including the physical contents of the two arrows establishing the bidirectional interaction between the robot mechanical system and the environment.

How to design robots that embed embodied intelligence remains an open question. How to set design parameters by exploring a large space of possible choices may go beyond traditional design methods used in robotics. Computational tools can come to rescue, like, for instance evolutionary algorithms that can find effective combinations out of the space of design parameters [10].

Full modelling of the internal and external interactions in a soft robot would give a boost to soft robot design.

5. Embodied Intelligence principles vs. abstraction into mathematical descriptions

Embodied intelligence is based on a receptacle of principles [9]. Soft robotics evolved rapidly with a basic trial-and-error design approach, taking those principles into account. While such an approach has been demonstrated to be valuable at the start of the new technological paradigm, soft robotics is now at a stage where an evolution towards a discipline is possible [11]. Are we able to capture the complex interaction of a soft robot body with the environment and describe it mathematically?

When principles can be abstracted and captured in mathematical descriptions, then we have tools for designing robots that can respond to given requirements, and still incorporating the starting principles. As an example, the reduced-order U-SLIP model [12] captures the principles for underwater legged locomotion learned from octopuses and crabs and enabled the design of the SILVER2 robot [13] responding to the need for seafloor exploration and sampling, with efficient self-stabilizing gaits that results from embodied intelligence. More extensive efforts to capture embodied intelligence principles into a theoretical framework are ongoing [14].

A possible path towards the challenge of a complete theory of embodied intelligence goes through mathematical modelling and computational tools that can fully describe the complex interaction of a soft robot with its environment.

Conclusions

The beneficial impact of embodied intelligence in robotics is bringing along a number of challenges that drive further research and technological progress. There are interesting open scientific problems still to investigate, that are interconnected. A full mathematical description of embodied intelligence principles, as discussed in Section 5, would be used for defining design methods that cope with the complexity of the design space, as discussed in Section 4, and consider the robot together with its ecological niche, as discussed in Section 2. Techniques for modelling the full spectrum of the deformations of a soft robot, as produced by its own actuators or by external forces, could be used in control systems that include external interactions in the control loop, as discussed in Section 3. The soft robotics community is meeting those challenges with a distinctive interdisciplinary approach, involving material science, physics, biology and mathematics. A leap forward can come from encouraging further such an interdisciplinary effort, to include computational modelling, control theory and AI towards a theory of embodied intelligence and a soft robotics discipline. Computational modelling techniques can be used to describe mathematically the internal and external interaction of a soft robot in its ecological niche. Control theory provides the tools for using such models of external interactions inside control loops. AI can contribute learning-based methods for building interaction models and control modules. This joint interdisciplinary effort can support the transition to a new stage of model-informed design in soft robotics, making the field evolve from a pioneering arena with trial-and-error design to a structured discipline.

References

[1] Berthoz A 2012 *Simplexity: Simplifying principles for a Complex World* (Yale University Press)

[2] Pfeifer R, Lungarella M and Iida F 2007 Self-organization, embodiment, and biologically inspired robotics *Science* **318**(1088)

[3] Albu-Schäffer A, Eiberger O, Grebenstein M, Haddadin S, Ott C, Wimböck T, Wolf S and Hirzinger G 2008 Soft robotics *IEEE Robot. Automat. Mag.* **15** 20–30

[4] Laschi C, Mazzolai B and Cianchetti M 2016 Soft robotics: technologies and systems pushing the boundaries of robot abilities *Science Robotics* **1**(1)

[5] McGeer T 1990 Passive Dynamic Walking *The International Journal of Robotics Research* **9**(2) 62–82

[6] Collins S H, Wisse M and Ruina A 2001 A 3-D Passive Dynamic Walking Robot with Two Legs

and Knees *International Journal of Robotics Research* **20**(7) 607–15

[7] George Thuruthel T, Ansari Y, Falotico E and Laschi C 2018 Control strategies for soft robotic manipulators: A survey *Soft Robotics* **5**(2) 149–63

[8] George Thuruthel T, Falotico E, Manti M and Laschi C 2018 Stable Open Loop Control of Soft Robotic Manipulators *IEEE Robotics and Automation Letters* *3*(2) 1292–8

[9] Pfeifer R and Bongard J C 2007 *How the body shapes the way we think: a new view of intelligence* (The MIT Press, Cambridge, MA)

[10] Lipson H 2014 Challenges and opportunities for design, simulation, and fabrication of soft robots *Soft Robotics* **1**(1) pp.21–7

[11] Hawkes E W, Majidi C and Tolley M T 2021 Hard questions for soft robotics *Science Robotics* **6**(53)

[12] Calisti M and Laschi C 2017 Morphological and control criteria for self-stable underwater hopping *Bioinspiration & Biomimetics* **13**

[13] Picardi G, Chellapurath M, Iacoponi S, Stefanni S, Laschi C and Calisti M 2020 Bioinspired underwater legged robot for seabed exploration with low environmental disturbance *Science Robotics* **5**(42)

[14] Hauser H, Ijspeert A J, Füchslin R M, Pfeifer R and Maass W 2011 Towards a theoretical foundation for morphological computation with compliant bodies *Biol Cybern* **105** 355–70

IOP Conf. Series: Materials Science and Engineering 1261 (2022) 012003 doi:10.1088/1757-899X/1261/1/012003

Embodied Intelligence in Plants

Barbara Mazzolai, Emanuela Del Dottore, Francesca Tramacere, Alessio Mondini and Laura Margheri

Istituto Italiano di Tecnologia, Bioinspired Soft Robotics Laboratory
Via Morego, 30 16163 Genova - Italy

barbara.mazzolai@iit.it

Abstract. Envisioning a rethink of the design of robotic systems is necessary for a step-change in developing more sustainable and efficient artificial machines. Recent trends in robotics have embraced the idea of taking inspiration from plants to create energy-efficient components, self-morphing growing robots, biodegradable robots, and the definition of novel models of embodied intelligence and morphological computation. Plants can move and grow in air, soil, and water. They can sense and explore the surrounding environment, continuously grow and adapt their shape, and even communicate with each other and with other organisms. Their role for us and our planet is fundamental: for the oxygen we breathe, the food we eat, and to preserve the equilibrium of biodiversity and global climate. Understanding their functioning is of paramount importance and represents an opportunity not only for scientific advancements but also for rethinking the design of artificial technologies that can better integrate with our ecosystems. With a specific focus on the aspects of plants' embodied intelligence, this contribution highlights some of the features of plants that have been investigated for engineering design and introduces new research lines currently at the forefront of the field. A perspective for innovation in science and robotics inspired by plants is also discussed, with a vision toward a new generation of sustainable robots.

1. Introduction

In nature, intelligent behaviors vastly emerge through the interaction of a body with its surrounding environment. This perspective of *intelligence* requires the embodiment, meaning delegating to the design and materials of a body the encoding of intelligent behaviors, or responses to external signals. This way, Embodied Intelligence as a field of research in engineering considers the body an enabler for cognition and thinking [1]. This becomes more evident when the complexity of the body increases, such as in the case of continuum, highly deformable structures with possibly infinite degrees of freedom, and when sensing and intelligence are distributed rather than centralized. Plants are peculiar examples of this type of architecture [2]. Their behaviors are linked to a complex and dynamic interaction between their morphology, distributed sensory-motor control at the level of their tissues, and the environment, expressing the basic principles of the embodiment.

The complexity of plants' morphology is increased by the fact that they grow for their entire life, into redundant structures, from apical extremities of their roots and shoots, overcoming in this way their characteristics of being sessile. This feature, also called indeterminate growth, confers to this organism the resilience to thrive, adapt and respond to changing conditions and survive under stress [2], by adapting their morphologies and physiological processes over time and space.

IOP Conf. Series: Materials Science and Engineering 1261 (2022) 012003 doi:10.1088/1757-899X/1261/1/012003

From materials to actuation principles, from sensing to control architecture, plants offer a variety of inspirational insights to design innovative, intelligent systems whose morphology, behavior, and interaction co-develop and can be more efficiently integrated into real-world environments [3].

The field of *plant-inspired robotics* is new but rapidly expanding and based on the interest in translating specific plant features into artificial solutions. This approach includes firstly their indeterminate growth [4,5], but also the capability to move without having muscles [6,7], their adaptive morphology [8], their structural materials with variable stiffness [9] and functional properties, such as distributed sensing and actuation [10], anchoring/attachment strategies [11], inter and intra-plant communication [12], or inspiring novel plant-hybrid energy generators [13].

The following sections discuss a series of selected examples of plant features where the intelligence reached through the embodiment is particularly evident. These include the analysis of the role of morphology and structural bistability for the fast movement of the carnivorous plant Venus flytrap, the role of hygromorphic properties and morphological characteristics of plant seeds for their passive movement and dispersion strategies, and the structural adaptability of climbing plants.

2. Catching prays: fast movements without muscles

Plants exploit instability and fluid-solid coupling together with hydraulic mechanisms to obtain fast movements: the elastic energy is first stored in the cell walls using a water flow released suddenly when a critical energy barrier is surpassed [14,15]. Following a change in the environmental conditions, the composition and the microstructure of the plants' tissue allow them to have a dynamic morphology that rapidly changes.

Dionaea muscipula, the Venus flytrap, and how the plant closes its leaves in a fraction of a second to capture prey (around 100 ms) represent a peculiar example of fast dynamic morphology. This phenomenon has fascinated scientists since the time of John Ellis [16]. The leaves close when at least two tactile hairs, positioned on their inner side, are touched within an interval of 0.75–20 s [17]. The trigger occurs thanks to a combination of morphological and structural characteristics that allow an accumulation of elastic energy, led by swelling and shrinkage, which is released by a snap-buckling mechanism when the triggering signal arrives. The storage of elastic energy occurs thanks to the particular geometry of the leaf that has two inverted curvatures, one convex/outward when the leaves are open and the other concave/inward when the leaves are closed [14]. By this double-curved shells modality, the bending and stretching deformations are coupled.

The mechanism of the Venus has inspired artificial flytrap-like systems. Some of them are based on an asymmetrically laminated carbon fiber prepreg (CFRP), which acts as a bistable artificial leaf, and a shape memory alloy (SMA), which acts as a triggering actuator to induce the snap motion [18]. Other systems use a combination of different actuation mechanisms (pneumatic, plus SMA spring) to respond to different environmental triggers (heat, moisture, or magnetic stimuli; [6]).

A deeper analysis of the microstructure of the leaf using polarized light microscopy [19] has been used to investigate the presence of reinforcing microfibrils running perpendicular to its midrib in the upper and lower epidermis (Figure 1). The arrangement of the microfibrils is different in the two areas and creates a preferential deformation layer during the snapping motion. This feature inspired the design of an artificial Venus based on a hygroscopic bistable system. With the use of microfabrication techniques, the system has been realized by bonding prestretched poly(dimethylsiloxane) – PDMS layers prior to depositing oriented electrospun polyethylene oxide (PEO) nanofibers that, upon water absorption/desorption, induce the snap in the artificial shell [19].

Like in natural Venus leaves, this artificial system exploits the synergistic integration of geometry, material properties, and bistability, taking advantage of the storage of elastic energy for a rapid, energy-efficient movement, overcoming the limits of actuation based only on the simple diffusion of water molecules. This actuator performed with a resulting snapping velocity of 0.5 s, one order of magnitude smaller than the same system without a pre-stretch.

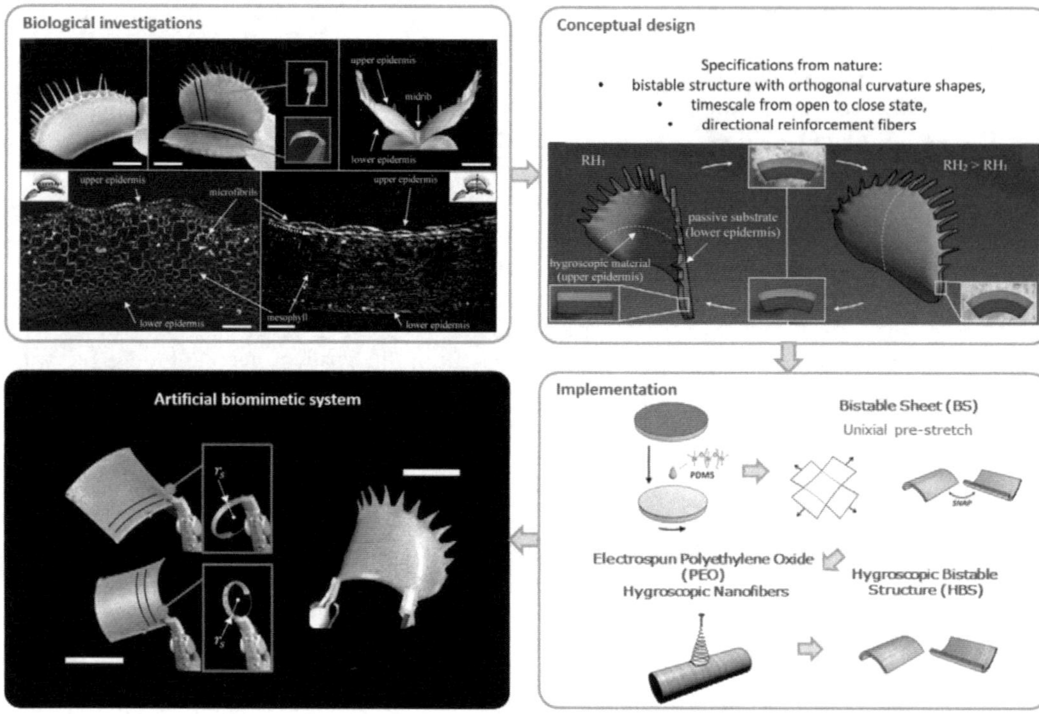

Figure 1 From biology to engineering: (biological investigations box) picture of the Venus flytrap (top) and the image with polarized light microscopy for the visualization of the microfibrils (bottom-); (conceptual design box) key specifications defined from biological investigations are grouped in a conceptual design for the artificial venus-like system; (implementation box) sequence of fabrication steps; (artificial biomimetic system box) results of the implementation for the artificial structures. Adapted from [19].

The model of the Venus underlines the importance of the synergy between structural and multi-functional properties of materials and the advantages in terms of movement speed and energy expenditure that can be gained in the design of environmentally responsive materials and smart structures extracting specifications from nature.

3. Plant seeds dispersion: passive mobility via morphological computation

Plants have evolved various strategies to disperse their seeds and overcome the limitations posed by their sessile nature [20]. They use vectors like wind, water, gravity, and animals or exploit structural environmental responsiveness allowing seeds to move away from their related and colonize new areas. Such characteristics are due to the intrinsic properties of materials [21–23] and individual structural features (e.g., hierarchical organization, presence of awns, hairs, wings, hooks, etc.; [24]) of seed tissues, allowing them to perform dynamic behaviors despite being passive.

Because of these features, seeds and their shells represent one of the most significant examples of *embodied intelligence* in the natural world, displaying a rich library of morphological and mechanical features optimized for passive release, take-off, flying, landing, and drilling [25].

Environmentally-driven movements in plants are linked to the capability of woody materials to absorb and desorb water and hence to swell or shrink, relying on their chemical composition and microstructures.

In pine cones, for example, under wet conditions –not favorable for seeds dispersal– the pine scales are curved upwards, the cone is closed, and the winged, airborne seeds are protected and hindered from being released (Figure 2, left). Under dry conditions, conversely, the scales bend down and liberate the seeds, which are free to fly away. Pine scales are constructed as functional bilayers with a highly swellable, lower sclereid layer (hard outer shell, with cells that are ~20% longer in their wet state as compared to their dry state) and a considerably less swellable upper layer composed of sclerenchymatous strands embedded in a matrix of 'brown tissue' [26,27].

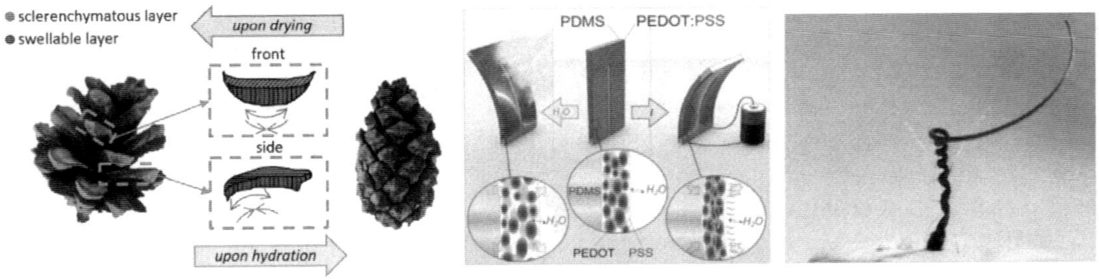

Figure 2 Left: illustration of the pine cone scales that are open in dry conditions and closed in wet conditions, with a schematic of the principle driving the motion; center: integration of the working principle displayed by the pine cone into an active material (PEDOT:PSS) able to respond to humidity changes [7]; right: example of self-burying seed (*Erodium cicutarium* (L.)) showing a coiling tail, thanks to a bi-layer arrangement of hygroscopically active and inactive layers.

The working principle of the pinecone, and other similar structures, has led to the design of materials that react to environmental parameters such as humidity. In Taccola et al., 2015 [7], for example, it has been developed an active material (based on poly(3,4-ethylenedioxythiophene) polystyrene sulfonate (PEDOT:PSS)) with a good combination of electrical conductivity, humidity responsiveness, mechanical properties, and chemical and thermal stability. These features give to the proposed actuator the potential to simultaneously combine active/passive actuation and sensing within a single-composite material (Figure 2, center).

Among various seeds that use wind dispersal strategy, particular attention has been posed to winged seeds or samaras. They disperse under windy conditions with the behavior of auto-rotators [28]. Their rotational mechanism occurs thanks to a heavy nut (their center of gravity, where the seed is enclosed) located at the base of a wing-shaped structure [29,30]. They have a small size and relatively low velocity but can generate very high lift forces thanks to the creation of leading-edge vortexes [31]. Analysis of winged seeds performances focused on surface roughness, the center of gravity location, leading-edge thickness, and planform design [32], in-flight characteristics such as terminal velocity [33], and on numerical simulations of the auto-rotation [34].

Drawing inspiration from biological analyses, examples of artificial samara seeds were fabricated via 3D printing (e.g., by [35,36]), showing experimental and theoretical models with an optimal wing fold angle to minimize the descent velocity. Also, a recent work [37] has presented an artificial single-winged samaras 3D printed using poly(lactic acid) (PLA), having similar morphological features to the natural one (shape, center of mass, and wing loading), and similar aerodynamics performances (rotation velocities, wing tip speeds and lab descent speeds).

In so-called "self-burying seeds", crawling and burying occur thanks to a helical unit, which has hygroscopic characteristics and responds to variations of external humidity by changing its configuration and spontaneously uncoiling when wet and re-coiling when dry [38,39]. The helical unit (Figure 2, right) is composed of a bilayer structure, one hygroscopically active with cellulose-based fibrils that adsorb water and the other hygroscopically inactive with ligneous cells. The actuation is driven by the coupling of the two differently structured tissue layers, which exhibit different elongations along a specific direction in the same moist conditions [40].

The mechanism of self-burial abilities has been investigated in biology, especially on *Erodium cicutarium* (L.), in terms of materials and structural fibers [24,40,41]; microstructure [42] mechanics and geometry [43,44]; and 3D micromechanics [45], offering a series of interesting functional insights from a biomimetic perspective for technological design. *Erodium cicutarium* (L.) self-burial mechanism provides characteristics to explore different solutions for artificial penetrators in unstructured and extreme environments, including space [46]. In soft robotics, the coiling and uncoiling motions of the

Erodium tail have been a source of inspiration for the design of novel soft pneumatic actuators [47]. Exploring the material characteristics, Abraham et al. (2012) [40] have reproduced the induced spiraling mechanism in a simplified physical model made from a thread embedded in an isotropic foam matrix.

4. Structural adaptability of climbing plants

Climbing plants are able to grow and move in cluttered environments, searching for support and attaching, anchoring, or cooping themselves onto it. They use stimulus-oriented growth (tropisms) and oscillatory growth patterns (circumnutations) to navigate their surroundings. They exploit their tactile perception, adhesion strategies, and capacity to morph their bodies, change their stiffness locally to extend over voids, grow vertically, pass physical constraints and reach higher exposure to light [48–50].

Tendrils of vines are particularly interesting from a robotic point of view for their morphological adaptation to differently shaped objects. Their coiling ability is associated with the presence of gelatinous fibers [51], in which a differential decrease of water content generates an asymmetric contractile force, from which the coiling derives [52]. Secondary coiling occurs afterward and allows the plant to get closer to the support by pulling it and creating an elastic spring-like anchorage to resist external loads or the wind. Then, the water loss leads to increased structural rigidity ("lignification") that prevents the tendril from uncoiling.

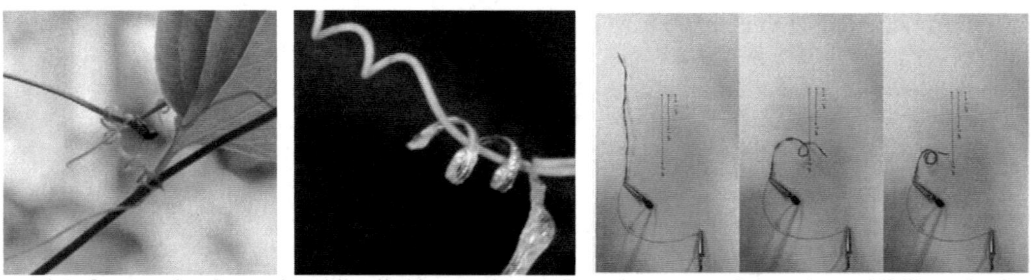

Figure 3 Left: example of natural tendril coiled in complex configuration; center: frame of a tendril-like soft robot based on reversible osmotic actuation [9]; right: sequence of three frames during coiling of a tendril-like structure based on SMA [53] (vertical scale bar drawn in the background on the paper, from longest to shortest lines: 10, 5, 1 cm).

Examples of climbing plant-inspired robots include systems imitating their searching abilities mediated by circumnutation movements [54–56], for an efficient environment exploration and contact; or robots that use tendril-like structures (e.g., SMA coil [53]) to improve stabilization by anchoring to existing structures or intertwine with other robots. Plant-inspired continuum structures are also adopted to extend exploration capability by exploiting varied robot topologies by extending tendril-like continuum arms [57].

When the support is unsuitable for the plant, the coiling is reversed, and the tendril uncoils. Such reversibility is a desired feature in soft robots and can be implemented in tendril-like soft arms using the same actuation strategy of plants based on the control of osmolyte gradients [9].

Inspired by the functional traits of a climbing cactus [50], 3D printable multi-material actuators (based on hydrogel and elastomeric materials) have been realized with similar morphology and material properties observable in the cactus tissue [58]. The artificial system has shown the ability to change its shape (from star to circular) upon absorption/desorption of water by its hydrogel component, showing a characteristic morphing similar to natural cactus across developmental stages [59].

5. Conclusions: what can plants offer to Embodied Intelligence

Plants offer a plethora of inspirational insights to design innovative technologies and robotic solutions energetically more -efficient thanks to embodied responsive abilities.

Abilities in plants result from a complex and dynamic interaction among their morphology, distributed sensory-motor architecture, and the surrounding environment. They have a structural re-programmability fundamental for exploiting environmental changes for sensing and mobility.

IOP Conf. Series: Materials Science and Engineering 1261 (2022) 012003 doi:10.1088/1757-899X/1261/1/012003

«Intelligence» in plants is inherent to the body structure, and the behavior emerges directly from the combination of structural units with changes in the environmental conditions. Several movements and behaviors of plants occur, therefore, without the need for any internal control or energy supply (e.g., a pine cone [60] or a drilling seed [22]).

In plants, a co-development of morphology and behavior (individual and collective) occurs, which is continuous, not preset. With their highly redundant and growing structures, centralized control is not evolutionarily advantageous. On the contrary, distributed control and the embodiment of intelligence into the structural material represent the most suitable form to better integrate into natural, changing environments and improve adaptation for survival. They indeed represent an excellent model in nature from which new concepts of artificial embodied intelligence can be extracted.

The field of plant-inspired robotics promises a series of technological and scientific innovation opportunities. The challenge ahead is researching new designs and new components for robots to make them more "sustainable". In this perspective, robots should be made of biodegradable or recyclable materials to avoid releasing further e-waste once the robot has come to an end of use. They also should use sustainable forms of energy to reduce their ecological footprint during their use. Fabrication processes also need to be revised to make them more sustainable. Therefore, the whole circle of robotic design, development, and disposal needs revisions in terms of an ecological transition.

A second challenge ahead looks for the potential use of bioinspired robots to test and validate scientific hypotheses. Several still unknown aspects in plant biology can be unveiled using a bioengineering approach and robots as physical tools. On the other way around, engineering tools and methodologies used to analyze an aspect of plants can reveal new information about them that would not have been possible with a classic methodology. This mutual benefit and the possibility to advance science and technology simultaneously represent a major aspect of innovation concerning the approach to scientific research.

References

[1] Hara F and Pfeifer R 2003 *Morpho-functional Machines: The New Species Designing Embodied Intelligence*

[2] Karban R 2008 Plant behaviour and communication *Ecology Letters* **11** 727–39

[3] Mazzolai B and Laschi C 2020 A vision for future bioinspired and biohybrid robots *Science Robotics* **5** eaba6893

[4] Sadeghi A, Mondini A and Mazzolai B 2017 Towards self-growing soft robots inspired by plant roots and based on additive manufacturing technologies *SoRo in press*

[5] Hawkes E W, Blumenschein L H, Greer J D and Okamura A M 2017 A soft robot that navigates its environment through growth *Science Robotics* **2** eaan3028

[6] Esser F J, Auth P and Speck T 2020 Artificial Venus Flytraps: A Research Review and Outlook on Their Importance for Novel Bioinspired Materials Systems *Front. Robot. AI* **7** 75

[7] Taccola S, Greco F, Sinibaldi E, Mondini A, Mazzolai B and Mattoli V 2015 Toward a new generation of electrically controllable hygromorphic soft actuators *Advanced Materials* **27** 1668–75

[8] Sadeghi A, Del Dottore E, Mondini A and Mazzolai B 2019 Passive Morphological Adaptation for Obstacle Avoidance in a Self-Growing Robot Produced by Additive Manufacturing *Soft Robotics*

[9] Must I, Sinibaldi E and Mazzolai B 2019 A variable-stiffness tendril-like soft robot based on reversible osmotic actuation *Nature Communications* **10**

[10] Cheng Y, Wang R, Chan K H, Lu X, Sun J and Ho G W 2018 A Biomimetic Conductive Tendril for Ultrastretchable and Integratable Electronics, Muscles, and Sensors *ACS Nano* **12** 3898–907

[11] Fiorello I, Tricinci O, Naselli G A, Mondini A, Filippeschi C, Tramacere F, Mishra A K and Mazzolai B 2020 Micropatterned Devices: Climbing Plant-Inspired Micropatterned Devices for Reversible Attachment *Adv. Funct. Mater.* **30** 2070256

[12] Emanuela Del Dottore, Alessio Mondini, Ali Sadeghi, and Barbara Mazzolai 2018 Swarming Behavior Emerging from the Uptake–Kinetics Feedback Control in a Plant-Root-Inspired Robot *Applied Sciences* **8** 47

[13] Meder F, Thielen M, Mondini A, Speck T and Mazzolai B 2020 Living Plant-Hybrid Generators for Multidirectional Wind Energy Conversion *Energy Technology* **8** 2000236

[14] Forterre Y, Skotheim J M, Dumais J and Mahadevan L 2005 How the Venus flytrap snaps *Nature* **433** 421–5

[15] Dumais J and Forterre Y 2012 “Vegetable Dynamicks”: The Role of Water in Plant Movements *Annu. Rev. Fluid Mech.* **44** 453–78

[16] Ellis J 1770 *Directions for Bringing Over Seeds and Plants, from the East-Indies and Other Distant Countries, in a State of Vegetation: Together with a Catalogue of Such Foreign Plants as are Worthy of Being Encouraged in Our American Colonies, for the Purposes of Medicine, Agriculture and Commerce : to which is Added, the Figure and Botanical Description of a New Sensitive Plant, Called Dionaea Muscipula : Or, Venus's Fly Trap* (London: L. Davis)

[17] Macfarlane J M 1892 Contributions to the history of Dionaea muscipula, Ellis *Contrib Bot Lab Penna* **1** 7–44

[18] Kim S-W, Koh J-S, Lee J-G, Ryu J, Cho M and Cho K-J 2014 Flytrap-inspired robot using structurally integrated actuation based on bistability and a developable surface *Bioinspir. Biomim.* **9** 036004

[19] Lunni D, Cianchetti M, Filippeschi C, Sinibaldi E and Mazzolai B 2020 Plant-Inspired Soft Bistable Structures Based on Hygroscopic Electrospun Nanofibers *Advanced Materials Interfaces* **7** 1901310

[20] Howe H F and Smallwood J 1982 Ecology of Seed Dispersal *Annu. Rev. Ecol. Syst.* **13** 201–28

[21] Chambers J C and MacMahon J A 1994 A DAY IN THE LIFE OF A SEED: Movements and Fates of Seeds and Their Implications for Natural and Managed Systems *Annu. Rev. Ecol. Syst.* **25** 263–92

[22] Burgert I and Fratzl P 2009 Plants control the properties and actuation of their organs through the orientation of cellulose fibrils in their cell walls *Integrative and Comparative Biology* **49** 69–79

[23] Sádlo J, Chytrý M, Pergl J and Pyšek P 2018 Plant dispersal strategies: a new classification based on the multiple dispersal modes of individual species *Preslia* **90** 1–22

[24] Abraham Y and Elbaum R 2013 Hygroscopic movements in Geraniaceae: the structural variations that are responsible for coiling or bending *New Phytol* **199** 584–94

[25] Fratzl P and Barth F G 2009 Biomaterial systems for mechanosensing and actuation *Nature* **462** 442–8

[26] Reyssat E and Mahadevan L 2009 Hygromorphs: from pine cones to biomimetic bilayers *Journal of The Royal Society Interface* **6** 951–7

[27] Dawson C, Vincent J F V and Rocca A-M 1997 How pine cones open *Nature* **390** 668–668

[28] Anon 2006 *Spinning Flight* (New York, NY: Springer New York)

[29] Norberg R Å 1973 AUTOROTATION, SELF-STABILITY, AND STRUCTURE OF SINGLE-WINGED FRUITS AND SEEDS (SAMARAS) WITH COMPARATIVE REMARKS ON ANIMAL FLIGHT *Biological Reviews* **48** 561–96

[30] Azuma A and Yasuda K 1989 Flight performance of rotary seeds *Journal of Theoretical Biology* **138** 23–53

[31] Lentink D, Dickson W B, van Leeuwen J L and Dickinson M H 2009 Leading-Edge Vortices Elevate Lift of Autorotating Plant Seeds *Science* **324** 1438–40

[32] Yasuda K and Azuma A 1997 The Autorotation Boundary in the Flight of Samaras *Journal of Theoretical Biology* **185** 313–20

[33] Seter D and Rosen A 1992 Stability of the Vertical Autorotation of a Single-Winged Samara *Journal of Applied Mechanics* **59** 1000–8

[34] Arranz G, Moriche M, Uhlmann M, Flores O and García-Villalba M 2018 Kinematics and dynamics of the auto-rotation of a model winged seed *Bioinspir. Biomim.* **13** 036011

[35] Rabault J, Fauli R A and Carlson A 2019 Curving to Fly: Synthetic Adaptation Unveils Optimal Flight Performance of Whirling Fruits *Phys. Rev. Lett.* **122** 024501

[36] Fauli R A, Rabault J and Carlson A 2019 Effect of wing fold angles on the terminal descent velocity of double-winged autorotating seeds, fruits, and other diaspores *Phys. Rev. E* **100** 013108

[37] Nave G K, Hall N, Somers K, Davis B, Gruszewski H, Powers C, Collver M, Schmale D G and Ross S D 2021 Wind Dispersal of Natural and Biomimetic Maple Samaras *Biomimetics* **6** 23

[38] Stamp N E 1984 Self-Burial Behaviour of Erodium Cicutarium Seeds *The Journal of Ecology* **72** 611

[39] Elbaum R, Zaltzman L, Burgert I and Fratzl P 2007 The Role of Wheat Awns in the Seed Dispersal Unit *Science* **316** 884–6

[40] Abraham Y, Tamburu C, Klein E, Dunlop J W C, Fratzl P, Raviv U and Elbaum R 2012 Tilted cellulose arrangement as a novel mechanism for hygroscopic coiling in the stork's bill awn *J. R. Soc. Interface.* **9** 640–7

[41] Aharoni H, Abraham Y, Elbaum R, Sharon E and Kupferman R 2012 Emergence of Spontaneous Twist and Curvature in Non-Euclidean Rods: Application to *Erodium* Plant Cells *Phys. Rev. Lett.* **108** 238106

[42] Elbaum R and Abraham Y 2014 Insights into the microstructures of hygroscopic movement in plant seed dispersal *Plant Science* **223** 124–33

[43] Evangelista D, Hotton S and Dumais J 2011 The mechanics of explosive dispersal and self-burial in the seeds of the filaree, *Erodium cicutarium* (Geraniaceae) *Journal of Experimental Biology* **214** 521–9

[44] Jung W, Kim W and Kim H-Y 2014 Self-burial Mechanics of Hygroscopically Responsive Awns *Integrative and Comparative Biology* **54** 1034–42

[45] Zhao C, Liu Q, Ren L, Song Z and Wang J 2017 A 3D micromechanical study of hygroscopic coiling deformation in Pelargonium seed: from material and mechanics perspective *J Mater Sci* **52** 415–30

[46] Pandolfi C, Comparini D and Mancuso S 2012 Self-burial Mechanism of Erodium cicutarium and Its Potential Application for Subsurface Exploration *Biomimetic and Biohybrid Systems* Lecture Notes in Computer Science vol 7375, ed T J Prescott, N F Lepora, A Mura and P F M J Verschure (Berlin, Heidelberg: Springer Berlin Heidelberg) pp 384–5

[47] Geer R, Iannucci S and Li S 2020 Pneumatic Coiling Actuator Inspired by the Awns of Erodium cicutarium *Front. Robot. AI* **7** 17

[48] Rowe N and Speck T 2005 Plant growth forms: an ecological and evolutionary perspective: Research review *New Phytologist* **166** 61–72

[49] Rowe N P and Speck T 2014 Stem biomechanics, strength of attachment, and developmental plasticity of vines and lianas *Ecology of Lianas* ed S A Schnitzer, F Bongers, R J Burnham and F E Putz (Chichester, UK: John Wiley & Sons, Ltd) pp 323–41

[50] Soffiatti P and Rowe N P 2020 Mechanical Innovations of a Climbing Cactus: Functional Insights for a New Generation of Growing Robots *Front. Robot. AI* **7** 64

[51] Bowling A J and Vaughn K C 2009 Gelatinous fibers are widespread in coiling tendrils and twining vines *American Journal of Botany* **96** 719–27

[52] Gerbode S J, Puzey J R, McCormick A G and Mahadevan L 2012 How the Cucumber Tendril Coils and Overwinds *Science* **337** 1087–91

[53] Gallentine J, Wooten M B, Thielen M, Walker I D, Speck T and Niklas K 2020 Searching and Intertwining: Climbing Plants and GrowBots *Front. Robot. AI* **7** 118

[54] Wooten M and Walker I 2018 Vine-Inspired Continuum Tendril Robots and Circumnutations *Robotics* **7** 58

[55] Wooten M B and Walker I D 2016 Circumnutation: From Plants to Robots *From Animals to Animats 14* Lecture Notes in Computer Science vol 9825, ed E Tuci, A Giagkos, M Wilson and J Hallam (Cham: Springer International Publishing) pp 1–11

[56] Visentin F, Naselli G A and Mazzolai B 2020 A New Exploration Strategy for Soft Robots Based on Proprioception *2020 3rd IEEE International Conference on Soft Robotics (RoboSoft)* 2020 3rd IEEE International Conference on Soft Robotics (RoboSoft) (New Haven, CT, USA: IEEE) pp 816–21

[57] Lastinger M C, Verma S, Kapadia A D and Walker I D 2019 TREE: A Variable Topology, Branching Continuum Robot *2019 International Conference on Robotics and Automation (ICRA)* 2019 International Conference on Robotics and Automation (ICRA) (Montreal, QC, Canada: IEEE) pp 5365–71

[58] Rodriguez N, Bastola A K, Behl M, Soffiatti P, Rowe N P and Lendlein A 2021 Approaches of combining a 3D-printed elastic structure and a hydrogel to create models for plant-inspired actuators *MRS Advances* **6** 625–30

[59] Bastola A K, Rodriguez N, Behl M, Soffiatti P, Rowe N P and Lendlein A 2021 Cactus-inspired design principles for soft robotics based on 3D printed hydrogel-elastomer systems *Materials & Design* **202** 109515

[60] Eger C J, Horstmann M, Poppinga S, Sachse R, Thierer R, Nestle N, Bruchmann B, Speck T, Bischoff M and Rühe J 2022 The Structural and Mechanical Basis for Passive-Hydraulic Pine Cone Actuation *Advanced Science* 2200458

IOP Conf. Series: Materials Science and Engineering 1261 (2022) 012004 doi:10.1088/1757-899X/1261/1/012004

From bioinspired to biological robots, and back again.

Josh Bongard

Department of Computer Science, University of Vermont, Burlington Vermont 05405, USA

E-mail: josh.bongard@uvm.edu

Abstract. The histories of the Artificial Intelligence and robotics communities have been primarily those of bioinspiration: observing, distilling, and instantiating a wealth of biological structures and mechanisms into machines. However, this transferal from biology to technology has been uneven: the machine learning community has focused on non-physical mechanisms, like synaptic plasticity, while the robotics community has focused on physical structures, like the quadrupedal body plan. Organisms however offer a wealth of physical mechanisms, like bones that strengthen in response to load or tissues that regenerate in response to amputation, that have yet to be instantiated in machines. Biological robots—machines built entirely from biological tissue–exhibit many of these morphological mechanisms 'for free', suggesting that biobots could be designed in future to exaggerate these adaptive mechanisms, or provide hints for how they can be incorporated into morphologically self-editing robots.

1. Introduction.

Machine learning researchers and roboticists have long looked to nature for ideas about how to create algorithms that learn or machines that adapt. However, bioinspired design presents several challenges. First, there are thousands of adaptive mechanisms described in the biology literature, some understood better than others, and each operates at a particular spatial and temporal scale, under certain environmental conditions, in a subset of species. Determining which among this vast array are likely to be useful if incorporated into algorithms and machines is far from obvious. Second, once a structure or mechanism is chosen, it must be abstracted, adapted, and incorporated into an algorithm or machine in such a way that it retains its adaptive properties, while operating at different scales and using, or being composed of, different materials. A third challenge arises when dealing with physical adaptive mechanisms: the mechanism may demand that the physical aspects of the machine change in a way that is difficult or impossible to achieve with today's materials.

The result of these challenges is that some adaptive structures and mechanisms have received much more attention than others. If we consider all products of biological evolution as either structures or mechanisms, one broad pattern that can be discerned is that machine learning researchers tend to focus on mechanism, while roboticists tend to focus on structure. If we further subdivide our classification of evolutionary results into neural or morphological structures or mechanisms, a further pattern can be observed: machine learning researchers tend to devote inordinate focus to neural mechanisms while roboticists tend to focus primarily on physical structures. This can be seen most clearly by the overwhelming attention that the neural mechanism of synaptic plasticity has received in the machine learning literature, while the most obvious example of a physical structure studied in robotics is that of the quadrupedal body plan.

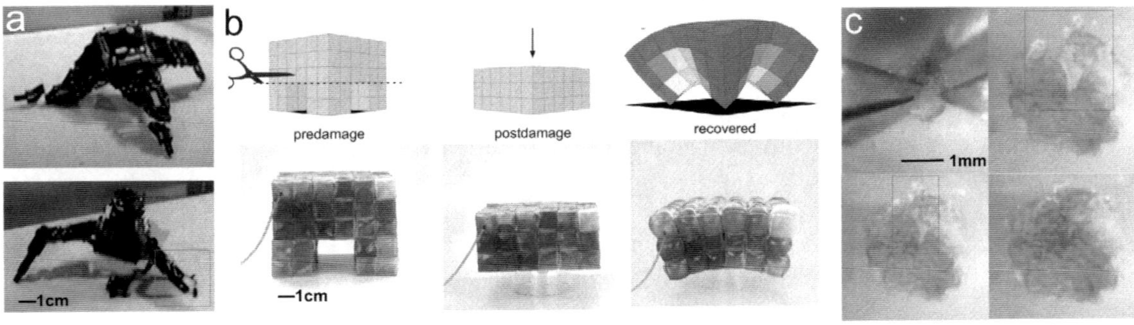

Figure 1. A rigid, soft, and biological robot recovering from injury. **(a):** A rigid robot's only option, after losing an appendage is to adapt behaviorally by learning a compensating policy [4]. **(b):** A soft robot may adopt behavioral and morphological changes to recover from injury, but it must learn how to do so [10]. **(c):** This particular biological robot has inherited adaptive potential from the tissues out of which it is constructed: it spontaneously recovers from injury, requiring no learning [8].

This means that two quadrants in this taxonomy of evolutionary products—neural structures and physical mechanisms—have been understudied, at least until recently. The rise of interest in neural architecture search [5] and neuromorphic computing [15] indicates that neural structure is becoming more topical within machine learning. Similarly, advances in soft robotics and materials science, both of which allow autonomous machines to alter their forms in ways that are difficult or impossible for traditional rigid machines, have provided exciting new opportunities for incorporating bio-inspired physical mechanisms into machines.

2. Bioinspired machines.

Embodied intelligence can be viewed as the enterprise of determining how to change not only one's behavior to adapt to changing circumstances but one's body as well. Soft materials incorporated into autonomous machines provides a diverse set of options for morphological change, including change in shape [16], material properties [18, 17, 19], numerosity [14, 12], and mass [13, 6, 7]. However, such plasticity provides avenues for incorporating physical change into machines. Questions of how best to recruit those processes to best instantiate a given physical adaptive mechanism observed in nature remains an open problem.

One particular type of adaptive mechanism ubiquitous throughout the living world is recovering from internal surprise, such as damage, injury, or amputation. It is instructive to consider how rigid, soft, and now biological robots have been built to instantiate adaptation to internal surprise. Rigid robots only have recourse to behavioral adaptation when damaged (Fig. 1a). As reported in [4], one way to accomplish this is to build and then update internal models before and after damage, and vet candidate control policies internally on self models before attempting them in reality, where inappropriate policies may cause further damage.

This line of inquiry was later continued using a soft robot composed of hollow silicone cubes [10]. This body plan allowed the robot, in principle, to 'regenerate' amputated limbs by expanding the inner volumes of neighboring cubes (Fig. 1b). Surprisingly, the robot often evolved alternative strategies, such as retracting its remaining limbs and regressing from walking to peristaltic motion. This provides an instructive example that robots may exhibit, at a high level, mechanisms observed in nature—changing body shape to recover from amputation—but do so in different ways, as robots are built from different materials as living systems, are not constrained by the historical accidents of biological evolution.

3. Biological machines.

Finally, a third step in this study of adapting to internal surprise in machines was recently taken by damaging a biological robot [8].

Biological robots are an emerging technology [8, 1, 11, 9, 2] in which genetically unmodified cells or tissues are liberated from the usual developmental environment of an embryo, rearranged into new patterns, which directs the rearranged tissue toward a stable adult form and function that differs markedly from that of the host organism. Previous work has demonstrated that organisms can not only survive drastic interventions during development, but exploit those interventions to aid adaptive behavior. For example, eye precursor cells were ectopically implanted in the tail of a developing tadpole, where they connected with the growing animals' nervous system and became functional eyes [3]. The adult organisms could even use these ectopic eyes for light-mediated learning.

Recently, we hypothesized that an evolutionary algorithm could be created that searches over the space of all possible candidate interventions, seeking those that rearrange virtual biological tissue in virtual environments into new biological forms. The search method could be instructed to seek 'reconfigured organisms' that exhibit some desired behavior *in silico*, even if the form of that organism is very far from that of the wild type organism from which its tissues are drawn. Good candidate designs should also be realizable using the techniques that will be employed to build these organisms *in vitro*. We have found that such a method can indeed produce AI-designed organisms that walk [8] or swim [2].

Unlike rigid or soft robots, biological robots are built from smart materials: cells themselves are sophisticated machines capable of non-trivial sensorimotor behavior and, given complex internal states that includes memory function, 'cognitive' function. This has important implications for considering such constructs as robots. For example, such robots come with useful capabilities built in: when damaged, we found that AI-designed biological robots could heal themselves (Fig. 1c), a capability beyond our previous work in exposing rigid [4] and soft [10] robots to injury.

4. Bioinspired machines revisited.

Usually, when designing bioinspired machines, we seek to imprint biological structures and mechanisms into otherwise inert materials, such as metal, plastics and electronics. When attempting to design biological machines, it seems useful to assume a different approach: it behooves us to create search methods that seek ways to exaggerate adaptive mechanisms already latent in the biological material, rather than attempting to build it into machines that lack it.

Comparing biological machines to bioinspired machines, and the ways we may best design each, also raises new research questions. For instance, one may ask how mm-sized 'xenobots'—biobots made from *Xenopus laevis* tissue—are able to heal themselves after damage, when they are already in a compromised state, far from the adult form of the donating organism? The answer may lie in a shift in perspective: from the point of the view of the organism, it is indeed in a very different form. However, at the cellular level, any given *X. laevis* skin cell may have the same neighbors it would have had in the wild type animal: other skin cells. Thus, the 'new' phenotype may not actually seem so novel from a cell's perspective, and thus it may not be in as disadvantaged a position as originally seemed when it must participate in a healing process to recover from a wound. In this way, biological robots challenge us to rethink adaptation in particular, such as recovery from injury, and intelligence in general. When we observe an animal immediately responding in an appropriate manner to a 'new' environment, that environment may not be as novel as it really appears.

Such new points of view also help us revisit bioinspired robots in new ways. For instance, our increased appreciation for organisms as machines made of intelligent machines (organs) made of intelligent machines (cells) led us to the creation of fractal robots: robots made of robots made of robots. As the fields of bioinspired and biological machines continue, it will be interesting to explore how increasingly sophisticated loops from technology to biology and back again will reinvigorate our long term goal of creating intelligent machiners.

References

[1] Philip Ball. Living robots. *Nature Materials*, 19(3):265–265, 2020.

[2] Douglas Blackiston, Emma Lederer, Sam Kriegman, Simon Garnier, Joshua Bongard, and Michael Levin. A cellular platform for the development of synthetic living machines. *Science Robotics*, 6(52):eabf1571, 2021.

[3] Douglas J Blackiston and Michael Levin. Ectopic eyes outside the head in xenopus tadpoles provide sensory data for light-mediated learning. *Journal of Experimental Biology*, 216(6):1031–1040, 2013.

[4] J. Bongard, V. Zykov, and H. Lipson. Resilient machines through continuous self-modeling. *Science*, 314:1118–1121, 2006.

[5] Thomas Elsken, Jan Hendrik Metzen, and Frank Hutter. Neural architecture search: A survey. *The Journal of Machine Learning Research*, 20(1):1997–2017, 2019.

[6] Elliot W Hawkes, Laura H Blumenschein, Joseph D Greer, and Allison M Okamura. A soft robot that navigates its environment through growth. *Science Robotics*, 2(8), 2017.

[7] Kazuya Horibe, Kathryn Walker, and Sebastian Risi. Regenerating soft robots through neural cellular automata. In *EuroGP*, pages 36–50, 2021.

[8] Sam Kriegman, Douglas Blackiston, Michael Levin, and Josh Bongard. A scalable pipeline for designing reconfigurable organisms. *Proceedings of the National Academy of Sciences*, 117(4):1853–1859, 2020.

[9] Sam Kriegman, Amir Mohammadi Nasab, Douglas Blackiston, Hannah Steele, Michael Levin, Rebecca Kramer-Bottiglio, and Josh Bongard. Scale invariant robot behavior with fractals. *Proceedings of the Robotics: Science and Systems Conference*, 2021.

[10] Sam Kriegman, Stephanie Walker, Dylan Shah, Michael Levin, Rebecca Kramer-Bottiglio, and Josh Bongard. Automated shapeshifting for function recovery in damaged robots. In *Proceedings of the Robotics: Science and Systems Conference*, pages 28–37, 2019.

[11] M Levin, J Bongard, and JE Lunshof. Applications and ethics of computer-designed organisms. *Nature Reviews Molecular Cell Biology*, 21(11):655–656, 2020.

[12] Shuguang Li, Richa Batra, David Brown, Hyun-Dong Chang, Nikhil Ranganathan, Chuck Hoberman, Daniela Rus, and Hod Lipson. Particle robotics based on statistical mechanics of loosely coupled components. *Nature*, 567(7748):361–365, 2019.

[13] Barbara Mazzolai. Plant-inspired growing robots. In *Soft Robotics: Trends, Applications and Challenges*, pages 57–63. Springer, 2017.

[14] William Savoie, Thomas A Berrueta, Zachary Jackson, Ana Pervan, Ross Warkentin, Shengkai Li, Todd D Murphey, Kurt Wiesenfeld, and Daniel I Goldman. A robot made of robots: Emergent transport and control of a smarticle ensemble. *Science Robotics*, 4(34), 2019.

[15] Catherine D Schuman, Thomas E Potok, Robert M Patton, J Douglas Birdwell, Mark E Dean, Garrett S Rose, and James S Plank. A survey of neuromorphic computing and neural networks in hardware. *arXiv preprint arXiv:1705.06963*, 2017.

[16] Dylan Shah, Bilige Yang, Sam Kriegman, Michael Levin, Josh Bongard, and Rebecca Kramer-Bottiglio. Shape changing robots: bioinspiration, simulation, and physical realization. *Advanced Materials*, 33(19):2002882, 2021.

[17] Jun Shintake, Bryan Schubert, Samuel Rosset, Herbert Shea, and Dario Floreano. Variable stiffness actuator for soft robotics using dielectric elastomer and low-melting-point alloy. In *2015 IEEE/RSJ International Conference on Intelligent Robots and Systems (IROS)*, pages 1097–1102. IEEE, 2015.

[18] Han Yuan, Eric Courteille, and Dominique Deblaise. Static and dynamic stiffness analyses of cable-driven parallel robots with non-negligible cable mass and elasticity. *Mechanism and Machine Theory*, 85:64–81, 2015.

[19] Michelle C Yuen, R Adam Bilodeau, and Rebecca K Kramer. Active variable stiffness fibers for multifunctional robotic fabrics. *IEEE Robotics and Automation Letters*, 1(2):708–715, 2016.

IOP Conf. Series: Materials Science and Engineering 1261 (2022) 012005 doi:10.1088/1757-899X/1261/1/012005

Embodied Intelligence & Morphological Computation in Soft Robotics Community: Collaborations, Coordination, and Perspective

S.M.Hadi Sadati[1†*], Marwa ElDiwiny[2†], Surya Nurzaman[3], Fumiya Iida[4], Thrishantha Nanayakkara[5]

[1]The Robotics and Vision in Medicine Lab, School of Biomedical Engineering & Imaging Sciences, King's College London, UK.
[2]Brubotics, Vrije Universiteit, Brussel, Belgium
[3]Mechanical Engineering Discipline, School of Engineering, Malaysia Campus, Monash University, Melbourne, Australia.
[4]Machine Intelligence Laboratory, Department of Engineering, Cambridge University, UK.
[5]Dyson School of Design Engineering, Imperial College London, UK.
[†]Equal contributions.

E-mail: [*]smh_sadati@kcl.ac.uk

Abstract. The agile nature of physical interactions in animal and plant species has inspired many recent advances in robotics and their control frameworks. However, they still face challenges in interaction with ever-changing unconstructured world that we live in. Intelligence is one of nature's survival solutions for biological creatures to adapt to and reshape their surroundings. Our robots are no different in these remits. An important key to their survival and effectiveness in the natural world is the concept of Artificial Intelligence (AI). In this chapter we provide a brief overview of the rapidly emerging Soft Robotics research community around AI relevant concepts such as Embodied intelligence and Morphological Computation. More specifically, we focused on the importance of setting such "*community goals*" to create a diverse interdisciplinary research environment, an enormously important element to keep up with our rapidly progressing world. To this end, we focused on the collaborations withing and between the communities, impacts of the recently established IEEE Soft Robotics Technical Committee on coordinating these efforts, and most important of all, the key ideas and perspectives based on 200 interviews with researchers across different fields.

1. Introduction

Animals can solve realtime computational problems to do with perception and action in a natural world that robots find it hard to solve. Realtime computation means solving computational problems to come up with good enough answers to meet deadlines imposed by the environment. i.e. If a mountain goat slips on a steep cliff, it has to brake before the momentum builds too much. The problem of braking before it is too late is imposed by the environment. The solution clearly cannot come from the nervous system alone due to its slow communication pathways. Then an embodied solution to use passive dynamics is needed [1]. Similar problems are faced by humans performing object examination such as checking if a fruit is ripe or not or a steak is well cooked or not using haptic feedback. They have to find a finger stiffness and movement

IOP Conf. Series: Materials Science and Engineering 1261 (2022) 012005 doi:10.1088/1757-899X/1261/1/012005

control strategy to perceive before it is too late to take the correct decision [2].

There is an increasing interest among the robotics community to use biological inspirations to exhibit efficient, robust, and compliant behavior [3]. In many notable bioinspired robotic examples [4], there is no clear boundary between biomechanical and sensor based control algorithms [5]. In other words, the "controller" and "controlled" are not explicitly detached from one another [6]. For instance, [7] addresses this integrated/embodied approach by use of "templates and anchors" where some of the control task is delegated to passive dynamics of the physical body. Related examples such as the RHex robot [8], the pure passive dynamic model of a goat's hoof with no sensors or actuators for slip reduction [1], and various models of passive dynamic walkers [9] suggest that the notion of control can take many forms that includes benefiting from passive dynamics of physical systems.

In other words, the brain and the body should work together to overcome challenges of our unstructured world to achieve extremely robust, energy efficient, and highly adaptive robots similar to biological systems [10]. This hints interesting paradigms for

- *Embodied intelligence* (EI), "investigating tight coupling between an agent body and brain"[11],
- *Physical intelligence* (PI), "physically encoding sensing, actuation, control, memory, logic, computation, adaptation, learning and decision-making into the body of an agent" [11], and
- *Morphological Computation* (MC), "outsourcing part of control, sensing and computation tasks to the robot body to be handled through emergence of functional behaviours due to properties such as architectured kinematics, compliance, natural resonance, damping and friction" [10].

The ultimate goal is to identify these elements in nature, both the abstract concepts and means of utilization, and to transfer similar level of perception and intelligence to man-made agents for intelligent autonomous operation in real world. More specifically, an agent that benefits from embodied intelligence for survival in a natural environment. Such bodies can be pre-designed, initiate in a general state to adapt to the task and environment, or even grow to a final robust shape while adapting to their environment, as for the living creatures. These are in contrast with the traditional robotics that the robot embodiment and natural behaviour are part of the problem to be overcome by high energy actuation and computationally expensive control units [10].

Research communities are the core to such progress and the great importance of EI has been well captured by the emerging research field of *Soft Robotics* (SoRo). In this chapter, we focus on the importance of setting "*community goals*", such as EI, MC, and SoRo. Such a community goal can put together researchers from different fields, generations, and geological regions, that are enormously important in our rapidly progressing world.

To this end, we present an overview of the research collaborations in the EI and MC communities based on co-authorship between these researchers in publications with these keywords. Then, we elaborate on one of the most important coordinated efforts to grow these research fields, the formation of IEEE Robotics and Automation Society (RAS) Technical Committee (TC) on Soft Robotics, and its achievements. Finally, a collection of ideas and opinions, some have never appeared in a publication before, are presented based on about 200 interviews with key researchers in the Soft Robotics community. We conclude this chapter by reflecting on opportunities and ideas for further collaborations in the community.

2. Communities & Collaborations: A Brainstorm Map

It is reasonable to assume that the research communities around EI and MC topics are highly inter-connected, given the similarities of concepts, their goals, and current target application.

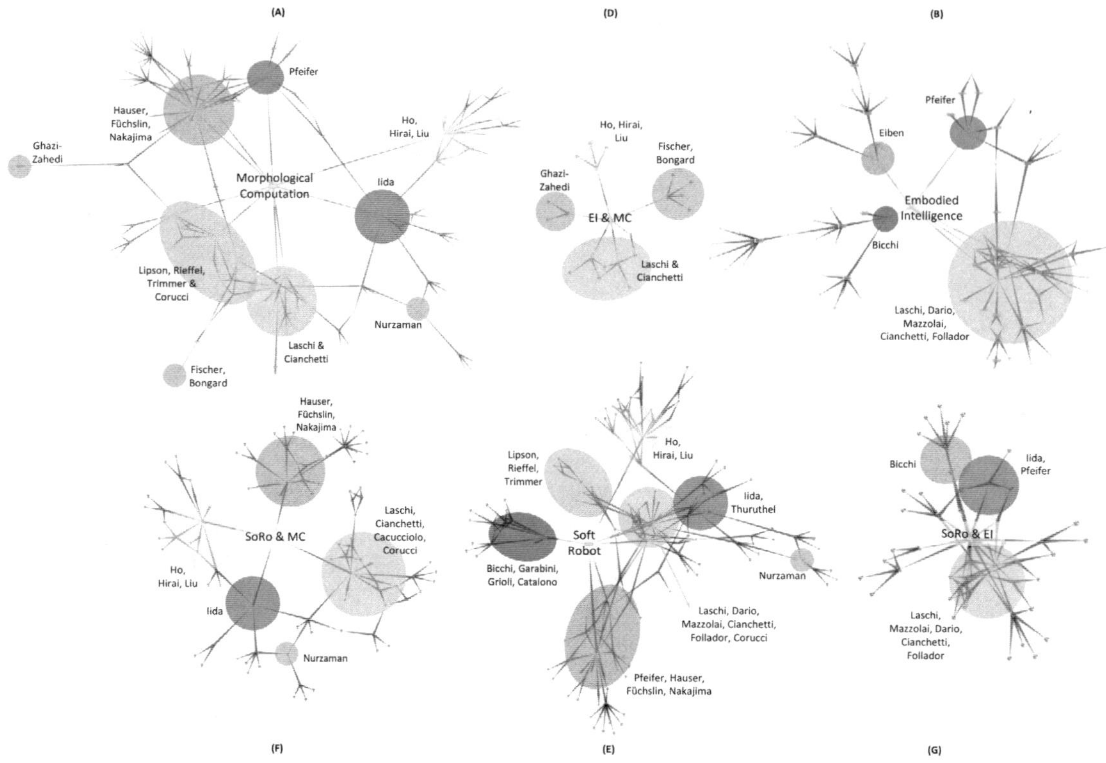

Figure 1. Brainstorm map of the researchers and their collaborations in publications with the keywords MC (A), EI (B), and both MC & EI (C) in the manuscripts' title, abstract, and list of keywords. Nodes are either researchers or publications linking the collaborators. The documents are extracted by searching the exact keywords in *www.scopus.com*. The same set of manuscripts are analysed based on SoRo (D), both SoRo & MC (E), and both SoRo & EI (F) keywords to highlight the contributions from Soft Robotics community to EI and MC research. The analysis and brainstorm map are done by *Qiqqa* software. The plots are ordered in a way that makes the comparison between the main topics and overlapping research easier.

However, that is not exactly the case. We conducted a statistical survey to better understand the collaborations formed around these topics.

A quick search of the exact terms, *Embodied Intelligence* and *Morphological Computation*, in the title, abstract, and keywords fields of accessible manuscripts through *www.scopous.com* results in 123 publications related to MC and 90 publications related to EI. The 36% higher number of publications referring to MC shows the rapid growth of the research around the topic compared to its slightly older companion topic, EI. MC owes it to its grasp within the Soft Robotics community and the communities rapid growth over the past decade. The list of top five highly cited authors are different between the two with only 6 publications with reference to both the terms in our search. hence, it is fair to refer to two to some extent distinct communities around the EI and MC topics. The more frequent appearance of the term *Artificial Intelligence* in combination with EI (28 manuscripts) and MC (14 manuscripts) terms shows wider acceptance in the two the communities.

We used Qiqqa software to analyse the collaborations inside and between the two communities based on the manuscripts' co-authorship. Fig. 1 shows the brainstorm map of the relevant publications to each topic (central node). The peripheral nodes are either the researchers or

IOP Conf. Series: Materials Science and Engineering 1261 (2022) 012005 doi:10.1088/1757-899X/1261/1/012005

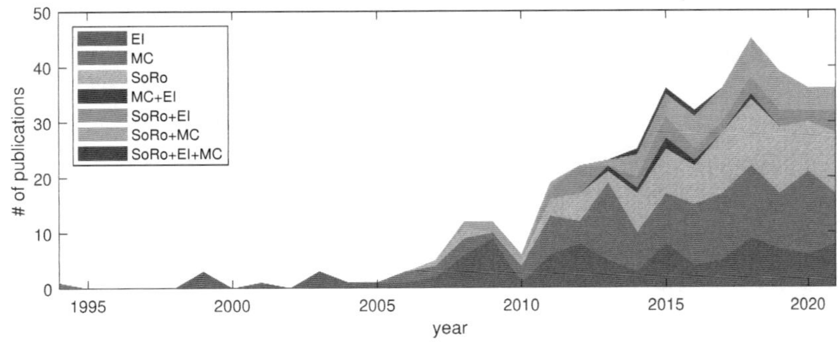

Figure 2. Histogram of the number of manuscripts containing each keyword or their combination in 1994-October 2021.

their manuscripts bridging them to the other researchers. The more interconnected nodes show the larger collaborations and closer partnerships between the researchers. The authors with five or more publications are directly connected to the central node as the most influential researchers in each group of interconnected nodes.

Fig. 1-A presents the manuscripts referring to *Morphological Computation.* Apart from R. Pfeifer who is one of the funding fathers of the field, five other main (coloured) and two emerging (grey) groups of researchers can be identified. The Wide reach of collaborations around F. Iida and the highly interconnected partnerships between the research teams in Italy (C. Laschi and M. Cianchetti) and in United States (H. Lipson, J. Rieffel, and B. Trimmer) are the highlights of this graph. It is worth mentioning that this is a map the interconnected part of the community. There are isolated research groups with multiple relevant publications that are not depicted here, for example the teams around T. Nanayakkara, J. Dambre, A. Bicchi, M. Garabini, L. Magnani, G. Dodig-Crnkovic, A. Adamatzky, K. Hosoda, P. Manoonpong, V. Muller, and P. Oudeyer.

Fig. 1-B presents a less inter-connected community around the EI topic. Four research groups are identifiable, two of which (A. Eiben and A. Bicchi) are not connected to the rest of the community. R. Pfeifer is a dominant figure in the EI community too. A strong research community is present in Italy and Switzerland. Despite the MC community, the EI community shows less activity in Asia and America.

Fig. 1-C analyses the 6 publications with reference to both the keywords in our search. The groups around C. Laschi and M. Cianchetti is the dominant figure in bringing the two concepts together, while most of the other main research researchers in the both communities (e.g. R. Pfeifer) avoided referring to both the topics simultaneously , However, a reference to both the terms is more common among the emerging nodes in the MC community (K. Ghazi-zahedi, J. Bongard, and M. Fischer).

A closer look at the contributions by the Soft Robotics community to these concepts helps with identifying the transitions from EI and MC to Soft Robotics, and how the trends have shifted from one concept after another. Fig. 1-D-F presents the interconnected part of the brainstorm map for the SoRo sub-community of MC and EI research (referenced in 91 manuscripts) and its overlap with EI (29 manuscripts) and MC (48 manuscripts) keywords. The SoRo map is more inter-connected and diverse with a strong foundation in Italian research centers. The overlap maps are very similar to the maps for MC and EI. There are only three not very recent manuscripts (2014-16) with reference to all the three keywords mostly by C. Laschi, M. Cianchetti, and V. Cacucciolo that may indicate the diverse interest of the new emerging Soft Robotics research teams when it comes to choosing between MC and EI concepts.

Finally, Fig. 2 presents a histogram for the number of manuscripts containing each keyword or their combination since the first publication on MC in 1994 by K. kamejima et al. [12] till Oct 2021. Apart from the early works in the computer science community [12], MC is a more recent concept which experienced faster growth since 2010. The correlation between the growth in MC, SoRo, and SoRo+MC compared to SoRo+EI shows the significant role of the Soft Robotics community in this growth. As mentioned earlier, there have been some efforts to reflect on the similarity of these concepts by referring to all the three terms (SoRo+EI+MC) in 2014-16. This chapter is a similar effort after 5 years to further promote such collaborations.

Once more, it is worth mentioning that these plots are based on the quick searches of the exact terms with a limited scope to a manuscript title, abstract, and keywords. Furthermore, our analysis only covers the SoRo sub-category of our EI and MC database, and does not reflect on the entire Soft Robotics community. The graphs would take a different shape if the entire content of a manuscript was analysed for similar terms (e.g. *Physical Intelligence*) around the EI and MC concepts. As a result, the provided analysis may not perfectly reflect the real impact and nature of the collaborations within and between the communities, but serves as a source of inspiration for promoting such collaborations.

3. Coordination Efforts: The IEEE RAS Soft Robotics Technical Committee

Embodied intelligence is fundamental for robotics research, and its great importance has been considered in an emerging field of robotics known as soft robotics. One of the most important coordinated efforts to grow the field is probably the formation of IEEE Robotics and Automation Society (RAS) Technical Committee (TC) on Soft Robotics, which was established in October 2012. Starting with approximately fifty members registered in its communication channel, the TC already had almost a thousand members in 2020. In order to push soft robotics field forward, since May 2013, the TC has been consistently published bimonthly newsletters to highlight scientific journal articles, important news and relevant popular articles on its official website (softrobotics.org/newsletters/), with its 48th edition published in March 2021. Due to the significantly increasing numbers of articles, the newsletter started to be published trimonthly in the 49th edition (April – June 2021 edition). The newsletter covers topics such as soft grasping and manipulation, soft locomotion, soft sensing, soft actuators, soft robot fabrication, embodied intelligence, and stretchable electronics. Up until the 49th edition, the newsletter has featured 1044 scientific articles and 387 popular articles related to soft robotics, excluding a large number of featured videos and news like conference announcement or even relevant vacancies. In 2013, the newsletter featured 64 scientific and six popular articles, while in 2020 it featured 189 scientific and 49 popular articles, demonstrating a significant increase of community size and the popularity of the field. Aside from the increasing number of feature articles, there are other interesting statistics that we can also glean from the newsletters. The first is the increasing number of 'categories' that are used to group the articles. In the first edition, for example, the scientific articles could be easily grouped into categories like sensing, actuation or manipulation. Along with the growth of the community, the featured articles become more diverse and therefore new categories like learning, modelling, control and state estimation were introduced. Another interesting statistic is the significant increase of the number of popular new articles. In the first year of the newsletter publication in 2013, there were merely six featured popular articles. In 2020, the newsletter featured 49 popular articles. As there were six editions of the bimonthly newsletter each year, it means an average of more than eight article per edition, exceeding the number of popular articles featured in the whole year of first year of the newsletter publication in 2013. Embodied intelligence is probably even more fundamentals than soft robotics itself but certainly there must be some lessons that can be taken from the growth of the soft robotics community. It is probably fair enough to say that the community cannot have such significant growth without the coordination provided by the IEEE TC on Soft Robotics, with the soft

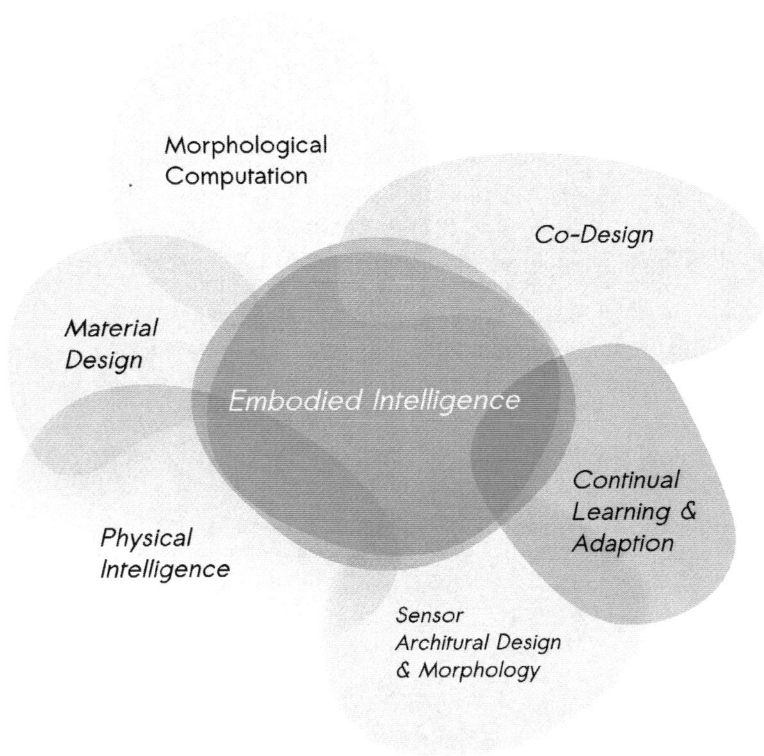

Figure 3. Key aspects of Embodied Intelligence.

robotics newsletter as one of the main driving factors.

4. Ideas & Perspectives: Reflecting on 200 Interviews

The most intriguing ideas and ground-breaking perspectives in a research field are the consequences of asking key questions. Questions such as, what is the limit of soft robots for embodied intelligence? Can soft robots be designed to adapt to environmental uncertainty e.g damage? In evolutionary terms, diverse creatures are evolved to adapt to their environment, and it seems impossible to evolve everything that could reduce fitness, which explains why humans do not have feathers.

How evolution can come up with these designs and put these materials together? There is a harmonic interplay between the brain, body, and environment. This interplay is not settled yet for example which one is more significant and whether intelligence can be manifested in the body without having a brain? Looking back to evolution, It seems the body-mind problem "embodied intelligence" according to Fumiya Iida is a hard question and a debate that is still not settled. It is mystical that humans start as sperm and evolve to the structure we have as a human. There is another deeper aspect that we still struggle to understand, what are soul, gut feelings, and intuition? All these concepts are mystical. According to Gusz Eiben "It is magical going from no-life to life, from no-intelligence to intelligence.

The question is do we need to figure out that in soft robotics? Short answer, maybe and maybe, not all creatures have a brain, for example, plants exhibit intelligence through their bodies which have known physical intelligence. Since soft robotics is based on soft materials, what kind of materials should be pushed forward to exhibit embodied intelligence? e.g. physical intelligence, effective sensing.

That leads us to the question: what is missing for an effective soft robot design? Is that developing new materials? Do we need this level of intelligence? According to Herbert Shea's thoughts regarding this point " What does it take for an artificial octopus tentacle to solve Rubik cubes, it would take several aspects that we do not master yet, like huge and efficient interaction among embedded sensing capabilities, actuating high forces to twist the Rubik's cube, local computation at tentacles tips, and main central controller to control the preferred tentacles and more unveiled question that we do not yet understand". How we can bring all these together is still challenging and it could happen when we have a defined and deep understanding of soft systems".

To understand the interplay between the brain and body in soft robotics, according to Jennifer Mather, there is the beautiful and mysterious interplay between the Octopus brain and its body "The octopus has eight flexible arms, with a huge neurological loads coordination which is necessary, Octopus are not conscious about the details of their arm, but it is believed that each arm has mini-brains which is considered lower-level subroutine. Interestingly, the octopus model is quite different from humans, we have a highly centralized controller, in contrast, they have a more distributed controller"

Designing soft robots that can adapt to damage is crucial. An interesting example, what happens if the octopus loses one of its arms, how the controller is re-distributed? it is believed that the arms are equally potentially, each arm can do what the other arm does, but even being equally potentially they tend to be not equal, they use four arms for walking, the other four arms for exploring. We do not know about the control in the arms. Another interesting engineering aspect is octopus gait analysis which could be inspiring for soft robots design. For example, human has left, right, for the horse even more interesting, but for octopus is more sophisticated gait analysis, what are the specific gait rules, which arms they should use, it is hard to imagine, which arm come first or next, but how they organize to walk, we do not know, that could be interesting to use soft robots locomotion design to explain how octopus gait analysis work.

Another important aspect of soft robotics is sensing design and how it can be embodied in soft bodies. A fascinating example from evolution is the eagle, which is considered to have an embodied sensing with low dimensional control space, according to Steven Brunton "the eagle is such a good example of a system that expertly interacting with very complicated fields and reflecting on that, there are many things are missing when it comes to robots design". The eagle and other biological systems have incredible embodied sensing, they have feathers, and inertial guidance that goes far beyond anything we build to sense their environment efficiently. Interestingly, the decision made by an eagle to hunt, fly, and avoid obstacles is based on low-dimensional control space. It does not care about billions of degrees of freedom, certain patterns matter that could be characterized and exploited. That leads us to the question of how we can leverage physical intelligence with minimal control in soft robots design which still needs more exploration.

Fig. 3 summarizes the more frequent referred to aspects of EI in our interviews. What could be still missing in soft robotics is how to co-design. According to Oliver Brock" how we can bring two paradigms: digital computation (e.g Turing machines, optimization) and morphological computation together to achieve clever designs?" Co-design is still not well established for complex problems e.g. how we bring different classes of materials to achieve interesting functionalities like toughness or leveraging physical intelligence with minimal control. An initial step towards cracking how to co-design is intuitive design with the ultimate goal of an adaptive morphology by growth. The way of how to co-design would be a new chapter in robotics

5. Conclusion

Biological examples show that adaptations spanning across neural and physical structures are important for realtime actions to maximise rewards. *Embodied intelligence*, *Physical intelligence*, *Morphological Computation*, and *Soft Robotics* concepts have paved the road toward identifying the abstract forms behind such adaptations. In this chapter, we highlighted the importance of setting such community goals in establishing a diverse and inclusive interdisciplinary research environment, which are extremely important in the rapidly progressing world that we live in.

6. Acknowledgment

SMH Sadati is supported by core funding from the Wellcome/EPSRC Centre for Medical Engineering, Wellcome Trust [WT203148/Z/16/Z], and an NIHR Cardiovascular MIC Grant.

7. References

[1] S.-A. Abad, N. Herzig, S. M. H. Sadati, and T. Nanayakkara, "Significance of the compliance of the joints on the dynamic slip resistance of a bioinspired hoof," *IEEE Transactions on Robotics*, vol. 35, no. 6, pp. 1450–1463, 2019.

[2] N. Sornkarn and T. Nanayakkara, "Can a soft robotic probe use stiffness control like a human finger to improve efficacy of haptic perception?" *IEEE transactions on haptics*, vol. 10, no. 2, pp. 183–195, 2017.

[3] D. Renjewski, A. Spröwitz, A. Peekema, M. Jones, and J. Hurst, "Exciting engineered passive dynamics in a bipedal robot," *IEEE Transactions on Robotics*, vol. 31, no. 5, pp. 1244–1251, 2015.

[4] D. Polani, "An informational perspective on how the embodiment can relieve cognitive burden," in *Artificial Life (ALIFE), 2011 IEEE Symposium on.* IEEE, 2011, pp. 78–85.

[5] R. Pfeifer, M. Lungarella, and F. Iida, "Self-organization, embodiment, and biologically inspired robotics," *science*, vol. 318, no. 5853, pp. 1088–1093, 2007.

[6] M. Hoffmann and V. C. Müller, "Trade-offs in exploiting body morphology for control: from simple bodies and model-based control to complex bodies with model-free distributed control schemes," *arXiv preprint arXiv:1411.2276*, 2014.

[7] R. J. Full and D. E. Koditschek, "Templates and anchors: neuromechanical hypotheses of legged locomotion on land," *Journal of experimental biology*, vol. 202, no. 23, pp. 3325–3332, 1999.

[8] J. C. Spagna, D. I. Goldman, P.-C. Lin, D. E. Koditschek, and R. J. Full, "Distributed mechanical feedback in arthropods and robots simplifies control of rapid running on challenging terrain," *Bioinspiration & biomimetics*, vol. 2, no. 1, p. 9, 2007.

[9] S. Collins, A. Ruina, R. Tedrake, and M. Wisse, "Efficient bipedal robots based on passive-dynamic walkers," *Science*, vol. 307, no. 5712, pp. 1082–1085, 2005.

[10] K. Ghazi-Zahedi, J. Rieffel, S. Schmitt, and H. Hauser, "Editorial: Recent Trends in Morphological Computation," *Frontiers in Robotics and AI*, vol. 8, p. 708206, May 2021. [Online]. Available: https://www.frontiersin.org/articles/10.3389/frobt.2021.708206/full

[11] M. Sitti, "Physical intelligence as a new paradigm," *Extreme Mechanics Letters*, vol. 46, p. 101340, Jul. 2021. [Online]. Available: https://www.sciencedirect.com/science/article/pii/S2352431621001012

[12] K. Kamejima, T. Aoki, and Y. C. Watanabe, "Existence and Implications of Fixed Points in Morphological Computation Processes," *Proceedings of the ISCIE International Symposium on Stochastic Systems Theory and its Applications*, vol. 1994, pp. 129–137, 1994.

Adopting Physical Artificial Intelligence in Soft Aerial Robots

Pham Huy Nguyen,[1] **Mirko Kovač**[1,2*]

[1]Aerial Robotics Laboratory, Department of Aeronautics, Imperial College London, South Kensington, London, UK

[2]Materials and Technology Centre of Robotics, Department for Functional Materials, Swiss Federal Laboratories of Materials Science and Technology (EMPA), Dübendorf, Switzerland.

E-mail: n.pham@imperial.ac.uk, m.kovac@imperial.ac.uk*

Abstract. In recent years, there has been a lot more attention towards the utilization of physically intelligent features in robotics. In this work, we provide a perspective on the physical artificial intelligence (PAI) paradigm and its impact on the conceptualization, design, and manufacturing of current and future aerial robots and infrastructure. We highlight the theory, enabling technologies, system features, and the tasks that the PAI paradigm will improve beyond the current approaches with conventional rigid aerial robots. We also discuss the multi-disciplinary effort required to collaborate with and educate researchers in the development of physically intelligent robots. PAI promises to lead the development of a new era of robust flying robotic organisms that are capable of adapting to and performing multi-functional tasks autonomously in a complex and unstructured environment. Aerial robotics is a great field of study to validate PAI as a development methodology.

1. Introduction

In recent years, there has been a significant multidisciplinary effort towards developing highly intelligent robots akin to biological organisms [1–7]. This new dawn towards the next generation of "robotic organisms" reinforces the need for robots with capabilities to autonomously adapt to the wild and unstructured environment [1, 8]. Thus, highlighting capabilities such as physical reconfigurability, robustness, multimodality and multifunctionality, through the utilization of smart structures. In addition, the potential to self-repair, self-grow, self-respond to internal sensors and external sensors, self-power, self-reconfigure, maintain homeostasis, and safely biodegrade at its end-of-life condition, are all interesting features that will further push bio-inspired robots into the future.

Although the "brain" of robots, the computational intelligence, has been in maturing over the last century [9, 10], gaps still remain in the development of the intelligence encoded in the physical body of robots, also known as physical artificial intelligence (PAI) [1]. A route to developing these physically intelligent robots is by first studying biology and then developing these physical agents through the co-evolution of the body's material make up, the body's morphology, its embedded actuation and sensing scheme, and its underlying control scheme [4]. By utilizing this PAI paradigm, research fields such as material science, biology, computer science, chemistry and mechatronics can be interconnected to evolve human-made autonomous systems towards becoming comparable to the biological systems observed in nature (Fig. 1).

In the context of intelligence embedded into the body of robots and their biological counterparts, there have been various paradigms highlighted such as morphological intelligence, morphological computation, embodied intelligence, and more recently physical intelligence and PAI [1–3, 11–14], as shown in Fig. 1a. In its essence, PAI is a synthesis method that enables the incorporation of PI features in "robotic" organisms, through the study of biological organisms and their functions. PAI tries to emphasize the design of the systems starting from considerations of its materials, body structure, self-perception and actuation, and embedded control, through interdisciplinary processes. Although each paradigm underlines various similar principles, there are slight differences between them. For example, PAI differs from morphological intelligence in that it involves the encoding of embedded decision-making, learning, and computation schemes. This is exemplified by the tentacles of some cephalopods, where there are neuron clusters distributed locally among its different tentacles, allowing them to make their own decisions without requiring direct input from the central nervous system [15, 16]. The PAI paradigm also slightly differentiates from morphological computation (MC) as it is independent of evolutionary body varying morphological processes. Grouped within embodied intelligence (EI), MC is often highlighted as a way to offload computation from the brain to the body [17]. PAI emphasizes the integration of biological, chemical, and material-related functionalities, while EI focuses on the interconnection between the body, the brain, and the environment [11, 14]. PAI also has an emphasis in the development of biohybrid systems and their processes. Thus, PAI couples closely with digital artificial intelligence and any form of computational intelligence which are related through EI. PAI also utilizes digital AI and other digital processes as a tool to co-evolve systems from the ground up. Finally, the PAI design paradigm enables the creation of physically intelligent (PI) features, in "robotic" organisms. Therefore, these different paradigms are strategically different yet complements each other in the effort to holistically advance robotics.

In this paper, we seek to provide insights into how various PAI methodologies can be utilized in the development of soft aerial robots, outlining instances where inspiration from biology is utilized to assist the development of these next generation systems. Figure 1b highlights the various system features, enabling technologies, and theories that are currently being researched as part of the PAI paradigm. Thus far, the progress in the adoption of PAI capabilities in aerial robots has been limited. Hence, there are ample opportunities and urgent needs for more advanced PAI implementations to create intelligent and adaptable aerial robots that can perform multi-capabilities, cross multi-terrains, and perform tasks autonomously in real-world conditions.

We analyze current and future PAI utilization in aerial robot related applications, such as: (1) improving maneuverability and flight efficiency (2) multi-modal mobility across terrain interfaces and fluid boundaries (3) robustness to landing and collision (4) manipulation, perching, resting, and energy management in complicated environments (5) bio-inspired aerial construction and nesting (6) bio-hybrid flying structures, biodegradability (8) self-regeneration and self-healing in aerial robot.

2. Nature inspires the adoption of the PAI paradigm

Currently, classical aerial robots broadly utilize rigid body structures, such as carbon fibre frames. Along with on-board sensing and dynamic loop control, rigid aerial robots have a defined, precisely controlled, and stable flight envelope. This makes their flights safe to operate and predict. However, when compared to their biological flying counterparts, rigid aerial robots still struggle to match their performance parameters. As illustrated in Fig. 1c, it is noted that flying animals, which have physically intelligent bodies, are able to achieve better mechanical robustness, longer flight times, better agility and maneuverability. Compounding the advantages of these animals are their capability to switch effortlessly between gliding and dynamic flying,

IOP Conf. Series: Materials Science and Engineering 1261 (2022) 012006 doi:10.1088/1757-899X/1261/1/012006

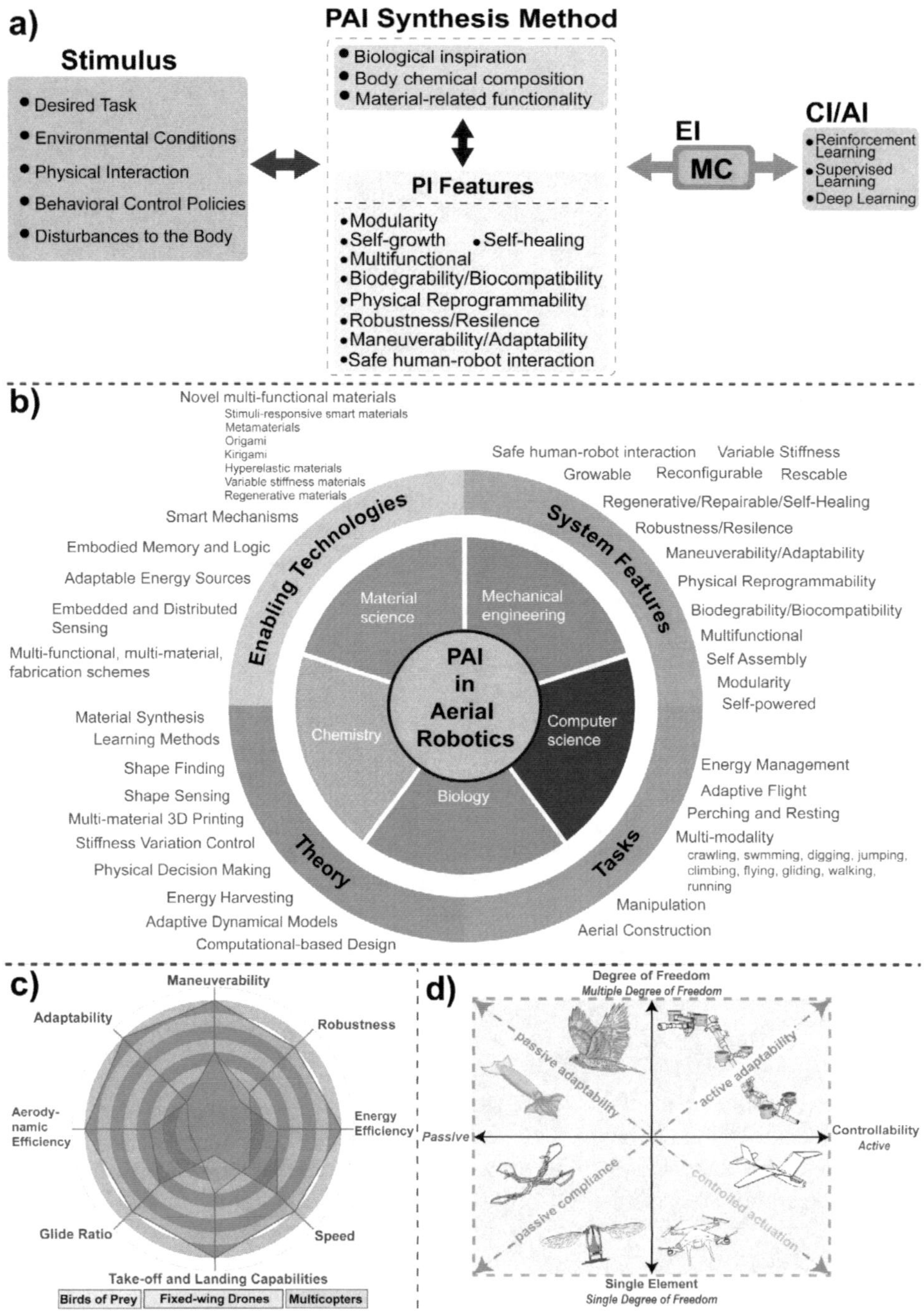

Figure 1. a) The relationship between the various intelligence paradigms. The PAI synthesis method enables the incorporation of embedded PI features in robotic organisms. b) PAI highlights various enabling technologies and theories that pushes the boundaries of the development of different systems features for soft aerial robots. c) Spider chart comparing aerial performances of bird of preys, multicopters, and fixed-winged drones. d) Design space of current aerial robots based on two key parameters: degree-of-freedom and controllability.

and being able to manipulate, land safely, and perch/de-perch. Birds, for example, utilize their morphological features and a neuroplastic-based control scheme, with redundant sensing and actuation, to exploit dynamic and biomechanical effects to lead to higher aerodynamic performances [3, 18].

The implementation of PAI within novel aerial robotic systems first involves examining the drone design space based on two main axes, the degree-of-freedom (DOF) of the system and its system controllability, as seen in Fig. 1d. This highlights design trade-offs seen in the directional arrows of each quadrant. In the bottom-right quadrant, aerial robots with a limited DOF and requires high precision control is representative of classical rigid multicopters or fixed-wing robots. In the top right quadrant, with an increasing number of DOFs that require high precision control, the design space shifts to highly adaptable, active, reconfigurable and foldable multi-rotors [19, 20]. On the other end of the spectrum, the bottom-right quadrant, a lower DOF and less requirement for active control, leads to design of drones with PAI embodied adaptable bodies, such as soft bodied drones and passive foldable multi-rotors [3, 21–23]. In the top left quadrant, aerial vehicles with multiple DOFs, utilizing redundant sensing and actuation as well as minimalistic neuroplastic-based control schemes, push towards the realm of biological flying animals and bio-inspired drones akin to their biological counterparts. Thus, in the left quadrants, PAI methodologies play a significant role in potentially reducing control and computation complexities, while increasing versatility, performance, and robustness and soft aerial robots.

Aerial animals highlight how they adapt their morphologies in order to extend their dynamic envelope, either to match the task they are trying to perform or to match the object or environment they are trying to interact with. The inspire-abstract-implement paradigm (IAI paradigm) [4], highlights how nature constantly inspires the implementation of current aerial robots that abstracts PAI features, from studying biomechanical or ecological features of biological animals. Thus, the IAI paradigm elucidates the fundamental knowledge in the functionalization of biological organisms, leading to successful implementations of highly adaptable aerial robots.

In order to adapt to multiple tasks and the ever changing environment, aerial robots and their biological counterparts need to adapt their body, utilizing a PAI method called metamorphosis. Metamorphosis here is derived from the evolutionary development method, where the biological agent innately changes its morphology and form on demand, adapting to the desired task, environment, or other stimulus [24]. In biology, metamorphosis is a transformation process in an organisms' physical body during its growth life cycle to adapt to different ecological habitats [25, 26]. This stems from the need to improve the survival rate of the specified species, so that the larval and adult stage of the species operate in different niches. This way, there is no necessity to compete for the same food sources [27]. Metamorphosis examples are often seen in insects such as grasshoppers, beetles, and butterflies that grow from a nymph or caterpillar stage, to then acquire the ability to fly in its final metamorphosis stage [28]. Metamorphosis is also prominent in some amphibians such as frogs and salamanders that transition from operating in aquatic to terrestrial environments, post metamorphosis [29].

Metamorphosis plays a key role in the development of multifunctional robotic designs that can perform different tasks and expand their operational range in different environments, be it aerial, terrestrial, or aquatic. Metamorphosis in robotics has also fallen under various research titles such as self-reconfigurable, cellular, foldable, and morphable robotics [30–32]. As we will see throughout this paper, metamorphosis augments the functionalities of soft aerial platforms. From the capabilities of adapting their size to go through apertures, or to absorb impact, or to manipulate objects [3, 19, 33]. Metamorphosis is also highlighted in multi-modal transitions, from aerial to terrestrial environments [34, 35] and aerial to aquatic environments [36, 37].

3. PAI Adaptation in Flight

Wings have been shaped from millions of years of natural selection for various flight animals and insects from their hollow bones, membrane-like wings, to feathers, to longer forelimbs [18, 50]. The evolution of wings have led to aerodynamic adaptations in varying atmospheric conditions for efficient flight and for several complex maneuvers such as diving and perching. The study of wings and wing membranes in gliding and flying animals and insects continues to be a source of inspiration for the development of aerial robots, as shown in Fig. 2.

Insect-inspired robots have been well studied within bio-inspired aerial robotics [51–53], with a trend towards the design of small-scale micro-flyers that utilize high power-dense soft actuators [54], as seen in Fig. 2d. However, insect-inspired robots still face challenges such as limited flight time and payload capacities due to their size [53]. By scaling up, membrane-based flyers and birds inspired aerial robots both open the opportunity for concepts that can carry more significant payloads in comparison to microflyers [50, 55], as shown in Fig. 2e and f.

Currently, several PAI adaptations for aerial robots have been towards the design of impact and collision resilient system [21, 56–58], morphological changes to extend of the systems flight envelope [18], and the extended capabilities for multi-tasking and operating in multiple environments [3, 33]. By utilizing the IAI concept, the function and morphological differences of the muscles and the skeletal system of the organisms would be studied, identifying their role in accomplishing flapping flight. This then leads to abstracting other PAI methods in flight animals, such as how reislin operates in insect wings to enable wing elasticity and folding [59] (highlighted in Fig. 2a with the earwig insect), how bird feathers operate to manipulate aerodynamic effects [39, 60] (highlighted in Fig. 2c with the hawk), or how membranes enable robust and efficient flight in membrane-based flyers [55] (highlighted in Fig. 2b with the grey headed flying fox). The IAI concept also studies the behavioural and ecological adaptations of flight animals, to further understand how flight animals interact with their food sources, preys, predators, and unstructured environment [4]. This thorough study of the physical and behavioral mechanisms of flying animals will provide novel insights to enhance the flight capabilities and adaptability of modern day flying robots.

There is another aspect to understanding how aerial robots measure up and compare to biological inspirations. The study of the muscular skeletal structures of animals towards flight is currently lacking. Specific muscular function during flight from in vivo tests would provide interesting insights in the development of bioinspired systems. Currently, membranes on gliding animals and bird feathers yield an amazing level of robustness in comparison to engineered wings. The adaptations enable a diverse range of kinematics to extend the systems' flight envelope, like speed and maneuverability. As engineers narrow the gap between animals capable of flight and flying robots, we not only develop more adaptable robots but also robots that can provide biologists with insights into further understanding animals and their full capabilities.

4. PAI Adaptation to Multi-modal Locomotion

In this section, we discuss how animals and their robotic counterparts utilize their adaptive morphologies to metamorphosize in order to complete tasks, protect themselves from collisions, and traverse and operate in multi-modalities [61]. We then discuss how the abstraction of these fundamental knowledge leads to applicable areas of PAI in order to develop more robust and versatile multi-modal aerial robots.

In multi-modal applications, some animals have the ability to adapt to different fluid mediums, using multi-functional locomotor skills. Some animals are able to walk, fly, and swim by adapting their morphology and structure [7, 33, 62]. When performing aerial to aquatic transitions, robots and animals need to take into account the fluid inertia and drag underwater. This capability is highlighted in various multi-modal animals, some seen in Fig. 2g and h, such as flying fishes (*Exocoetidae*), flying squids(*Ommastrephidae*), insects, and birds [63–66]. The

IOP Conf. Series: Materials Science and Engineering 1261 (2022) 012006 doi:10.1088/1757-899X/1261/1/012006

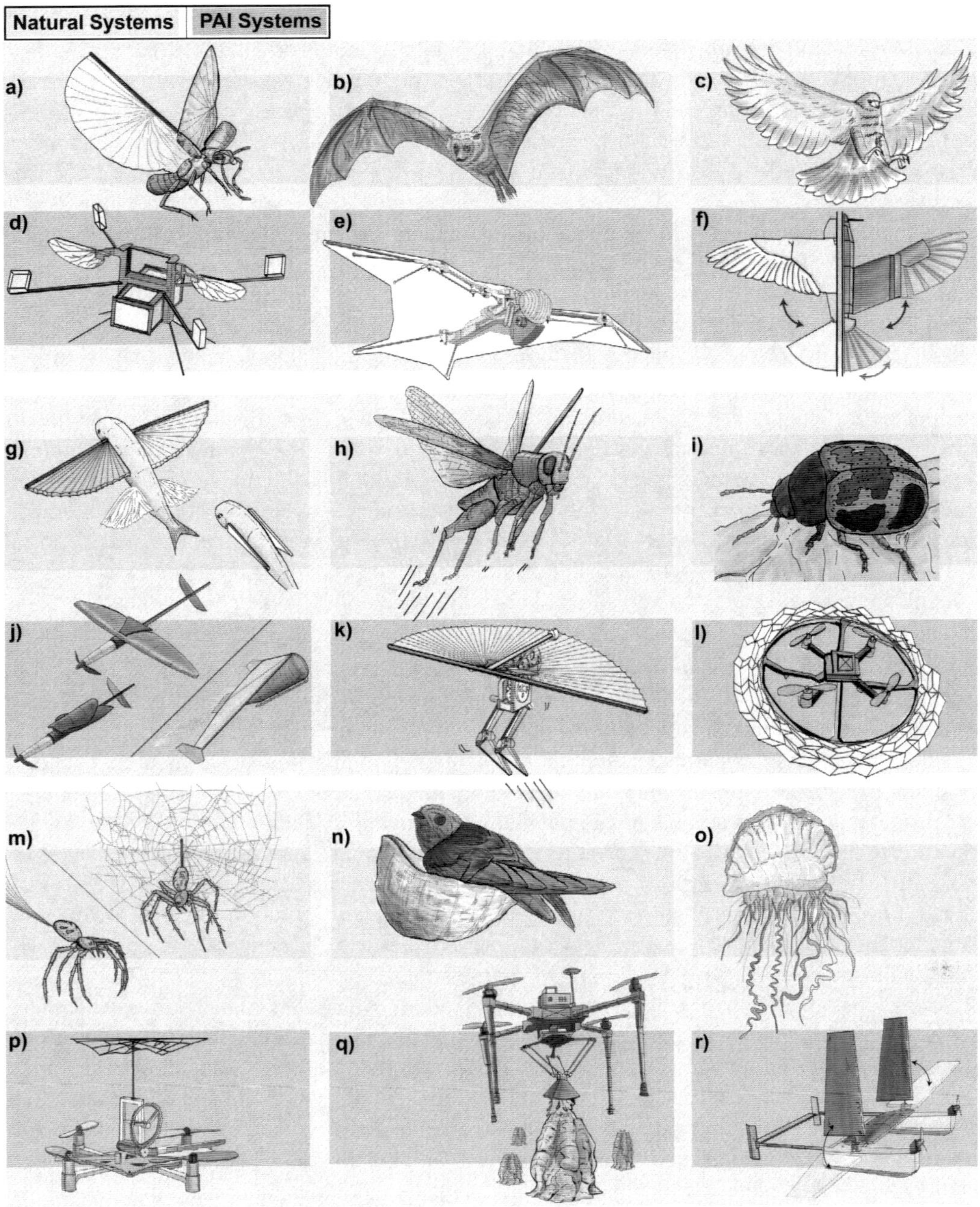

Figure 2. Illustration of natural biological systems and their PAI counterparts. a) Earwig Insect (*Eauricularia*) b) Grey Headed Flying Fox (*Pteropus poliocephalus*) c) Hawk (*Accipitridae*) d) Robobee [37] e) RoboBat [38] f) [Left] PigeonBot [39] [Right] RaptoBot [40] g) Flying Fish (*Exocoetidae*) h) Locust (*Acrididae*) i) Exoskeleton of a milkweed leaf beetle (*Labidomera clivicollis*) j) Two versions of AquaMAVs [41, 42] k) Jump-gliders [43–45] l) Rotorigami [21] m) [Left] Balloning Spider [Right] Perching and construction of spiderweb (*Erigone atra*) n) Swiflet and construction of nest (*Aerodramus fuciphagus*) o) Portguese man o' war (*Physalia physalis*) p) SpiderMAV [46] q) Aerial additive manufacturing [47, 48] r) SailMAV [49]

Aquatic Micro Air Vehicle (AquaMAV), is an example that highlights capabilities to perform aerial and aquatic locomotion, as seen in Fig. 2j. The AquaMAV utilizes PAI principles, from the flying squid [41, 42] that jets out of the water into a gliding phase by utilizing a pressurized jet of water. The AquaMAV on the other hand, utilizes the expansion of compressed gas [67] or chemical reaction with water [42] to generate that water jet to propel itself out of water. Similarly, other cephalopods are known to also utilize water jet propulsion underwater to locomote [68].

Another common multi-modal example often seen in nature is jumping-gliding or jumping-flying. Jumping-gliding is often seen in various arboreal animals such flying lizards, flying squirrels, flying fish, flying frogs, and flying snakes, pelagic flying squid and different insects [69–73]. Jumping-flying is also most often seen in insects [74, 75], as seen in Fig. 2h. Most of these animals utilize a foldable or deployable wing mechanism that is deployed after the jump, and is restored right before landing to protect their membrane or wings from damage [43–45, 76, 77], as seen in Fig. 2k. Similarly, birds like the *Guillemot Uria Aalge* are able to fly, swim on and underwater utilizing a flapping mechanism [78]. Other birds, like the gannet from *Sulidae* family, fold their wings to dive at high speed to enter water. They utilize soft sack filled structures in their face and neck to absorb impact when hitting the water. The aerial-aquatic locomotion of pelagic flying squids [64], that spread their tentacles wide after escaping the water boundary, have further been highlighted as a method to saving energy while gliding through the air, in comparison to swimming.

5. PAI Adaptation in Impact and Collision Resiliency

Aerial robotic platforms often have to handle high-impact interaction with highly cluttered environments during flight. In such a complicated environment, along with the lack of abundant sensing, dodging obstacles become increasingly difficult. This makes collisions with external disturbances inevitable. From the study of biological organisms, insect bodies are noted to absorb impact by being highly compliant [52, 79, 80]. While some utilize exoskeletons that provide external protection during collisions [79], seen in Fig. 2i, smaller insects, such as flies, rely on their compliant legs to dampen the collision impact when landing or perching [81]. Whereas in larger flying animals, elastic components in the joints generally assist with absorbing impact, but collision mitigation schemes, such as a high angle of attack stall maneuver to induce high drag, are still required [82, 83].

Collisions with an aerial robotic platform often lead to unwanted disturbances that end with crashes and/or structural damages to the platform. PAI enabled methods can play a role towards the development of highly collision resilient aerial robots, that do not require complicated onboard sensors or planning schemes. The goal would be to design PAI systems that are able to sustain collisions without resulting in loss of control or structural damage due to contact or crashes. Currently, aerial platforms either have an external cage-like structure to absorb impact [21, 22, 84–86] or embedded compliance within their bodies to absorb impact [23, 87, 88], as seen in Fig. 2l.

PAI enabled capabilities, seen in soft robotics, have emerged as interesting candidates to handle the collision of aerial systems. Some interesting work have been the utilization of deformable rotors in multi-rotors [89], compliant joints and/or drone arms [21, 38, 57, 87, 90, 91], and physically reprogrammable structures such as origami, kirigami, or tensegrity [3, 53, 92–95]. Soft manipulators, due to their compliances have also been utilized as landing gears as well as grippers [96–104]. In the future, we project other PAI enabled methods such as the utilization of active variable stiffness structures to maintain compliance when absorbing collision with the environment, but remain stiff while flying. PAI enabled collision-safe aerial robots will also outperform rigid systems during aggressive and dynamic landing or perching situations, by absorbing the impact in tandem with performing the task.

IOP Conf. Series: Materials Science and Engineering 1261 (2022) 012006 doi:10.1088/1757-899X/1261/1/012006

6. PAI Adaptation in Manipulation and Perching

Manipulation in aerial robots and their biological counterparts is one of the most attractive topics due to its multi-functionality and versatility. Manipulation in aerial-based organisms allows the carrying and delivery of payload, capturing of food sources, perching and resting, and even burrowing or digging.

6.1. Perching

Perching is often used by animals in nature to maintain a resting position, hide from errant weather, and be placed in an advantageous position for surveillance. Nature has highlighted various energy-saving perching mechanisms, often combined with coordinated flight dynamics. For example, Birds decelerate in the air through the high angle of attack maneuver placing them them in an upright position at which point they can orientate their claws on their feet towards the target to grab [105, 106]. Both birds and bats utilize a PAI enabled grasping mechanism called digital tendon locking, when perching, where no additional energy is utilized to maintain a fully grasped claw orientation [107, 108]. Bees, on the other hand, approach and hover in front of the perch before soft attaching legs to the perching surface [109]. Flies and other insects, utilize their compliant legs to dampen collision and perch at the same time [110]. Another interesting insect inspiration is seen with spiders, that utilize their silk threads as tensile anchoring elements, so they can attach on different perches and also create nests [46, 111], as shown in Fig. 2m. Beyond just animals, plant seeds utilize various bristles and microspines to hitchhike on host creatures for pollination purposes [112].

Perching aerial robots have utilized various bio-inspired mechanisms as well. From the use of dry adhesion [113, 114], electrostatic adhesion [115], microspines [116–118], suction cups [119], threads as anchoring points [46, 120], and claw-like graspers as well [102–104, 121–125]. Figure 2p highlights our work on tensile based perching, as seen in spiders [46].

6.2. Aerial Manipulators

Beyond perching, other desired aerial manipulation tasks include handling, sampling, transporting, and interacting with the environment [126–128]. Currently, this field is dominated by the use of rigid robot arms, single-DOF [129], multi-jointed [130, 131], or even multi-armed systems [132] that hang from the base of these multi-rotor systems. For these systems, questions still remain about their operational capabilities in cluttered and unstructured environments. They also generally face limitations due to their rigid designs, such as their weight, bulk, and lack of adaptability, compliance, and maneuverability. Here, PAI developed methods, such as soft manipulators, can emerge as one of the low-cost solutions to tackle these aforementioned challenges [133, 134]. Although soft actuators and grippers have been utilized on drones [101, 103, 104, 135], there are opportunities to explore the utilization of full-length, multi-DOF, soft manipulators integrated within drones as a flight-critical subsystem. Currently, we have only seen an origami based soft aerial manipulator modules realized in aerial robots [96]. In the future, the introduction of soft continuum manipulators [136] and soft growing manipulators [137, 138], would highlight lightweight, highly adaptable and robust manipulators that are operable in multi-mediums.

Another interesting form of manipulation is the utilization of an active tail that would assist with aerial stability or maneuverability [139], as well as other tasks like multi-modal dynamic perching, climbing, and jumping [140].

7. PAI Adaptation in Construction and Nesting

In nature, construction has always played a key role in the survival of an individual, their offspring, or a collective. Construction of nests, traps, or underground structures is highlighted in insects, birds, and mammals alike [141]. Construction in nature often highlights coordinated collective construction efforts of a group of animals [142], as seen in ants and termites that utilize a decentralized and stochastic construction approach for construction [143]. In this section, we focus on the PAI capabilities of the individual in constructing shelter or traps. For example, arachnids can create soft silk structures that have variable strength, elasticity, and adhesion [144], as shown in Fig. 2m. The weaver bird, *Philetairus socius*, is capable of weaving branches, twigs, and muds to create nests that last decades [145]. The *Aerodramus fuciphagus*, the swiftlet, utilizes solidified saliva, as adhesion and structure, to create these intricate nests made of amorphous materials [146], illustrated in Fig. 2n. The *Apis mellifera*, honey bees, also utilize their saliva to improve the rigidity and resistance to heat of their bee hives. On the other hand, some flight animals utilize burrowing as a subtractive method to construct shelter underground, such as the burrowing owl, (*Athene cunicularia*), the ground woodpeckers (*Geocolaptes olivaceus*), or digger bees (*Anthophorini*), for example.

A few research teams have utilized PAI inspired methods for construction with aerial robots. For example, the creation of tensile web-like structures using drones, for perching [120, 147, 148] and the utilization of 3D printing polyurethane foam for repairing gaps and creating structures during flight [47, 48], as highlighted in Fig. 2p and q. As the interest in ad-hoc construction increases, future developments will relate to aerial 3D printing and curing of smart building materials. Construction related to burrowing and digging highlights PAI methods inspired by plant roots [149] and soft bodied digging animals, like earthworms, that perform growing and digging at the same time [150, 151].

8. Discussions and Future Outlook

In this work, we highlighted the interesting interconnections between morphology, the environment, the tasks, and the brain. There is a significant interest and advancement in the embodiment of the PAI paradigm in robots that need to perform multiple tasks and operate in multiple complex environments. The PAI paradigm enhances the capabilities of the system's body to offset the need to process interaction complexities with the mind or computational intelligence. As we have highlighted, biological organisms have dominant PAI traits that help them deal with the natural environment. In even harsher extreme environments, like the bottom of the ocean, on-board computing power is further lacking and PAI traits would be the main option of intelligence. As the interest in developing more advanced PAI skills continues to rise, there are still many topics and tasks that can be evolved and developed further, especially in relation to aerial robots. The PAI paradigm will play a viable solution to encode robots with additional functions, such as adapting its body based on the energetic requirements, multi-modal hitchhiking, collision absorption and protection, fast-rapid movement utilizing elastic instabilities, embedded circuitry, logic, and memory, capabilities of shape sensing, capabilities of regeneration and self-healing, and biodegradability.

8.1. Energy Management

One of the largest bottlenecks faced by most aerial vehicles is the limited flight time. As seen in biological inspirations, birds' morphable wings are capable of adapting to hovering flight, but also able to transition to a more gliding/forward flight in order to travel long distances [70]. Another way we have seen animals improve energetic autonomy is by gliding, drifting, or sailing from one place to another [152, 153]. Animals also perch and rest as a form of energy saving method [118]. Finally, as often seen in plants [154], insects [155, 156], and other animals [157–159], hitchhiking or piggybacking is highly effective strategy in low energy locomotion. Animals and plants, by attaching themselves onto a moving object, such as adhesion onto animal fur/skin or sitting inside a mother's pouch as seen with marsupials, they can travel in multi-mediums (air, water, and land) whiles aided to minimize the use of energy. Utilizing the different PAI enabling technologies, there are various unexplored and nascent approaches that could be further developed.

8.1.1. Multi-modal perching and manipulation A multi-modal perching mechanism for a multi-functional aerial robot highlights an interesting application to improve the energy consumption of systems that would want to perch and rest or hitchhike to travel long distances without additional energy consumption, on land or in water. The ability to perch in either medium would also allow the system to perform surveillance too. As previously highlighted, bio-inspired approaches to perching, such as electroadhesives [115], dry adhesives [113, 114], vacuum suction cups [119, 160], microspines and pre-loaded spikes [102, 116, 117, 161], and even claw-like graspers [102, 121–125], work well in a on land but struggles in aquatic use cases because of the shear flow faced underwater.

Currently, the study of multi-modal biological attachment systems is lacking. Perhaps the animal that has been studied the most for its cross medium attachment capabilities has been the octopus [162], where the octopus suctorial organs have been an inspiration to many robotic attachment systems [163–165]. There have been other instances, although not well documented, where sea animals have shown capabilities of multi-modal perching, underwater and on land. Sea creatures such as limpets, leeches, clingfishes, waterfall climbing gobies, and remora fishes, highlight successful underwater perching mechanisms [166]. Even insects such as the net-wing midge larvae and diving beetle highlight the use of suction adhesion in multi-mediums [65, 167]. In gliding animals, the Sunda flying lemur, have been known to cusp their hands to form a suction cup to assist attachment to surfaces along with their claws [168]. Biologically-inspired suction adhesion shows great potential for cross-medium attachment on variable surfaces, far superior to current standard suction cups, without the need for constant activation to maintain attachment. Some of these animals, such as the waterfall climbing gobies, highlight a novel ability to move while still maintaining suction to climb [169].

Currently, avian-inspired grippers for perching are generally an add-on gripper module that increases the weight of the robot and also requires additional control. Thus, there is an aerial robotic solution here that is capable of metamorphosis from a flight to a perching mode. Thus, there would be a shared functionality between flying and perching, without an additional penalty an added gripper. This could perhaps be achieved by utilizing smart structures, such as multistable structures, metamaterials, origami, kirigami, or even tensegrity structures [2, 170].

Another interesting form of manipulation is shown with animals with multiple tentacles, like the Portugese man o' war (Physalia physalis), that have tentacles that can grow up to $30m$ into the water, as illustrated in Fig. 2o. Their active tentacles are lined with numerous venomous microscopic nematocysts to sting and paralyze prey while they pull the prey towards their digestive polyps [153]. The use of this form of grasping strategy has been called entanglement [171]. Entanglement is also highlighted in the passive PAI features of human hair [172], utilizing collective soft slender structures that randomly tangle themselves around or inside objects to create soft grasping. This grasping method would be highly adaptable to the environment or object it is interacting with, and because of its numerous strands will be capable of high strain payloads, and be adaptable in aquatic, aerial, and terrestrial environments for perching, grasping, and possibly locomotion.

Growth as a means of manipulation for search and rescue for aerial robots, through metamorphosis, is also a possibility. Growth in nature can often be related to growth in volume, length or surface area [137, 138]. Here, we label growth as highly-extensible and deformable systems that would enable manipulation, perching, or search and rescue tasks. Origami structures are highly extensible, but still capable of maintaining rigidity depending on their current state of operation, therefore they have been utilized as growing drone manipulators [137]. Another approach to highly extensible growing structures have been through the use of braiding or weaving. Based on the braiding orientation and materials, the structure can have multi-DOF and be highly compressible/extensible, yet remain lightweight [173]. An interesting approach to growing has been through tip-localized growth, now known as eversion soft systems [137]. These everting soft robots, highlight their capabilities to grow and maneuver just at the tip, seen with plant roots, cells, fungal hyphae [137]. Regarding a growing structure within an aerial robotic system, we can reimagine a multi-rotor system with arms that can extend or contract to modify its morphology during flight through clustered environments. Similarly, extended growth can be used for perching or sensor placement or even exploration. The same arms can also extend to provide landing or collision support.

8.1.2. Coupled collision mitigation for perching and landing As previously mentioned, aerial robots are prone to collisions and crashes in a cluttered environment due to inaccuracies in the onboard obstacle avoidance systems. Thus to ensure mission safety, PAI enabled structures that can mitigate collisions are needed. However, currently, researchers have taken perching and collisions are two separate problems in aerial robotics. This is because, often than not, they are complicated enough problems to solve in isolation. However, biological animals can dynamically perch by utilizing the compliance of its body and its limb upon impact [3, 81]. Thus, by studying nature, it can lead to aerial robots with compliant bodies and joints that are capable of absorbing high impact forces while performing adaptive and dynamic perching strategies [174]. Here, PAI enabled technologies like variable stiffness structures, enables aerial robots to modulate their stiffness depending on the collision scenario.

8.1.3. Gliding, Drifting, and Sailing Gliding, as previously mentioned, is a fluid-based form of flight that allows animals, plants, or insects to travel a large horizontal distance, by exchanging potential energy in altitude for kinetic energy in movement [70, 175]. Gliding has always been motivated by the need for injury prevention from falling, anti-predatory reaction, or horizontal transit to seek food or mate, or genetic seed dispersal. As previously highlighted, vertebrates such as flying squirrels, gliding frogs, mobula rays, and gliding snakes have been able to successfully navigate aerodynamically utilizing this morphology [81]. Invertebrates such as the ballooning spider, gliding ants and gliding squids, have also been seen to glide through the air [176]. With a unique approach, the ballooning spider spins small silk threads to "catch" air in order to quickly glide away [177]. Mobula rays, flying fishes, and flying squids have shown their gliding capabilities to transition between aquatic and aerial environments as well [63, 175]. Plants seeds, like the asian climb gourd (Alsomitra macrocarpa), box elder (Acer negundo, Aceraceae), western salsify (Tragopogon dubius), etc, also have morphological adaptations to glide or parachute, in various fashions, for long distances [178, 179]. The investigation of gliding and the mechanics of gliding enables the design of highly multi-modal and morphologically adaptive flyers with unique flight capabilities and improved flight efficiencies. An example is seen with vertebrate-based gliders, where their angle of attack is up to four times the angle of attack of general fixed-winged aerial robots, to assist their lift capabilities at reduced airspeeds. In addition, they arch their spines to vary the camber created by their wings to adapt to different aerodynamic conditions.

Similar to gliding, drifting and sailing are methods utilized to travel long distances without the expending of additional energy. Drifting is often found in countless tropical drift seeds and fruits made of different morphological adaptations. They are physically intelligent in the way they are able to withstand perilous journeys across the oceans, that sometimes can last up to years at sea [180, 181]. Their PAI enabled features include being buoyant while having a thick exoskeleton that is impervious to salt water penetration. The coconut is a well known drift fruit made of a hard endocarp that is covered by a thick husk [182]. Another interesting example is known as the seabean or seapod that grows into the giant legume, called liana [183]. The liana is a well-known climbing plant that often hangs on trees, up to $30m$ above the ground, in tropical rainforests [183]. The liana seapod can grow up to $2m$ long, made of small seabeans at each of its segments, as the seabeans grow, they break off the individual segments of the seapod and fall into the river or sea to begin their journey, maintaining buoyancy through the air cavity designed within the seed.

Finally, sailing is considered an improved form of drifting, where a mast and mainsail is utilized to maneuver and capture aerodynamic lifts to move forward in the ocean. Although nature boasts a few animals that have a sail-like fin, like sharks or sailfishes, these fins are not utilized for sailing. The by-the-wind sailor, also known as the velella velella, is a well known jellyfish that exhibits sailing capabilities. However, their sail is not active, so they do not have the capabilities to control locomotion using their on-board sail [152]. The Portugese man o' war (Physalia physalis) also deploys an inflatable sail called pneumatophores [153], as seen in Fig. 2o. Previously, we have presented a new class of aerial-aquatic robots, called SailMAV, that metamorphosizes its wings into sails [49] for long duration cross-medium locomotion, as highlighted in Fig. 2r. Gliding, drifting, and sailing also highlights future capabilities in environmental sensor deployment across various locations and biospheres.

8.2. Utilizing Elastic Instability

A common bottleneck in the performances of soft robotic actuators applied to aerial tasks has been their limited speed of actuation. As we look towards biology for solutions, the venus fly traps, mantis shrimp, chameleon or toad tongue, red kangaroos, etc., highlight the use of energy storage to lead to power amplification for high powered movements [184–193]. Shape shifting utilizing elastic instabilities in structures like buckling, wrinkling, folding, creasing, and snapping has been looked to as ways to enable shape change and create highly responsive soft actuation systems [194]. Examples of these structures (origami, kirigami, mechanical metamaterials, and growing structures) have been highlighted in soft robotic examples of jumping [195–197], running [198, 199], swimming [200], perching for micro-aerial vehicles [201], and manipulating [202–206]. In aerial applications, digital tendon locking seen in bats and birds have a snapping mechanism utilized during perching that does not utilize additional energy to keep hold of the grasp [187]. Some flapping-wing insects also utilize an elastic flight thorax that is capable of elastic energy storage to enable their wings to passively rotate through its stroke cycle without requiring additional energy [207]. The utilization of storing and releasing elastic strain energy through buckling or snapping of bistable structures, can possibly also enable stable, rapid, cross-medium escape between water and air.

8.3. Embedded Fluidic Circuitry

Currently, the field of soft robotic actuation is dominated by the use of fluidic-based actuation, either utilizing pneumatics or hydraulics. As the complexities and degree-of-freedom of these systems increase, there is a need for multiple actuators to operate together. Standardized pneumatic components such as tubing, valves, and pumps, add substantial weight, bulk, and rigidity to the system, which is not ideal for future soft aerial robotic applications. To address these issues, there has been burgeoning interest in embedding onboard mechanical control-based hardware directly as part of the system, such as soft valves and pressure-driven digital circuits, utilizing principles from microfluidics and digital logic [204, 208–210]. There have been also been an effort towards onboard pump systems [211, 212]. This push towards onboard embedded controls will lead the charge towards variable-stiffness compliant aerial robots that have embedded circuitry to control various complexities.

8.4. PAI Adaptation in Self-healing and Biodegradability

Sustainability in aerial robotics is not often discussed as a PAI skill the real deployment on the environment [6, 213–215]. Current aerial robots do not have the ability to recover or repair themselves from being damaged. They are limited to one life-cycle with broken metal or composite components that are potentially hazardous to the natural environment. In recent years, the study of self-healing materials have allowed soft robots to be more robust to damage [216–218]. There have been uses of Diel-Alder materials [217], bio-degradable polymers [219–221], and even bio-hybrid materials [222, 223]. These structures recover from mechanical damages by various triggers such as UV light or temperature [216, 217]. Self-healing aerial robots of the future will enable longer life-cycles of highly-adaptable, soft-bodied flapping wings or MAV systems [215, 224].

As robots reach the end of their functional life, they often provide a negative environmental impact, because of the traditional metal and rigid composites they are made with. To improve post-life effects of aerial robots and soft robots alike, biodegradability is a possible solution. Akin to nature, organic material that makes up a living organism degrades harmlessly into the environment. Thus, for aerial vehicles, there needs to be a constant push towards the development of biodegradable bodies [213, 224], actuators [216, 217], and sensors [8, 214, 225, 226]. There have been various recently developed biodegradable bio-polymers that are pneumatically or electrically actuated. In the field of soft aerial robotics, gliding drones and concepts of biodegradable structure and sensing in MAVs have been showcased [215, 224]. With the active components such as motors, propellers, and batteries still made of non-biodegradable material and electronics, there is still a lot of research opportunities to develop eco-friendly and biodegradable flying systems [227]. Eco-sustainability in environmentally-friendly soft aerial robotics must demonstrate a net positive environmental impact following the circular economy paradigm, considering its whole life cycle [6].

9. Concluding Remarks

In this paper, the PAI paradigm highlights the potential of encoding physically intelligent features to aerial robots to push the boundaries of conventional design strategies seen with rigid-drones. These design strategies break away from highly precise, single-task rigid robot architectures to aerial robots, akin to their biological counterpart, that are capable of developing new adaptable functionalities. Highlighted in this paper have been a few promising and interesting implementations of PAI in extending the flight envelope and maneuverability, enhancing multi-modal capabilities, collision resiliency, perching abilities of aerial robots. PAI enforces the use of new soft yet robust materials and design strategies for the development of novel, robust, and versatile aerial robots by studying and understanding the encoded physical intelligence and adaptive morphology in biological inspirations.

10. Acknowledgements

We would like to thank Peter Zheng, Basaran Bahadir Kocer, Feng Xiao, Fabian Wiesemüller for their helpful suggestions during the paper editing process. We would also like to thank the members of the Aerial Robotics Lab of Imperial College London for their support and numerous stimulating discussions on this topic. This work was partially supported by funding from EPSRC, and the EU H2020 AeroTwin project. Mirko Kovac is supported by the RoyalSociety Wolfson fellowship (RSWF/R1/18003).

References

[1] Miriyev A and Kovač M 2020 *Nature Machine Intelligence* **2** 658–660
[2] Sitti M 2021 *Extreme Mechanics Letters* **46** 101340
[3] Mintchev S and Floreano D 2016 *IEEE Robotics & Automation Magazine* **23** 42–54

[4] Kovač M 2014 *Soft Robotics* **1** 28–37
[5] Kim S, Laschi C and Trimmer B 2013 *Trends in biotechnology* **31** 287–294
[6] Mazzolai B and Laschi C 2020 *Science Robotics* **5** eaba6893
[7] Shah D, Yang B, Kriegman S, Levin M, Bongard J and Kramer-Bottiglio R 2021 *Advanced Materials* **33** 2002882
[8] Rossiter J 2021 *Artificial Life and Robotics* **26** 269–274
[9] Poole D, Mackworth A and Goebel R 1998
[10] Kacprzyk J and Pedrycz W 2015 *Springer handbook of computational intelligence* (Springer)
[11] Howard D, Eiben A E, Kennedy D F, Mouret J B, Valencia P and Winkler D 2019 *Nature Machine Intelligence* **1** 12–19
[12] Pfeifer R, Iida F and Gómez G 2006 Morphological computation for adaptive behavior and cognition *International Congress Series* vol 1291 (Elsevier) pp 22–29
[13] Hauser H, Ijspeert A J, Füchslin R M, Pfeifer R and Maass W 2011 *Biological cybernetics* **105** 355–370
[14] Floreano D, Mondada F, Perez-Uribe A and Roggen D 2004 Evolution of embodied intelligence *Embodied artificial intelligence* (Springer) pp 293–311
[15] Hochner B 2012 *Current biology* **22** R887–R892
[16] Shigeno S, Andrews P L, Ponte G and Fiorito G 2018 *Frontiers in Physiology* **9** 952
[17] Ghazi-Zahedi K, Haeufle D F, Montúfar G, Schmitt S and Ay N 2016 *Frontiers in Robotics and AI* **3** 42
[18] Chin D D, Matloff L Y, Stowers A K, Tucci E R and Lentink D 2017 *Journal of The Royal Society Interface* **14** 20170240
[19] Patnaik K and Zhang W 2021 *International Journal of Intelligent Robotics and Applications* 1–16
[20] Derrouaoui S H, Bouzid Y, Guiatni M and Dib I 2021 *Unmanned Systems* 1–27
[21] Sareh P, Chermprayong P, Emmanuelli M, Nadeem H and Kovac M 2018 *Science Robotics* **5228** 1–13
[22] Hedayati H, Suzuki R, Leithinger D and Szafir D 2020 *arXiv preprint arXiv:2008.07615*
[23] Dilaveroğlu L and Özcan O 2020 Minicore: A miniature, foldable, collision resilient quadcopter *2020 3rd IEEE International Conference on Soft Robotics (RoboSoft)* (IEEE) pp 176–181
[24] Joachimczak M, Suzuki R and Arita T 2016 *Artificial Life* **22** 271–298
[25] Sims K 1994 Evolving virtual creatures *Proceedings of the 21st annual conference on Computer graphics and interactive techniques - SIGGRAPH '94* (ACM Press)
[26] Bongard J 2011 *Proceedings of the National Academy of Sciences* **108** 1234–1239
[27] Truman J W and Riddiford L M 1999 *Nature* **401** 447–452 URL `https://doi.org/10.1038/46737`
[28] Campbell N A, Reece J B, Urry L A, Cain M L, Wasserman S A, Minorsky P V and Jackson R B 2018 *Biology: a global approach* (Pearson New York, NY)
[29] Fritzsch B 1990 *Journal of Neurobiology* **21** 1011–1021
[30] Dickson J D and Clark J E 2012 *IEEE/ASME Transactions On Mechatronics* **18** 494–505
[31] Alattas R J, Patel S and Sobh T M 2018 *Journal of Intelligent & Robotic Systems* **95** 815–828
[32] Cheung K C, Demaine E D, Bachrach J R and Griffith S 2011 *IEEE Transactions on Robotics* **27** 718–729
[33] Sareh S, Siddall R, Alhinai T and Kovac M 2017 Bio-inspired soft aerial robots: adaptive morphology for high-performance flight *Soft Robotics: Trends, Applications and Challenges* (Springer) pp 65–74
[34] Daler L, Mintchev S, Stefanini C and Floreano D 2015 *Bioinspiration and Biomimetics* **10** ISSN 17483190
[35] Kovač M, Wassim-Hraiz, Fauria O, Zufferey J C and Floreano D 2011 *2011 IEEE International Conference on Robotics and Biomimetics, ROBIO 2011* 1503–1508
[36] Siddall R, Ortega Ancel A and Kovač M 2017 *Interface focus* **7** 20160085 ISSN 20428901
[37] Chen Y, Wang H, Helbling E F, Jafferis N T, Zufferey R, Ong A, Ma K, Gravish N, Chirarattananon P, Kovac M *et al.* 2017 *Science Robotics* **2**
[38] Ramezani A, Chung S J and Hutchinson S 2017 *Science Robotics* **2** Art–No
[39] Chang E, Matloff L Y, Stowers A K and Lentink D 2020 *Science Robotics* **5**
[40] Ajanic E, Feroskhan M, Mintchev S, Noca F and Floreano D 2020 *Sci. Robot.* **5** eabc2897
[41] Siddall R and Kovač M 2014 *Bioinspiration & biomimetics* **9** 031001
[42] Zufferey R, Ancel A O, Farinha A, Siddall R, Armanini S F, Nasr M, Brahmal R, Kennedy G and Kovac M 2019 *Science Robotics* **4**
[43] Kovač M, Fauria O, Zufferey J C, Floreano D *et al.* 2011 The epfl jumpglider: A hybrid jumping and gliding robot with rigid or folding wings *2011 IEEE International Conference on Robotics and Biomimetics* (IEEE) pp 1503–1508
[44] Woodward M A and Sitti M 2014 *The International Journal of Robotics Research* **33** 1511–1529
[45] Baek S M, Yim S, Chae S H, Lee D Y and Cho K J 2020 *Science Robotics* **5**
[46] Zhang K, Chermprayong P, Alhinai T M, Siddall R and Kovac M 2017 Spidermav: Perching and stabilizing micro aerial vehicles with bio-inspired tensile anchoring systems *2017 IEEE/RSJ International Conference on Intelligent Robots and Systems (IROS)* pp 6849–6854

[47] Hunt G, Mitzalis F, Alhinai T, Hooper P A and Kovac M 2014 3D printing with flying robots *2014 IEEE International Conference on Robotics and Automation (ICRA)* pp 4493–4499 ISSN 1050-4729
[48] Chermprayong P, Zhang K, Xiao F and Kovac M 2019 *IEEE Robotics and Automation Magazine* **26** 54–66 ISSN 1558223X
[49] Zufferey R, Ancel A O, Raposo C, Armanini S F, Farinha A, Siddall R, Berasaluce I, Zhu H and Kovac M 2019 *IEEE Robotics and Automation Letters* **4** 2894–2901 ISSN 23773766
[50] Altshuler D L, Bahlman J W, Dakin R, Gaede A H, Goller B, Lentink D, Segre P S and Skandalis D A 2015 *Canadian Journal of Zoology* **93** 961–975
[51] Phan H V and Park H C 2020 *Current Opinion in Insect Science*
[52] Mintchev S, de Rivaz S and Floreano D 2017 *IEEE Robotics and automation letters* **2** 1248–1255
[53] Floreano D and Wood R J 2015 *Nature* **521** 460–466
[54] Chen Y, Zhao H, Mao J, Chirarattananon P, Helbling E F, Hyun N s P, Clarke D R and Wood R J 2019 *Nature* **575** 324–329
[55] Socha J J, Jafari F, Munk Y and Byrnes G 2015 *Canadian Journal of Zoology* **93** 901–924
[56] Mintchev S, Daler L, Eplattenier G L, Floreano D and Member S 2015 *Proc. of the IEEE Conference on Robotics and Automation* 2190–2195
[57] Mintchev S, Shintake J and Floreano D 2018 *Science Robotics* **0275** 1–8 ISSN 2470-9476
[58] Bucki N and Mueller M W 2019 Design and control of a passively morphing quadcopter *IEEE International Conference on Robotics and Automation* vol 2019-May pp 9116–9122 ISBN 9781538660263 ISSN 10504729
[59] Yazawa K, Numata K and Norma-Rashid Y 2018 *PloS one* **13** e0193147
[60] Matloff L Y, Chang E, Feo T J, Jeffries L, Stowers A K, Thomson C and Lentink D 2020 *Science* **367** 293–297
[61] Low K, Hu T, Mohammed S, Tangorra J and Kovac M 2015 *Bioinspiration & biomimetics* **10** 020301
[62] Garrad M, Rossiter J and Hauser H 2018 *IEEE Robotics and Automation Letters* **3** 2056–2062
[63] Breder Jr C M 1930 *American Society of Ichthyologists and Herpetologists (ASIH)* **4** 114–121
[64] Muramatsu K, Yamamoto J, Abe T, Sekiguchi K, Hoshi N and Sakurai Y 2013 *Marine biology* **160** 1171–1175
[65] Chen Y, Shih M C, Wu M H, Yang E C and Chi K J 2014 *Journal of the Royal Society Interface* **11** 20140273
[66] Butler P J and Jones D R 1997 *Physiological reviews* **77** 837–899
[67] Siddall R and Kovač M 2015 A water jet thruster for an aquatic micro air vehicle *2015 IEEE International Conference on Robotics and Automation (ICRA)* (IEEE) pp 3979–3985
[68] Luo Y, Xiao Q, Zhu Q and Pan G 2020 *Physics of Fluids* **32** 111901
[69] Norberg U M 1985 *The American Naturalist* **126** 303–327
[70] Socha J J 2011 *Integrative and Comparative Biology* **51** 969–982
[71] Vidyasagar A, Zufferey J C, Floreano D and Kovač M 2015 *Bioinspiration & biomimetics* **10** 025006
[72] Emerson S B and Koehl M 1990 *Evolution* **44** 1931–1946
[73] Zhao F, Wang W, Wyrwa J, Zhang J, Du W and Zhong P 2021 *Applied Sciences* **11** 3362
[74] Wei D, Gao T, Li Z, Mo X, Zheng S and Zhou C 2019 *Frontiers in neurorobotics* **13** 87
[75] Floreano D, Zufferey J C, Srinivasan M V and Ellington C 2009 *Flying insects and robots* (Springer)
[76] Desbiens A L, Pope M, Berg F, Teoh Z E, Lee J and Cutkosky M 2013 Efficient jumpgliding: Theory and design considerations *2013 IEEE International Conference on Robotics and Automation* (IEEE) pp 4451–4458
[77] Armour R, Paskins K, Bowyer A, Vincent J and Megill W 2007 *Bioinspiration & biomimetics* **2** S65
[78] Thaxter C, Wanless S, Daunt F, Harris M, Benvenuti S, Watanuki Y, Grémillet D and Hamer K 2010 *Journal of Experimental Biology* **213** 1018–1025
[79] Xing Y and Yang J 2020 *Journal of Biomechanics* **109** 109952
[80] Chen Y, Xu S, Ren Z and Chirarattananon P 2021 *IEEE Transactions on Robotics*
[81] Balebail S, Raja S K and Sane S P 2019 *PloS one* **14** e0219861
[82] Doyle C E, Bird J J, Isom T A, Kallman J C, Bareiss D F, Dunlop D J, King R J, Abbott J J and Minor M A 2012 *IEEE/ASME Transactions On Mechatronics* **18** 506–517
[83] Moore J, Cory R and Tedrake R 2014 *Bioinspiration & biomimetics* **9** 025013
[84] Klaptocz A, Briod A, Daler L, Zufferey J C and Floreano D 2013 Euler spring collision protection for flying robots *2013 IEEE/RSJ International Conference on Intelligent Robots and Systems* (IEEE) pp 1886–1892
[85] Liu Z and Karydis K 2021 Toward impact-resilient quadrotor design, collision characterization and recovery control to sustain flight after collisions *2021 IEEE International Conference on Robotics and Automation (ICRA)* pp 183–189
[86] Briod A, Kornatowski P, Zufferey J C and Floreano D 2014 *Journal of Field Robotics* **31** 496–509
[87] Mintchev S, de Rivaz S and Floreano D 2017 *IEEE Robotics and Automation Letters* ISSN 2377-3766

[88] Shu J and Chirarattananon P 2019 *IEEE Robotics and Automation Letters* **4** 3820–3827
[89] Nguyen D Q, Loianno G *et al.* 2020 Towards design of a deformable propeller for drone safety *2020 3rd IEEE International Conference on Soft Robotics (RoboSoft)* (IEEE) pp 464–469
[90] Shintake J, Rosset S, Schubert B E, Floreano D and Shea H R 2015 *IEEE/ASME Transactions on Mechatronics* **20** 1997–2008
[91] Tonazzini A, Mintchev S, Schubert B, Mazzolai B, Shintake J and Floreano D 2016 *Advanced Materials* **28** 10142–10148
[92] Dufour L, Owen K, Mintchev S and Floreano D 2016 A drone with insect-inspired folding wings *2016 IEEE/RSJ International Conference on Intelligent Robots and Systems (IROS)* pp 1576–1581
[93] Di Luca M, Mintchev S, Heitz G, Noca F and Floreano D 2017 *Interface focus* **7** 20160092
[94] Daler L, Mintchev S, Stefanini C and Floreano D 2015 *Bioinspiration & biomimetics* **10** 016005
[95] Jafferis N T, Helbling E F, Karpelson M and Wood R J 2019 *Nature* **570** 491–495
[96] Kim S J, Lee D Y, Jung G P and Cho K J 2018 *Science Robotics* **3** eaar2915
[97] Miron G, Bédard B and Plante J S 2018 *Actuators* **7** ISSN 2076-0825
[98] Lee D J and Jung G P 2020 *IEEE Robotics and Automation Letters* **5** 6097–6104
[99] Li H, Yao J, Wei C, Zhou P, Xu Y and Zhao Y 2021 *Mechanism and Machine Theory* **158** 104226
[100] Kovac M 2016 *Science* **352** 895–896
[101] Zhang K, Zhu Y, Lou C, Zheng P and Kovac M 2019 *RoboSoft 2019 - 2019 IEEE International Conference on Soft Robotics* 821–827
[102] Nguyen H N, Siddall R, Stephens B, Navarro-Rubio A and Kovač M 2019 A passively adaptive microspine grapple for robust, controllable perching *2019 2nd IEEE International Conference on Soft Robotics (RoboSoft)* (IEEE) pp 80–87
[103] Garcia Rubiales F J, Ramon Soria P, Arrue B C and Ollero A 2021 *Sensors* **21** ISSN 1424-8220
[104] Fishman J, Ubellacker S, Hughes N and Carlone L 2021 Dynamic grasping with a "soft" drone: From theory to practice *2021 IEEE/RSJ International Conference on Intelligent Robots and Systems (IROS)* pp 4214–4221
[105] Watson M 1869 *Journal of anatomy and physiology* **3** 379
[106] Chi W, Low K H, Hoon K H and Tang J 2014 An optimized perching mechanism for autonomous perching with a quadrotor *2014 IEEE International Conference on Robotics and Automation (ICRA)* pp 3109–3115
[107] Quinn Thomas Hand Baumel J J 1990 *Zoomorphology* **109** 281–293 ISSN 1432-234X
[108] Simmons N B and Quinn T H 1994 *Journal of Mammalian Evolution* **2** 231–254
[109] Evangelista C, Kraft P, Dacke M, Reinhard J and Srinivasan M V 2010 *Journal of Experimental Biology* **213** 262–270
[110] Hyzer W G 1962 *Science* **137** 609–610
[111] Humphrey J 1987 *Oecologia* **73** 469–477
[112] Van Leeuwen C H 2018 *Frontiers in plant science* **9** 153
[113] Thomas J, Pope M, Loianno G, Hawkes E W, Estrada M A, Jiang H, Cutkosky M R and Kumar V 2016 *Journal of Mechanisms and Robotics* **8** 051007 ISSN 1942-4302
[114] Daler L, Klaptocz A, Briod A, Sitti M and Floreano D 2013 A perching mechanism for flying robots using a fibre-based adhesive *2013 IEEE International Conference on Robotics and Automation* (IEEE) pp 4433–4438
[115] Graule M, Chirarattananon P, Fuller S, Jafferis N, Ma K, Spenko M, Kornbluh R and Wood R 2016 *Science* **352** 978–982
[116] Pope M T, Kimes C W, Jiang H, Hawkes E W, Estrada M A, Kerst C F, Roderick W R, Han A K, Christensen D L and Cutkosky M R 2017 *IEEE Transactions on Robotics* **33** 38–48 ISSN 15523098
[117] Kovač M, Germann J, Hürzeler C, Siegwart R Y and Floreano D 2009 *Journal of Micro-Nano Mechatronics* **5** 77–91
[118] Hang K, Lyu X, Song H, Stork J A, Dollar A M, Kragic D and Zhang F 2019 **6637** 1–11
[119] Liu S, Dong W, Ma Z and Sheng X 2020 *IEEE Robotics and Automation Letters* **5** 4766–4773
[120] Braithwaite A, Alhinai T, Haas-Heger M, McFarlane E and Kovač M 2018 *Tensile Web Construction and Perching with Nano Aerial Vehicles* (Cham: Springer International Publishing) pp 71–88 ISBN 978-3-319-51532-8
[121] Roderick W R, Jiang H, Wang S, Lentink D and Cutkosky M R 2017 Bioinspired grippers for natural curved surface perching *Conference on Biomimetic and Biohybrid Systems* (Springer) pp 604–610
[122] Zhang Z, Xie P and Ma O 2013 Bio-inspired trajectory generation for uav perching *2013 IEEE/ASME International Conference on Advanced Intelligent Mechatronics* (IEEE) pp 997–1002
[123] Doyle C E, Bird J J, Isom T A, Kallman J C, Bareiss D F, Dunlop D J, King R J, Abbott J J and Minor M A 2013 *IEEE/ASME Transactions on Mechatronics* **18** 506–517
[124] Erbil M A, Prior S D and Keane A J 2013 *International Journal of Micro Air Vehicles* **5** 207–228

[125] Chi W, Low K H, Hoon K H and Tang J 2014 An optimized perching mechanism for autonomous perching with a quadrotor *IEEE International Conference on Robotics and Automation* pp 3109–3115 ISBN 978-1-4799-3685-4 ISSN 10504729
[126] Ruggiero F, Lippiello V and Ollero A 2018 *IEEE Robotics and Automation Letters* **3** 1957–1964
[127] Khamseh H B, Janabi-Sharifi F and Abdessameud A 2018 *Robotics and Autonomous Systems* **107** 221–235
[128] Bartelds T, Capra A, Hamaza S, Stramigioli S and Fumagalli M 2016 *IEEE Robotics and Automation Letters* **1** 477–483
[129] Hamaza S, Georgilas I, Heredia G, Ollero A and Richardson T 2020 *Journal of Field Robotics* **37** 1224–1245
[130] Ikeda T, Minamiyama S, Yasui S, Ohara K, Ichikawa A, Ashizawa S, Okino A, Oomichi T and Fukuda T 2019 *Journal of Field Robotics* **36** 1212–1221
[131] Sheng X, Ma Z, Zhang N and Dong W 2021 *Journal of Mechanisms and Robotics* **13** 011023
[132] Suarez A, Jimenez-Cano A E, Vega V M, Heredia G, Rodriguez-Castaño A and Ollero A 2018 *Mechatronics* **50** 30–44
[133] Fishman J and Carlone L 2021 Control and trajectory optimization for soft aerial manipulation *2021 IEEE Aerospace Conference (50100)* (IEEE) pp 1–17
[134] Gomez-Tamm A, Ramon-Soria P, Arrue B and Ollero A 2019 Current state and trends on bioinspired actuators for aerial manipulation *2019 Workshop on Research, Education and Development of Unmanned Aerial Systems (RED UAS)* (IEEE) pp 352–361
[135] Mishra S, Yang D, Thalman C, Polygerinos P and Zhang W 2018 Design and control of a hexacopter with soft grasper for autonomous object detection and grasping *Dynamic Systems and Control Conference* vol 51913 (American Society of Mechanical Engineers) p V003T36A003
[136] Hughes J, Culha U, Giardina F, Guenther F, Rosendo A and Iida F 2016 *Frontiers in Robotics and AI* **3** ISSN 2296-9144
[137] Blumenschein L H, Coad M M, Haggerty D A, Okamura A M and Hawkes E W 2020 *Frontiers in Robotics and AI* **7**
[138] Sadeghi A, Mondini A and Mazzolai B 2017 *Soft robotics* **4** 211–223
[139] Siddall R, Byrnes G, Full R J and Jusufi A 2021 *Communications Biology* **4** 1020 ISSN 2399-3642
[140] Shield S, Jericevich R, Patel A and Jusufi A 2021 *Integrative and comparative biology*
[141] Petersen K H, Napp N, Stuart-Smith R, Rus D and Kovac M 2019 *Science Robotics* **4** eaau8479
[142] Castree N and Braun B 1998 *Remaking reality: Nature at the millenium* 3–42
[143] Franks N R, Wilby A, Silverman B W and Tofts C 1992 *Animal behaviour* **44** 357–375
[144] Craig C L 1997 *Annual review of entomology* **42** 231–267
[145] Collias E C and Collias N E 1964 *The Auk* **81** 42–52
[146] Kang N, Hails C and Sigurdsson J 1991 *Ibis* **133** 170–177
[147] Augugliaro F, Mirjan A, Gramazio F, Kohler M and D'Andrea R 2013 Building tensile structures with flying machines *2013 IEEE/RSJ International Conference on Intelligent Robots and Systems* (IEEE) pp 3487–3492
[148] Zhang K, Chermprayong P, Alhinai T M, Siddall R and Kovac M 2017 SpiderMAV: Perching and Stabilizing Micro Aerial Vehicles with Bio-inspired Tensile Anchoring Systems *International Conference on Intelligent Robots and Systems*
[149] Martinez A, DeJong J, Akin I, Aleali A, Arson C, Atkinson J, Bandini P, Baser T, Borela R, Boulanger R *et al.* 2021 *Géotechnique* 1–19
[150] Niu S, Luo Y, Shen Y and Kim K J 2015 Enabling earthworm-like soft robot development using bioinspired ipmc-scissor lift actuation structures: Design, locomotion simulation and experimental validation *2015 IEEE International Conference on Robotics and Biomimetics (ROBIO)* (IEEE) pp 499–504
[151] Ortiz D, Gravish N and Tolley M T 2019 *IEEE Robotics and Automation Letters* **4** 2630–2636
[152] Francis L 1991 *Journal of Experimental Biology* **158** 117–132
[153] Munro C, Vue Z, Behringer R R and Dunn C W 2019 *Scientific reports* **9** 1–12
[154] Sorensen A E 1986 *Annual Review of Ecology and Systematics* **17** 443–463
[155] Vieira-Neto E, Mundim F and Vasconcelos H 2006 *Insectes sociaux* **53** 326–332
[156] Dunlop J A, Wirth S, Penney D, McNeil A, Bradley R S, Withers P J and Preziosi R F 2012 *Biology letters* **8** 457–460
[157] Marliave J B and Mills C E 1993 *Canadian journal of zoology* **71** 257–263
[158] Xu Y, Shi W, Arredondo-Galeana A, Mei L and Demirel Y K 2021 *Scientific Reports* **11** 1–20
[159] Tyndale-Biscoe H 2005 *Life of marsupials* (Melbourne, VIC, Australia: CSIRO Publishing)
[160] Tsukagoshi H and Osada Y 2021 Soft hybrid suction cup capable of sticking to various objects and environments *Actuators* vol 10 (Multidisciplinary Digital Publishing Institute) p 50
[161] Lussier Desbiens A, Asbeck A T and Cutkosky M R 2011 *The International Journal of Robotics Research* **30** 355–370

[162] Wilker J J 2017 *Nature* **546** 358–359
[163] Tramacere F, Beccai L, Mattioli F, Sinibaldi E and Mazzolai B 2012 Artificial adhesion mechanisms inspired by octopus suckers *2012 IEEE International Conference on Robotics and Automation* (IEEE) pp 3846–3851
[164] Mazzolai B, Mondini A, Tramacere F, Riccomi G, Sadeghi A, Giordano G, Del Dottore E, Scaccia M, Zampato M and Carminati S 2019 *Advanced Intelligent Systems* **1** 1900041
[165] Xie Z, Domel A G, An N, Green C, Gong Z, Wang T, Knubben E M, Weaver J C, Bertoldi K and Wen L 2020 *Soft robotics* **7** 639–648
[166] Federle W and Labonte D 2019 *Philosophical Transactions of the Royal Society B* **374** 20190199
[167] Frutiger A 1998 *Journal of the North American Benthological Society* **17** 104–120
[168] Chong S M, Oh P Y, Chua D, Sekar S D and Hsu C D 2020 *Journal of wildlife diseases* **56** 713–716
[169] Schoenfuss H L and Blob R W 2003 *Journal of Zoology* **261** 191–205
[170] Ilievski F, Mazzeo A D, Shepherd R F, Chen X and Whitesides G M 2011 *Angewandte Chemie* **123** 1930–1935
[171] Wood R, Becker K, Teeple C, Charles N and Mahadevan L 2021
[172] Hughes J, Plumb-Reyes T, Charles N, Mahadevan L and Rus D 2021 Detangling hair using feedback-driven robotic brushing *2021 IEEE 4th International Conference on Soft Robotics (RoboSoft)* (IEEE) pp 487–494
[173] Stoy K, Walker K, Nielsen S, Ayres P, Heinrich M, Leon D and Cheheltan A 2021 A large-scale, light-weight, and soft braided robot manipulator with rapid expansion capabilities *2021 IEEE 4th International Conference on Soft Robotics (RoboSoft)* (IEEE) pp 495–500
[174] Zhang K, Chermprayong P, Tzoumanikas D, Li W, Grimm M, Smentoch M, Leutenegger S and Kovac M 2019 *Journal of Field Robotics* **36** 230–251
[175] Mohanty P K and Nayak D 2018
[176] Yanoviak S P, Kaspari M and Dudley R 2009 *Biology letters* **5** 510–512
[177] Sheldon K S, Zhao L, Chuang A, Panayotova I N, Miller L A and Bourouiba L 2017 Revisiting the physics of spider ballooning *Women in mathematical biology* (Springer) pp 163–178
[178] Nathan R, Katul G G, Horn H S, Thomas S M, Oren R, Avissar R, Pacala S W and Levin S A 2002 *Nature* **418** 409–413
[179] Jones K 1995 *Pau d'Arco: Immune power from the rain forest* (Inner Traditions/Bear & Co)
[180] Armstrong W 1990 *Pacific Discovery* **43** 32–39
[181] Nelson E 1978 *Watsonia* **12** 103–112
[182] Ward R G 1977 *The Dispersal of the Coconut: Did it Float Or was it Carried?*
[183] Zhang Y, Xin K, Liao B, Ai X and Sheng N 2021 *Forests* **12** 1375
[184] Müller U K and Kranenbarg S 2004 *Science* **304** 217–219
[185] Zack T, Claverie T and Patek S 2009 *Journal of Experimental Biology* **212** 4002–4009
[186] Anderson E J and Grosenbaugh M A 2005 *Journal of Experimental Biology* **208** 1125–1146
[187] Quinn T H and Baumel J J 1990 *Zoomorphology* **109** 281–293
[188] Bennett M and Taylor G 1995 *Nature* **378** 56–59
[189] McGowan C P, Skinner J and Biewener A A 2008 *Journal of Anatomy* **212** 153–163
[190] Lappin A K, Monroy J A, Pilarski J Q, Zepnewski E D, Pierotti D J and Nishikawa K C 2006 *Journal of Experimental Biology* **209** 2535–2553
[191] Deban S M, O'Reilly J C, Dicke U and Van Leeuwen J L 2007 *Journal of experimental biology* **210** 655–667
[192] Forterre Y, Skotheim J M, Dumais J and Mahadevan L 2005 *Nature* **433** 421–425
[193] Volkov A G, Murphy V A, Clemmons J I, Curley M J and Markin V S 2012 *Journal of plant physiology* **169** 55–64
[194] Holmes D P 2019 *Current opinion in colloid & interface science* **40** 118–137
[195] Zhang C, Zou W, Ma L and Wang Z 2020 *Robotics and Autonomous Systems* **124** 103362
[196] Carlson J, Friedman J, Kim C and Sung C 2020 Rebound: Untethered origami jumping robot with controllable jump height *IEEE International Conference on Robotics and Automation*
[197] Gorissen B, Melancon D, Vasios N, Torbati M and Bertoldi K 2020 *Science Robotics* **5**
[198] Scarfogliero U, Stefanini C and Dario P 2009 *Mechanism and Machine Theory* **44** 580–590
[199] Chi Y, Tang Y, Liu H and Yin J 2020 *Advanced Materials Technologies* **5** 2000370
[200] Chen T, Bilal O R, Shea K and Daraio C 2018 *Proceedings of the National Academy of Sciences* **115** 5698–5702
[201] Zhang H, Lerner E, Cheng B and Zhao J 2020 *IEEE/ASME Transactions on Mechatronics* 1–1
[202] Baumgartner R, Kogler A, Stadlbauer J M, Foo C C, Kaltseis R, Baumgartner M, Mao G, Keplinger C, Koh S J A, Arnold N *et al.* 2020 *Advanced Science* **7** 1903391
[203] Zhang Z, Ni X, Wu H, Sun M, Bao G, Wu H and Jiang S 2021 *Soft Robotics*
[204] Pal A, Restrepo V, Goswami D and Martinez R V 2021 *Advanced Materials* **33** 2006939

[205] Wang X, Khara A and Chen C 2020 *Bioinspiration & Biomimetics* **15** 056017
[206] Thuruthel T G, Haider Abidi S, Cianchetti M, Laschi C and Falotico E 2020 A bistable soft gripper with mechanically embedded sensing and actuation for fast grasping *2020 29th IEEE International Conference on Robot and Human Interactive Communication (RO-MAN)* pp 1049–1054
[207] Lau G K, Chin Y W, Goh J T W and Wood R J 2014 *IEEE Transactions on Robotics* **30** 1187–1197
[208] Mahon S T, Buchoux A, Sayed M E, Teng L and Stokes A A 2019 Soft robots for extreme environments: Removing electronic control *2019 2nd IEEE International Conference on Soft Robotics (RoboSoft)* (IEEE) pp 782–787
[209] Jin L, Forte A E and Bertoldi K 2021 *Advanced Science* 2101941
[210] Roinila T, Messo T, Luhtala R, Scharrenberg R, de Jong E C, Fabian A and Sun Y 2018 *IEEE Transactions on Industrial Electronics* **66** 5769–5777
[211] Wehner M, Tolley M T, Mengüç Y, Park Y L, Mozeika A, Ding Y, Onal C, Shepherd R F, Whitesides G M and Wood R J 2014 *Soft robotics* **1** 263–274
[212] Sridar S, Poddar S, Tong Y, Polygerinos P and Zhang W 2020 *IEEE Robotics and Automation Letters* **5** 4062–4069
[213] Hartmann F, Baumgartner M and Kaltenbrunner M 2021 *Advanced Materials* **33** 2004413
[214] McEvoy M A and Correll N 2015 *Science* **347**
[215] Wiesemüller F, Miriyev F and Kovac M 2021 Zero-footprint eco-robotics: A new perspective on biodegradable robots *2021 Workshop on Aerial Robotic Systems Physically Interacting with the Environment (AIRPHARO)*
[216] Bilodeau R A and Kramer R K 2017 *Frontiers in Robotics and AI* **4** 48
[217] Terryn S, Langenbach J, Roels E, Brancart J, Bakkali-Hassani C, Poutrel Q A, Georgopoulou A, Thuruthel T G, Safaei A, Ferrentino P *et al.* 2021 *Materials Today*
[218] Roels E, Terryn S, Brancart J, Verhelle R, Van Assche G and Vanderborght B 2020 *Soft robotics* **7** 711–723
[219] Gariya N and Kumar P 2021 *Materials Today: Proceedings*
[220] Winfield J, Rossiter J and Ieropoulos I 2019 *American Chemical Society*
[221] Yamada Y 2021 *IEEE Robotics and Automation Letters* **6** 3777–3784
[222] Won P, Ko S H, Majidi C, W Feinberg A and A Webster-Wood V 2020 Biohybrid actuators for soft robotics: Challenges in scaling up *Actuators* vol 9 (Multidisciplinary Digital Publishing Institute) p 96
[223] Ricotti L, Trimmer B, Feinberg A W, Raman R, Parker K K, Bashir R, Sitti M, Martel S, Dario P and Menciassi A 2017 *Science Robotics* **2**
[224] Wiesemüller F, Winston C, Poulin A, Aeby X, Miriyev A, Geiger T, Nyström G and Kovač M 2021 *IEEE Robotics and Automation Letters* **6** 4017–4024
[225] Rossiter J, Winfield J and Ieropoulos I 2016 Here today, gone tomorrow: biodegradable soft robots *Electroactive polymer actuators and devices (EAPAD) 2016* vol 9798 (International Society for Optics and Photonics) p 97981S
[226] Curry E J, Ke K, Chorsi M T, Wrobel K S, Miller A N, Patel A, Kim I, Feng J, Yue L, Wu Q *et al.* 2018 *Proceedings of the National Academy of Sciences* **115** 909–914
[227] Yin L, Huang X, Xu H, Zhang Y, Lam J, Cheng J and Rogers J A 2014 *Advanced Materials* **26** 3879–3884

IOP Conf. Series: Materials Science and Engineering 1261 (2022) 012007 doi:10.1088/1757-899X/1261/1/012007

Bio-Inspired Robots Imitating Human Organs with Embodied Intelligence Behaviour

Ryman Hashem, Weiliang Xu and Fumiya Iida

R. Hashem and Fumiya Iida are with the Department of Engineering, University of Cambridge, Cambridge CB2 1PZ, UK (e-mails: (rh805) and (fi224) @cam.ac.uk). W. Xu is with the Department of Mechanical and Mechatronics Engineering, University of Auckland, Auckland 1010, New Zealand (e-mail: p.xu@auckland.ac.nz)

Abstract. Soft robotics is an emerging field that introduces promising engineering methods that replicate biological behaviours. Soft robotics aims to obtain a delicate interaction with their environment and be adaptable in different situations. Using the morphology and materials in robotics design is recognised as an embodied intelligence of the system. This method provides new ideas other than classic engineering strategies; it can translate biological behaviour into an engineering context. Embodied intelligence introduces potential ways to replicate human organs' motor activities with soft-bodied simulators. Researchers are looking for a test environment that imitates the complex human organs functionalities to advance the knowledge of the human body. Many recent diseases were discovered, such as stomach dysrhythmia. It is believed that a test environment that can replicate such illnesses can introduce a faster solution to patients suffering from those illnesses. This chapter will discuss soft robots that emulate human organs using embodied intelligence in their morphology for simpler control systems and continuous actuation behaviour.

1. Introduction

Embodied intelligence is a term that was introduced to look at the robotics system from a different perspective. Generally, the robotics system is viewed as the robot body where the motion occurs, and the controller where all the programming and signal interaction with the robot body; These two sections are usually separate. However, the embodied intelligence concept combines both sections as an integrated system to achieve a task [1]. A robot body can be intelligent to manage some tasks without needing a signal from a controller. In this method, the controller of a robot can be simplified dramatically to achieve complex tasks. For example, the adaptation of the robot body to a surrounding environment can be considered an intelligent behaviour. In nature, the structure of a living organism (brain and body) is usually coupled to perform mobilities and other essential chores [2]. Human organs are an example of an embodiment in a human body, which they have complex behaviour that is still under research. For example, the interaction of the esophagus and the stomach between their body and the food shows a high level of adaptation with embodied intelligence behaviour. In order to develop robots that can simulate the adaptation, compliant materials with adequate morphologies are needed.

Soft robotics is a sub-field of robotics that proposes robots composed of soft materials that are comparable with biological forms [3]. The ability of the biological responses to their heterogeneous environment stimulates soft roboticists' inspiration [4]. This ability is observed in the living organisms' elasticity and manoeuvre to their surroundings as a daily activity. The observations of the living organisms' behaviour can be interpreted and applied in the engineering domain to obtain a method that classical robotics cannot solve. Soft roboticists are investigating the behaviour of living organisms, where identifying nature's functions can be efficient and effective in engineering. Soft robotics techniques aim to recognise the living system's perpetual and compliant functionalities by engineering procedures and communicate the abstracted knowledge into the robotic domain.

Biomimicry presents inspiring approaches to the engineering field through the quantitative examination of biological functions (adaptation, actuation, and sensation) [5]. These processes can be represented by mimicking a living system, such as human organs' responses. By linking concepts from diverse frameworks, multidisciplinary soft robotics can help researchers increase their biological functions' knowledge through robotic systems (see figure 1 (a)). Soft robots inherit unique features relative to the human organ's properties. Soft robots can be formed in a complex configuration that presents more pragmatic geometry than inflexible robots. Additionally, soft materials are analogous to biological behaviour regarding elasticity and adaptation to the surrounding conditions, essential for achieving embodied intelligence in the robotic context. It is possible to develop soft robots that offer bio-mimicking human organs' motor functionalities with soft robotics techniques.

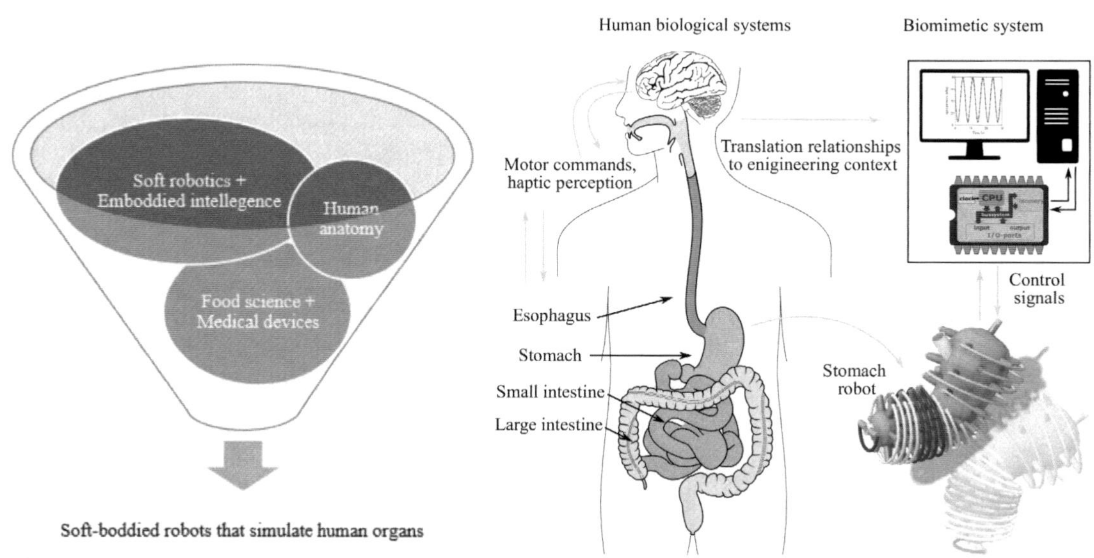

(a) (b)

Figure 1. (a) The embodiment occurs in the human biological systems and the translation of the related relationship toward the engineering systems. The brain generates command signals to the biological systems that stimulate the motor functionalities in the smooth muscles of the gastrointestinal tract. Haptic perceptions send signals back to the brain. A PC represents the translation of the biological system in the engineering context for generating control architecture that corresponds bidirectionally with a controller. For example, a stomach robot that can mimic the behaviour of a biological stomach is shown with a set of multiple actuators to perform the motor functionality. These actuators can be in a pneumatic form which communicates with the controller. To achieve the haptic perception in the soft robot, it is required to have a sensory system capable of measuring the deformation of the motor functionality and investigating the rheological behaviour. (b) The soft robotic human organs simulators links and integrates concepts from three areas: stomach anatomy, soft robotics, and food science.

Embodied intelligence assists in promoting different techniques to robotics that can be applied to simulate biological motor functionalities such as human organs. Manufacturing human organs in physical devices aid in the exploration of the organ's behaviour and related illnesses. Several studies have attempted to mimic human organs with the biomimetic method [6]. Computer-based studies on the properties of stomach functionalities have been carried out utilising numerous procedures, such as finite element analysis simulations [7, 8]. Assumptions were used to overcome many limitations that occurred in the simulation studies. Ferrua et al. advised that a dynamic soft robot can be a practical method for experimenting and validating the conduct theories [7]. For example, the swallowing robot was constructed from compliant materials and actuated by a pneumatic system to emulate a human esophagus [9]. Recent soft robots present inspiration and a step forward toward embodied intelligent robots that are adaptive to the unknown or variant environment, such as the swallowing robot [10], stomach robot [11, 12], and heart actuator [6]. The interaction between these robots and their environment is challenging to control. Nevertheless, what if the morphology of the robots can handle the interaction? In this case, we can see the benefit of embodied intelligence in reducing the control complexity and allowing a closer step for developing complex robots that simulate human organs.

In this chapter, we discuss three examples of soft robots that possess embodied intelligence through their body responses and simple control system: esophagus and stomach robots that bio-mimicking human organs functionalities and a ring actuator that simulates the contraction of the gastrointestinal tract (GI). A brief introduction from the medical perspective is communicated and followed by the relevant robotic systems. The discussion presents what has been done, the current state, and what will we need to move forward concerning a biologically inspired robot that simulates human organs.

2. Bio-inspired esophagus robots

In this section, we will introduce the swallowing functionality in a human body with the related muscular behaviour of the organ. The second half of this section illustrates the robotic system that simulates the swallowing functionalities.

2.1. Swallowing process in medicine

The feeding process is initiated to obtain nutrition for growth and body functions. It is a conventional manner and is essential to the continuance of living. Feeding includes the ingestion of solid and liquid foods conveyed to the digestive system where the nutrients can be derived.

Swallowing is the process between chewing and gastric digestion by allowing food to pass through the pharynx [13]. Swallowing or deglutition is classified physiologically into four phases: oral preparation, oral, pharyngeal, and esophageal phase [14]. The oral preparation and oral phases are under deliberate control to chew food to boluses [15]. The pharyngeal and esophageal phases are under involuntary control that continuously progresses when initiated [16]. These phases are coordinated to successfully transport boluses from the mouth to the stomach (see Table 1). The esophagus is a tubular hollow organ that connects the pharynx to the stomach and locates in front of the spine.

Dysphagia is a common disorder that affects the swallowing process [17]. It results from many major diseases, including stroke and cancer, usually shown among the elderly [18]. Patients with dysphagia struggle to ingest solid food and thick liquids. A practical treatment suggests modifying patients' diet to adjust bolus viscosity, which can be challenging to maintain. The investigation of dysphagia is limited by the complexity of testing on patients and the limited number of in-vitro devices. Food scientists usually initiate the simulation of the swallowing process with in-vitro devices to investigate the response of boluses in the esophagus, which aid in developing a solution for patients.

Table 1. Digestive process.

Gastrointestinal tract	Mechanical movement	Digestive juices
Mouth	Chewing	Saliva
Esophagus	Swallowing/Peristalsis	N/A
Stomach	Upper muscles & lower muscles peristalsis	Acids
Small intestine	Peristalsis	digestive juices
Large intestine	Peristalsis/bowel	N/A

2.2. Swallowing process in robotics

Biologically inspired swallowing robots were introduced as an in vitro solution for investigating food boluses and their response to dysphagia disorders [19]. These robots aim to mechanically simulate the peristaltic wave in the esophagus by empirically validating the motion. Therefore, the specifications of the esophagus are used to provide a correct esophagus-like robot. Specifying

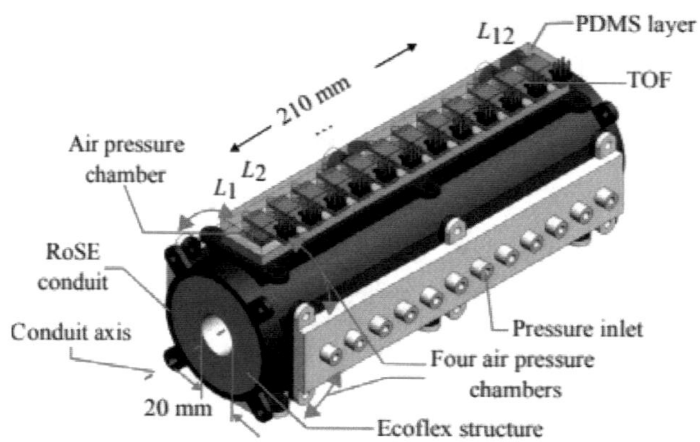

Figure 2. RoSEv2.0: a swallowing robot for simulating the esophageal behaviour and testing stent migration [20, 21].

the esophagus contractions and peristaltic waves is essential for developing an apparatus that simulates these biological behaviours. Anatomically, the esophagus can be considered a tubular organ extending 200 mm in length with a bore diameter between 18 to 23 mm [22]. The peristaltic wave in the esophags usually transfers boluses to the stomach within 3-10 s [14]. This period can be varied depending on the viscosity of the intake bolus, such as water; it can be transferred faster between 2-5 s with the help of gravity. The peristaltic wave is usually started slower in velocity, shallower in amplitude, and progresses faster and higher in the human organs. The amplitude and velocity of the peristaltic wave over the conduit are varied and divided into two sections: the first third of the tube length and the rest two-thirds [23]. The average amplitude in the first third is 53.4 ± 9.0 mmHg, while a larger amplitude occurs in the two-thirds at 69.5 ± 12.1 mmHg. The velocity also changes over the esophagus, starting at 2.92 cm/s and traversing down to 4.98 cm/s in the lower part.

Clinical trials on patients might cause discomfort and ethical issues, limiting medical studies on the swallowing process. Unlike medical practices, a swallowing robot can test the behaviour of dysphagia without including patients. A robot simulating the esophagus should determine the specification of the contraction waves reported above. In an esophageal stent method, a

IOP Conf. Series: Materials Science and Engineering 1261 (2022) 012007 doi:10.1088/1757-899X/1261/1/012007

meshed tube is installed in an esophagus suffering from dysphagia to remain a blocked region open. This method allows patients to swallow solids and liquids food. Stents usually migrate in the esophagus within time. Researchers who work on stent development are interested in swallowing robots as an experimentation method for the stent migration (see Fig. 2) [21]. The interaction between a stent and a robot presents an embodiment intelligent environment, and understanding the migration of the stent is highly related to the embodiment of the robot body. The effect of continuous peristaltic waves can also influence the stent state. This is still under research, and the analysis of this behaviour is yet to come.

2.3. Peristaltic wave in esophagus robot

How can we create a peristaltic wave that simulates GI behaviour, and how is it possible in the engineering context? Figure 3 shows a segment of the swallowing robot with a travelling peristaltic wave at the inner surface of the conduit. The conduit has four embedded chambers in each layer (12 layers in total). As the conduit is soft, applying pressure to inflate these chambers cause the conduit to contract. The peristaltic wave in the esophagus is a sinewave-like that has been already specified in the literature [20]. Inflating three layers in the robot conduit produces a sine wave that can be controllable. The compliance of material allows this simple-control complex-response relationship. By altering the inflation between the chambers, the robot can produce contractions. For example, simulation of the peristaltic wave in the stomach (see fig. 3, or the segmental contractions in the small intestine. The sinewave characteristics such as amplitude and wavelength are controllable. Different control methods can be applied to achieve a desired sine wave inside the conduit either by open-loop control for generating spatiotemporal waves [21] or with closed-loop through nonlinear model predictive control [20]. The ability to control the peristaltic wave opens the door to simulate many behaviours, such as dysphagia. Also, swallowing robots can be used for other applications such as industrial pumps, and by varying the peristaltic wave parameters, the robot can change the speed of pumping various contents.

In a robot that simulates the GI tract, adaptive behaviour is critical for generating the

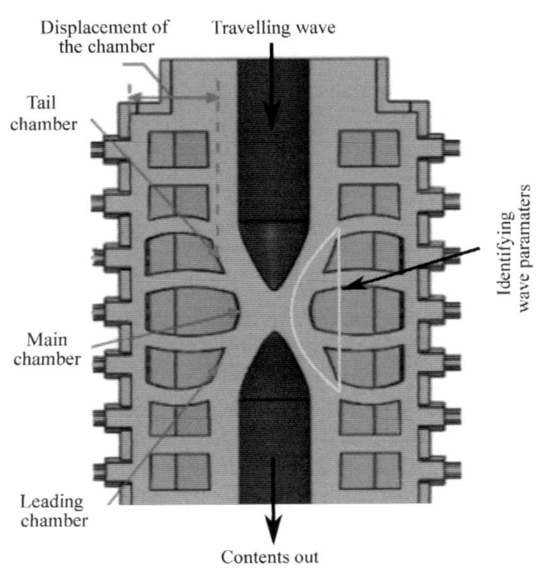

Figure 3. A section of RoSEv2.0 in a cross-section view that illustrate a travelling peristaltic wave from top to bottom (adapted from [20, 21]).

IOP Conf. Series: Materials Science and Engineering 1261 (2022) 012007 doi:10.1088/1757-899X/1261/1/012007

peristaltic wave. The morphology and material properties should inherit specifications similar to the biological example, such as stretchability and compliance. Adaptive behaviour considers intelligence behaviour as seen in nature. It is possible to generate a complex motion such as the peristaltic waves in a robot with simple control methods. Developing robots that mimic biological behaviour is essential for a set of applications, but also, the process of development provokes many questions about how biological motion exists. By developing such robots, we discover exciting and inspiring behaviours from nature [24], advancing knowledge in many disciplines.

3. Technology of Stomach Robots

In this section, we will present the behaviour of the human stomach and its essential role in the digestion process. After that, we depict the robotic system that mimics the peristaltic waves of a human stomach.

3.1. Medical background of a stomach

The digestion of intake food is a routine activity in various species to sustain living and growth. Inside the human body, regular digestion involves the rule of a churning movement and chemical breakdown of the food to chyme in a stomach [13]. Gastric motility provides mechanical motor functionalities. Related fields specialists declared that the motility cycles and motion strongly impact the digestion rule [8, 11]. The antral contraction wave (ACW) is the motor peristalsis activity that appears in the smooth muscles of a human gut [25]. More investigation on the ACW is required for analysing the performance, but few apparatuses are available as test environments that simulate the ACW [26, 27]. The state-of-the-art instruments are rigid forms, and inflexible actuators perform the actuation. These robots' motion and anatomy are unrealistic when contrasted with the perceived biological stomach. Acknowledging the gaps mentioned above has resulted in several soft-bodied stomach emulators that mimic human stomach motility and anatomy.

A significant known disease in the gut is dysrhythmias. This condition causes disorder in the stomach ACW rhythmic behaviour such as the direction frequency of the peristaltic wave [28]. Dysrhythmias cause difficulties in the normal digestion of food and liquids in a patient. A clinical solution that can identify a solution for dysrhythmias requires the examination of a specialist with an invasive technique.

The medical understanding of the digestion rule from the perspective of anatomical and physiological characteristics has improved lately, which has assisted in examining dysrhythmias. The knowledge of normal digestive processes can be known from healthy participants, but the diversity in participants' stomachs anatomy and situation introduce uncertainties for forming standardised measurements. Another challenge in the medical field is to provide ongoing nourishment to patients with dysrhythmia. The in vivo testing of the digestive process remain restricted through related complexity in the invasiveness methods [15].

Dysrhythmia can be examined in a test environment; a soft robot mimics the human stomach motility and can replicate dysrhythmia. The replication of GI motor functionality (healthy and ill) explains the causes of unhealthy behaviour. Furthermore, the Electrogastrogram (EGG) is the abstracted signal from the stomach, an example of a GI organ, which can also be used to simulate any irregular behaviour. Biomedical engineers have studied the EGG signals to identify stomach behaviour in healthy and unhealthy situations. It is proposed that the investigation of the EGG signals from a sick stomach can be implemented in a computer simulation, and signals can be examined with a stomach robot [29].

3.2. In-vitro stomach simulator

An in vitro soft-bodied robot mimicking the ACW of a stomach is needed for related research [30, 8, 26]. The motivation that drives this research can be exhibited in advancing the exploration of the digestion behaviour in both healthy and unhealthy stomachs. This exploration is plausible by an in vitro stomach robot. With an in vitro device, many studies can be conducted, such as the investigation of newly developed food and drugs that are related to the fields of food science [31]. Another application is represented by testing newly developed medical devices, such as micro-endoscopic capsules [11]. From these examples, we can see that a soft-bodied stomach robot inherits multi-disciplinary nature combining medical, engineering and food technology fields (Fig. 1).

Medical literature benefits the development of the in vitro device with the fundamental knowledge of the human stomach's anatomy and motility. The engineering input is correlated to the soft robotics methods, which can achieve a behaviour compared to biological responses. The material's behaviour introduces an adaptation capacity to the working environment. Another associated field in this study is food technology. Researchers from this field are highly interested in stomach robots to carry out their research, and a realistic stomach model will significantly impact the quality of quantitative food research. A stomach robot provides a test environment with repetitive, safe, and animal-free testing. Currently, the state-of-the-art rigid stomach robot is the backbone of stomach robot research and a source of inspiration to move the current state forward. The multi-disciplinary character of bio-inspired robots that simulate human organs proposes new opportunities for looking further than robots tailored for specific tasks.

It is challenging to develop soft robots that replicate the human stomach's motility, and state-of-the-art stomach robots perform ACW with rigid components. In such robots, there is no embodiment in the robot body, which either makes the controller very challenging or the overall functionality of the robot very simple. Therefore, an imitation of the human stomach motility by the soft robotics actuation techniques is necessitated for more versatile behaviour to generate peristaltic waves that can present embodied intelligent behaviour. The peristaltic wave data obtained from the medical field must be applied to the controller of the stomach simulator to provide a realistic stomach-like robot.

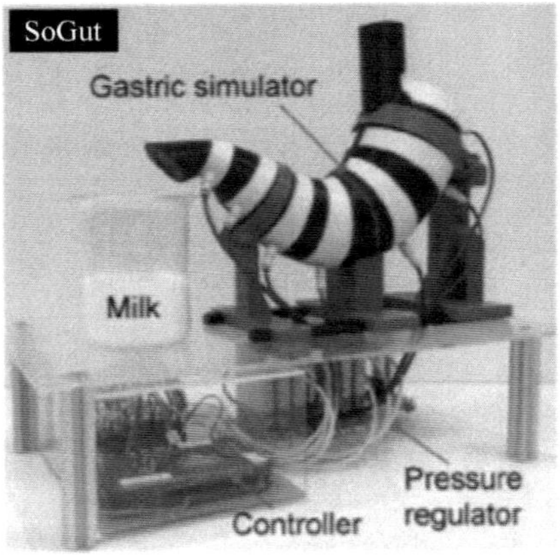

Figure 4. SoGut; soft robot stomach simulator with discrete soft actuators [12]

4. Soft ring actuator

4.1. Background

Bio-mimicry of the GI tract is essential to improve medical knowledge by testing on a physical robot, such as investigating dysphagia. As discussed in the stomach and esophagus sections, these organs' contraction waves are challenging to emulate. Other organs in the GI tract have a typical tubular shape and motor behaviour, a peristaltic wave of rhythmic contractions. It is the radial contraction that is generated by the smooth muscles. These contractions differ between organs in sequence, frequency, and velocity. Rather than designing a specific soft robot to mimic a targeted organ, we can create a generic soft ring actuator as a small section that can be modified to simulate any organ in the GI tract.

There are very few ring actuators that have been built for this purpose (see Fig. 5). Soft ring actuators are usually soft-bodied with embedded chambers that are pneumatically actuated. They emulate the contraction of a segment in the GI tract, tubular organs with different anatomy. Dang et al. presented a ring actuator to simulate a segment of the stomach [30]. It consists of one material with a single cylindrical chamber that inflates to produce a contraction. The same concept of the actuator is later used to construct a soft stomach robot. This ring actuator is simple to design, fabricate and control. However, the resulting contraction was not satisfactory for the significant contraction of the stomach, where a complete occlusion inside the soft ring is required. We previously presented an FEA study on a soft ring with multiple chambers with different designs to identify how the design parameters influence the contraction behaviour. Using a single material to construct the entire ring actuator presents unsatisfactory results compared to the biological body. Another ring actuator was developed from composite silicone to solve the large deformation issue in a soft ring actuator by emulating a segment of GI [32]. The same actuator was also used to build a full soft robotic stomach simulator with a continuous robot conduit to perform a mechanical peristaltic wave. Such an actuator can be employed to simulate other GI organs such as the esophagus and intestine.

(a) (b) (c)

Figure 5. Four examples of soft ring robots, (a) Finite element analysis for ring actuator of the same material with multiple chambers [33], (b) Bellows-driven soft pneumatic actuator that can be formed as a ring shape to simulate the peristaltic contractions and also serves as a gripper (gripping a strawberry) [34], (c) Entirely soft ring actuator that simulates a segment of the gastrointestinal tract with soft fingers [32].

4.2. The functionality of ring actuators

The ring actuator can be considered a fundamental design to develop more functional soft robotics that mimics an organ in the GI tract. In this method, more investigation on the specification of the organ behaviour can be structured with a small section for simplification. Another aspect of using this method is investigating the embedded sensory system for measuring the contraction. A sensory system can be a vital perspective for analysing the interactions between the surface and other elements that might influence the contraction. Though, it is very challenging to design an embedded sensor that can deform largely to accommodate the contraction of the soft ring actuators [35]—solving this problem by introducing sensors that adapt to the large deformation without influencing the performance of the robots. Embedded sensors in robots can present the forces that occur from the interactions, which aid in understanding the embodied intelligence and quantifying its performance.

Although soft ring actuators are constructed to mimic the contraction of the GI organ, they can be used in other applications such as gripping [30], in-hand manipulation [36], and industrial pumps [37]. The morphology of these actuators illustrates the adaptation to different environments and the applicability of using the same model for various applications, which is considered embodied intelligence. The self-organisation concept can also be seen in the in-hand manipulation application. The ring actuator consists of multiple fingers and can achieve manipulation without directly controlling the finger's bending motion. The break of symmetry concept determines the direction of the manipulation. Changing the frequency can influence the rotation of the manipulated object.

5. Discussion and conclusion

5.1. Embodied intelligence in organ simulators

The embodied intelligence can be seen in organ simulators by looking at the soft robots' behaviour that simulates peristaltic waves. The peristaltic waves require a self-organisation concept to be accomplished with the soft actuator. The material properties of soft robots, such as Ecoflex, achieve high strain/stress cycles with a simple control system. a series of embedded chambers in a tubular structure can be actuated in a sequence to achieve peristaltic waves. The deformation of a chamber influence the neighbouring chambers, which provide a continuous deformable surface rather than a discrete contraction. This aspect simplifies the control strategies and provides a more realistic deformable surface than biological behaviour. Self-organisation is one of the essential advantages of bio-inspired robotics, where rigid robots fail to mechanically simulate a movement like the smooth muscles in a human body. The three examples mentioned above used this technique (see Table 2).

The embodied intelligence can also be found in various applications that a soft robot can achieve. In the swallowing robot example, such a robot can serve as a test environment for simulating the swallowing process and an industrial pump. This example implies that the morphology of the swallowing robot is adaptive to a different environment by only altering the control schematic. The morphology of soft robotics can simplify the control complexity of required motion by employing the multi-degree of freedom inherited in their body. Another example is the soft ring actuator. The morphology of such an actuator helps these actuators to behave in many forms, such as simulating organ contractions, a gripper and an in-hand manipulation for industrial robotics applications. The in-hand manipulation method presented a break of symmetry and self-organisation concepts [36]. The symmetry break occurs in manipulating an object to rotate either clockwise or counterclockwise. The bifurcation can be determined by changing the frequency and duty cycles of the applied signals. The self-organisation behaviour is used by the soft fingers in the ring actuator (see Fig. 5) as the rotation of the object is also determined by the touchdown angle of a leader's finger. The other fingers always follow the same direction without an external control signal. This behaviour is specified as an underactuated system with no control over the bending angle of the soft fingers that rotate the contacted manipulated object. The self-organisation method is very beneficial for soft robotics applications. The morphology of the soft actuator is considered intelligent and can be included in the control strategies rather than cancel the useful multi-degree of freedom.

The current state of soft robotics with embodied intelligence presented various techniques and applications that benefited from the embodiment, such as self-organisation. Many robots were designed to accomplish a targeted task. However, improving the design of soft robots allow them to perform tasks in an unknown environment. The morphology plays a significant role in self-organisation. Soft actuators can achieve more adaptable behaviour in a complex work environment when the multi-degree of freedom can be employed in the design and control

Table 2. Summary of the current esophagus, stomach, and ring actuators that simulate the peristaltic waves in a human body.

Name	Body type	Actuation mode
Swallowing robot (RoSE) [21]	Soft body with rigid shell	Peristaltic contractions with pressurised air
Pump robot [38]	Foam conduit	Peristaltic with pressurised air
Artificial-esophagus [39]	Rigid body with soft conduit	Peristaltic contractions with SMA wires
TNO Intestinal Model (TIM-1)[40, 27]	Rigid body with flexible wall	Contractions with pressurised water
Human Gastric Simulator (HGS) [26]	Latex vessel with rigid rollers	Peristaltic with rigid rollers
Dynamic Gastric Model (DGM) [41, 42]	Rigid body	Rigid cylinder
Stomach simulator, [11]	Soft body with rigid actuators	Peristaltic with rigid wires
Soft robotic gastric simulator (SoGut) [12]	Soft body with rigid shell	Peristaltic with soft actuators
Soft robotic stomach simulator (SoRSS) [43]	Entirely soft-bodied	Peristaltic with soft actuators
Ring-shaped actuator [30]	Soft body with rigid shell	Shallow contractions with pressurised air
Ring soft pneumatic actuator (RiSPA) [32]	Entirely soft composite silicone	Large contractions with pressurised air
Soft ring actuator [33]	Entirely soft same material	Shallow contractions with pressurised air

strategies. The biological living organism couple the brain and body as one entity [2]. Following the biological example of embodied intelligence will result in an attractive behaviour in terms of adaptability, simple control, fewer actuation components, and overall intelligent soft robots that can achieve more than a simple task.

5.2. The challenges and future direction

The struggle to move forward in soft robotics can be narrowed down to sensing and actuation problems. Although material science provides methods that can be used in soft robotics, soft sensors are still a significant drawback that limits the controllability of soft robotics. Soft sensors suffer from reliable high stress/strain profiles that occur in soft materials. Soft materials stress/strain relationship can deviate with time, and the hysteresis curve might also get affected by materials properties changing with time. Another impediment is soft sensor discontinuity of measurement in large deformable robots. This issue occurs when the sensor stretchability reaches the maximum limit. Improving soft sensor design will solve many challenges the soft robotics control method faces.

The actuation challenge lies in simplifying soft robotics's rigid and bulky tethered system. A soft robot can be a small form but tethered with a sizeable rigid actuation system such as a pneumatic system. Developing a soft actuation system can simplify and provide an entirely soft robot that is untethered. The new generation of soft robotics is expected to employ these techniques to move forward in the soft robotics field. Self-organisation can be an excellent direction with actuation systems that communicate together to achieve a task. In a pneumatically actuated system, a discrete actuation method, a separate pneumatic valve controls each chamber, limiting the possibilities of soft robots' behaviour. A solution to this method is employing self-organisation in the actuation system by controlling the multi-chamber system with a design that communicates to each other by using the material properties and how the material responds to the same applied pressure [44]. We can develop a single input multi-output system with an entirely soft robot's body and actuators for more versatile applications by employing such techniques.

It is challenging to identify and quantify the intelligence of robots simulating organs. It is acknowledged that embodied intelligence exists in the material and morphology used to construct these robots. However, can we call these robots intelligent?. A significant question that researchers are looking to solve. Concepts have been established, such as the Turning test and the Chinese room. Although the robot body can entirely rely on the material embodied intelligence, today's actuation methods are still premature for obtaining such intelligence. An embodied intelligent soft actuator and sensors are an essential research direction that advances this field. Suppose we can develop a fully embodied intelligent soft robot that simulates a human organ (including soft actuators and sensors) with a central pattern generator control method. In that case, one can argue that we can investigate implant strategies and not only use these robots for a test environment. This can be done by closing the human-robot reality gap with embodied intelligent behaviour, which is still in an early stage but is feasible as a research direction.

In this chapter, we discussed embodied intelligence from the perspective of soft robots that simulate human organs. We briefly presented three examples of how soft robots can simulate an organ and why we need such a device. The swallowing robot, stomach robot, and ring actuators simulated the gastrointestinal tract were presented. We discuss the benefit of embodied intelligence to achieve an advanced robot that simulates organs. The current state and the future direction were also communicated.

References

[1] Pfeifer R, Lungarella M and Iida F 2012 *Communications of the ACM* **55** 76–87
[2] Pfeifer R, Lungarella M and Iida F 2007 *science* **318** 1088–1093

[3] Rus D and Tolley M T 2015 *Nature* **521** 467–475
[4] Majidi C 2013 *Soft Robotics* **1** 5–11
[5] Rus D and Tolley M T 2018 *Nature Reviews Materials* **3** 101
[6] Roche E T, Horvath M A, Alazmani A, Galloway K C, Vasilyev N V, Mooney D J, Pigula F A and Walsh C J 2015 *ASME IDETC-CIE* vol 57120 p V05AT08A042
[7] Ferrua M J and Singh R P 2010 *Journal of Food Science* **75** R151–R162
[8] Pal A, Indireshkumar K, Schwizer W, Abrahamsson B, Fried M and Brasseur J G 2004 *Proceedings.Biological sciences / The Royal Society* **271** 2587–2594
[9] Dirven S, Xu W, Cheng L K, Allen J and Bronlund J 2013 *International Journal of Biomechatronics and Biomedical Robotics* **2** 163–171
[10] Dirven S, Chen F, Xu W, Bronlund J E, Allen J and Cheng L K 2013 *Mechatronics, IEEE/ASME Transactions on* **PP** 1–9
[11] Condino S, Harada K, Pak N, Piccigallo M, Menciassi A and Dario P 2011 *Applied Bionics and Biomechanics* **8** 267–277
[12] Dang Y, Liu Y, Hashem R, Bhattacharya D, Allen J, Stommel M, Cheng L K and Xu W 2020 *Soft Robotics*
[13] Gray H 1918 *Anatomy of the human body* 20th ed (Philadelphia: Lea and Febiger)
[14] Dodds W J 1989 *Dysphagia* **3** 171–178
[15] Chen J 2009 *Food Hydrocolloids* **23** 1–25
[16] Hiiemae K 2004 *Journal of Texture Studies* **35** 171–200
[17] Logemann J A 1998 *Pro-ed Austin, TX* 395–400
[18] Dantas R O, Alves L M T, Dalmazo J, Santos C M d, Cassiani R d A and Nascimento W V d 2010 *Arquivos de gastroenterologia* **47** 339–343
[19] Chen F, Dirven S, Xu W, Bronlund J, Li X and Pullan A 2012 *Mechatronics* **22** 556–567
[20] Bhattacharya D, Hashem R, Cheng L K and Xu W 2021 *IEEE Transactions on Industrial Electronics*
[21] Bhattacharya D, Ali S J, Cheng L K and Xu W 2021 *Soft Robotics* **8** 397–415
[22] Orvar K B, Gregersen H and Christensen J 1993 *Digestive diseases and sciences* **38** 197–205
[23] Humphries T J and Castell D O 1977 *The American journal of digestive diseases* **22** 641–645
[24] Pfeifer R and Scheier C 2001 *Understanding intelligence* (MIT press)
[25] Ehrlein H J and Schemann M 2005 *Technische Universität München: Munich* 1–26
[26] Kong F and Singh R P 2010 *Journal of food science* **75** E627–E635
[27] Blanquet S, Zeijdner E, Beyssac E, Meunier J, Denis S, Havenaar R and Alric M 2004 *Pharmaceutical research* **21** 585–591
[28] Chen J, Pan J and McCallum R 1995 *Digestive Diseases* **13** 275–290
[29] O'Grady G, Du P, Cheng L K, Egbuji J U, Lammers W J, Windsor J A and Pullan A J 2010 *American journal of physiology-Gastrointestinal and liver physiology* **299** G585–G592
[30] Dang Y, Stommel M, Cheng L K and Xu W 2019 *Soft Robotics* **6** 444–454
[31] Li Y, Fortner L and Kong F 2019 *Food Research International* **125** 108598
[32] Hashem R, Kazimi S, Stommel M, Cheng L K and Xu W 2021 *Soft Robotics*
[33] Hashem R, Xu W, Stommel M and Cheng L K 2017 *5th International Conference, RITA* (Springer) pp 475–487
[34] Hashem R, Stomme M, Cheng L K and Xu W 2020 *IEEE/ASME Transactions on Mechatronics* **26**(5) 2327–2338
[35] Dang Y, Devaraj H, Stommel M, Cheng L K, McDaid A J and Xu W 2020 *Soft robotics* **7** 478–490
[36] Hashem R and Iida F 2022 *RoboSoft 2022, IEEE* (Accepted) p https://doi.org/10.48550/arXiv.2202.12525
[37] Ashigaki K, Iwasaki A, Hagiwara D, Negishi K, Matsumoto K, Yasuyuki Y, Habu H and Nakamura T 2018 *2018 7th IEEE Biorob* (IEEE) pp 1291–1296
[38] Esser F, Steger T, Bach D, Masselter T and Speck T 2017 *Conference on Biomimetic and Biohybrid Systems* (Springer) pp 138–147
[39] Miki H, Okuyama T, Kodaira S, Luo Y, Takagi T, Yambe T and Sato T 2010 *International Journal of Applied Electromagnetics and Mechanics* **33** 705–711
[40] Barker R, Abrahamsson B and Kruusmägi M 2014 *Journal of pharmaceutical sciences* **103** 3704–3712
[41] Mercuri A, Passalacqua A, Wickham M S, Faulks R M, Craig D Q and Barker S A 2011 *Pharmaceutical research* **28** 1540–1551
[42] Vardakou M, Mercuri A, Barker S A, Craig D Q, Faulks R M and Wickham M S 2011 *Aaps Pharmscitech* **12** 620–626
[43] Hashem R, Kazimi S, Stommel M, Cheng L K and Xu W 2021 *Soft Robotics. Under review*
[44] Gorissen B, Milana E, Baeyens A, Broeders E, Christiaens J, Collin K, Reynaerts D and De Volder M 2019 *Advanced Materials* **31** 1804598

IOP Conf. Series: Materials Science and Engineering 1261 (2022) 012008 doi:10.1088/1757-899X/1261/1/012008

Multifunctional Underwater Soft Robots: A Simulation Essay

Anup Teejo Mathew[1], Costanza Armanini[1], Aysha Ali Samra Ali Alshehhi[1], Ikhlas Mohamed Ben Hmida[1], Federico Renda[*1,2]

[1] Department of Mechanical Engineering, Khalifa University of Science and Technology, Abu Dhabi, UAE.
[2] Khalifa University Center for Autonomous Robotic Systems (KUCARS), Khalifa University, Abu Dhabi, UAE.

E-mail: * federico.renda@ku.ac.ae

Abstract. Underwater soft robotics is receiving growing popularity within the scientific community, thanks to its prospective capability of tackling challenges that are hard to deal with using traditional rigid technologies, especially while interacting with an unstructured environment. Recently, we proposed a multi-module underwater robotic system with deformable propellers, inspired by bacteria morphology [1]. Here, the same bio-inspired modular structure is employed to perform manipulation tasks, in order to design a multi-functional integrated system. Employing the Geometric Variable Strain Approach, we simulate a scenario where the flagellated robot moves towards a preferred target and, using the same soft appendages, it hooks to it, simulating a monitoring task. The modeling approach and the design allow the Embodied Intelligence principles to exploit the robot's surrounding environment (water), the shape of the grip-target and the robot's compliant nature to mediate effective navigation and safe interaction with the target, using few control inputs.

1. Introduction

The world's oceans cover two thirds of the Earth's surface and they have been an essential element of human welfare throughout history, representing a critical fount of resources which are crucial to economic development around the world. Remotely Operated (ROVs) and Unmanned Underwater Vehicles (UUVs) have been introduced to replace human divers in underwater operations, with incomplete success. While they have been successfully applied to safely patrol open stretches of sea, they suffer from non-negligible limitations when navigating close to the seabed or near submerged structures where unintended impacts must be prevented consistently [2]. Embodied intelligence principles can be exploited to allow the continuous adaptability of an underwater ROV's components to fulfill multiple functionalities. In this scenario, soft robotics appears as a promising solution to apply this idea. In recent times, the soft robotics community has grown exponentially, providing solutions that can't be obtained by traditional rigid technologies. In particular, underwater robotics represents a promising application field for soft robots, as demonstrated by the vast number of soft underwater robots recently proposed [3–11]. Unlike their rigid counterparts, soft robots can handle significant pressure variations, interact safely with the marine life and with delicate and uneven structures [12–15].

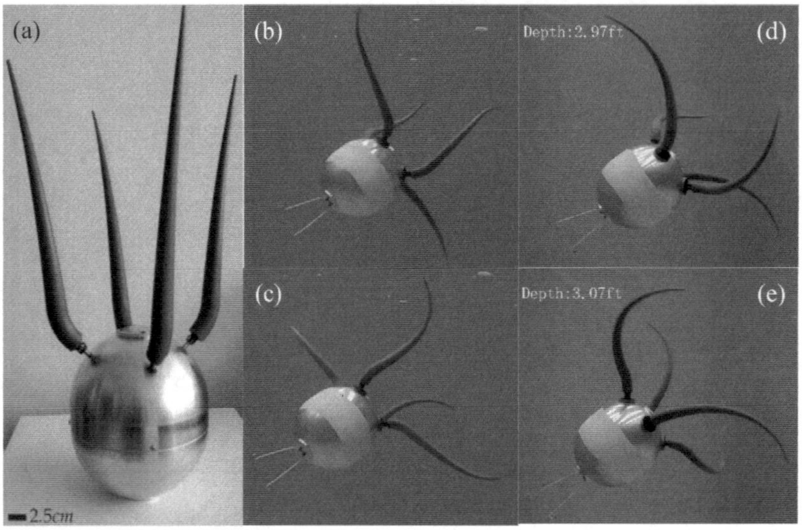

Figure 1. The multi module robot propelled by flagellar appendages previously presented in [1]. (Left) The robot consists of a dome shell containing the mechatronics components and four soft flagellar modules providing the propulsion capabilities. (Right) The robot navigating in a swimming pool.

In a recent work [1], we presented a soft propeller that is inspired by prokaryotic bacterial flagella. The soft module is actuated by a motor that continuously rotates, so that the propeller, interacting with the surrounding fluid, deforms to resemble a helical shape that finally provides a propulsive thrust. Leveraging on these results, we also built a complete untethered vehicle which consists of a dome shell containing all the mechatronics components and which is propelled by four soft modules. The four motors operate similarly to a quadrotor, by setting the speed of the four motors it is possible to obtain different motion directions, Fig. 1.

Extending these previous results, we investigate a multi-functional module for underwater intervention, which is capable of overcoming some of the most stringent limitations of the current state-of-the-art. We aim to do that by providing the same soft flagellate appendages that were used as propellers in [1] with grasping capabilities. We propose a modular structure where the soft propellers' geometry, compliance, and environmental interactions are leveraged to simplify locomotion and grasping tasks. Each module will combines propulsion and manipulation tools to obtain a redundant system able to adapt to different tasks. In our approach, instead of trying to solve directly the underwater mobile manipulation problem, fighting against the intrinsic and adverse conditions of the environment, we exploit the natural compliance and redundancy of the proposed design to wrap around and firmly cage the object or structure of interest. The flexibility and adaptability of the module structure, as well as the possibility of physically interacting with the surroundings, ensure a safer and more efficient operation in a cluttered environment, drastically reducing the entanglement problem. These features, coupled with peculiar noiseless locomotion, make this kind of robots suitable for a broad range of tasks currently unsolved, such as the non-invasive monitoring of water quality and marine habitats, easing the mapping of species and their distribution. The remaining parts of the paper are organized as follows. In Section 2 we tackle the design of the actuation mechanism to provide the grasping capabilities to the soft modules. In Section 3 we introduce the modelling of both the locomotion and the manipulation tasks, while in 4 we present the results of the simulations. Finally, in Section 5 we draw some conclusion and we discuss the possible future prospects of this work.

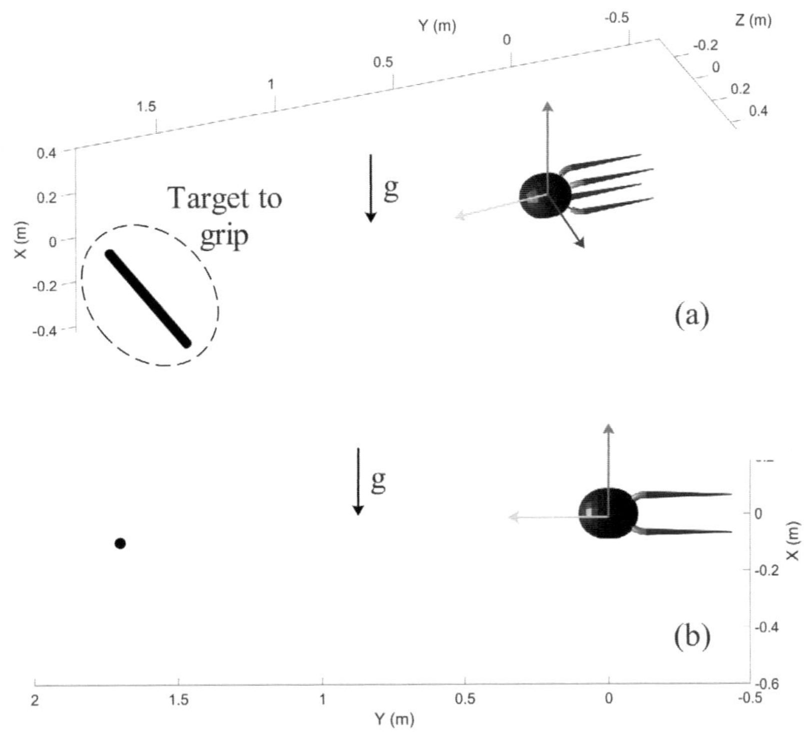

Figure 2. The envisaged scenario: the robot, propelled by the soft flagellar modules, reaches a target and, using the same multifunctional modules, anchors to it. The 3D (a) and the X-Y plane view (b) are shown.

2. Active Soft Filament

As discussed in the introduction, the aim of this work is to simulate a scenario where the multi- modular robot presented in [1] reaches a specific target, propelled by the soft flagellar modules, and finally anchors to it. In particular, envisaging a monitoring task, for example for the detection of a leakage in an underwater pipeline, we consider that the robot must attach to a rigid tube, Fig. 2. In the original design, the propulsive actuation is provided by a motor that continuously rotates. The soft propeller consists of two parts: a 50 mm long cylindrical *hook*, having radius 12.5 mm and a 45° pre-curvature, and a conical 300mm long *filament* having a radius ranging from 12.5mm to 2mm [16]. The *hook* transmits the continuous rotation of the motor to the filament which, interacting with the water, forms a helical shape providing the necessary propulsive thrust. While in the original design the hook was designed as a soft torsion-less body, here we consider it as rigid.

In order to provide the flagellar module with grasping capabilities, an appropriate actuation mechanism, such as pneumatic or cable-driven mechanisms [17, 18], should be designed. In this work, we consider the presence of embedded cable actuators along the soft *filament*. In particular, we simulate the presence of off-centered cables subject to a tension force, to obtain the required bending behaviour. The tension force can be provided by a motor (M2) pulley system as shown in Fig. 3. We consider an oblique cable present throughout the *filament* and having initial coordinates $(y, z) = (10, 0)$ mm and reaching the tip at $(y, z) = (2, 0)$ mm, where y and z are the local y and z coordinates. In this way, applying a tension to the cable results in a bending about the z-axis. Another motor (M1) and a gear system can supply the torque that rotates the *filament*. We can use a swivel mechanism to decouple the cable rotation and avoid

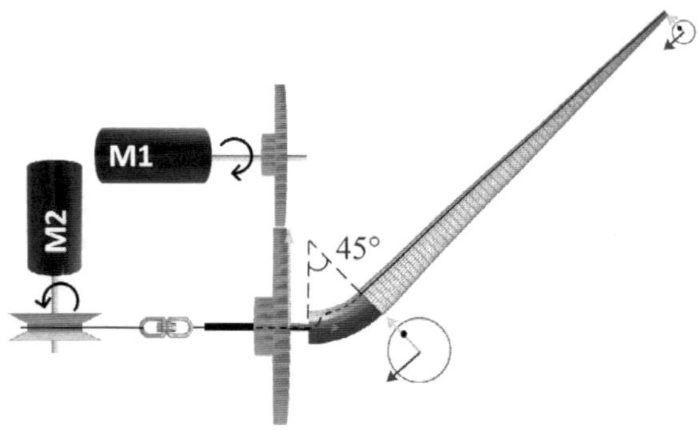

Figure 3. Schematics of the proposed design. An off-centered cable, actuated by a motor (M2), is described by the black dot inside the red circles (cross-section images of the filament), provides a bending actuation. Another motor (M1) rotates the filament providing propulsion to the robot. Red, green, and blue arrows represent local x, y, and z axes, respectively.

coiling of the cable.

3. Mathematical Modelling

The hybrid soft-rigid robot is modelled using the Geometric Variable Strain Approach (GVS), [19, 20]. In particular, a soft body is treated as a Cosserat rod, represented by a center line and a continuous stack of local reference frames representing the cross section of the body. In particular, in the GVS approach, the configuration is parametrized by the rod's strain fields, which are discretized on a truncated basis of strain functions. When the strains are treated as constants, the GVS approach simplifies to the Piecewise Constant Strain approach [21]. This modelling technique can be applied to model rigid, soft and hybrid soft-rigid robots, including internally actuated soft bodies [19] and mobile robots (free joints). In particular, the dynamic equations of the system take the form:

$$\boldsymbol{M}\ddot{\boldsymbol{q}} + (\boldsymbol{C} + \boldsymbol{D})\dot{\boldsymbol{q}} + \boldsymbol{K}\boldsymbol{q} = \boldsymbol{B}\boldsymbol{u} + \boldsymbol{F}, \quad (1)$$

where $\boldsymbol{q}$ is the joint coordinates vector of the system, $\boldsymbol{M}(\boldsymbol{q})$ is the generalized mass matrix, $\boldsymbol{C}(\boldsymbol{q}, \dot{\boldsymbol{q}})$ is the generalized Coriolis matrix, $\boldsymbol{D}$ is the generalized damping matrix, $\boldsymbol{K}$ is the generalized stiffness matrix, $\boldsymbol{B}(\boldsymbol{q})$ is the generalized actuation matrix, $\boldsymbol{F}(\boldsymbol{q}, \dot{\boldsymbol{q}})$ is the vector of generalized external forces and $\boldsymbol{u}$ is the vector of applied actuation forces. The joints coordinates of the system ($\boldsymbol{q}$) correspond to the 6 degrees of freedoms (DoF) of the body and the DoFs of each of the arms. In particular, for the multi-functional soft filaments, the actuation matrix $\boldsymbol{B}$ accounts for both the propulsive actuation provided by the motor's rotation and the tendon driven actuation. Similarly, the actuation force vector $\boldsymbol{u}$ accounts for both the motor's torque and the cable actuator strength.

3.1. SoRoSim Toolbox

To simulate the dynamic motion of the robot, we employ the MATLAB-based toolbox SoRoSim [22, 23]. The SoRoSim toolbox can solve static and dynamic simulations of open-chain, closed-chain, and branched soft, rigid, or hybrid robots. The user can interactively create soft and rigid links and combine them to form various linkage structures. The toolbox saves the links and linkages as MATLAB class elements, with their physical properties stored as class properties and the mathematical operations as class methods.

The design consists of a rigid link, represented by the spheroidal shell, the motors shafts and the four flagellar modules. The joint at the beginning of the shell is a passive free joint having six degrees of freedom. Similarly, the flagellar modules are created combining a rigid shaft with a revolute joint, the rigid *hook* and the soft *filament*, Fig. 3. The soft *filament* is modeled as an inextensibile Kirchhoff rod, allowing the torsion about the x-axis and the bending about the

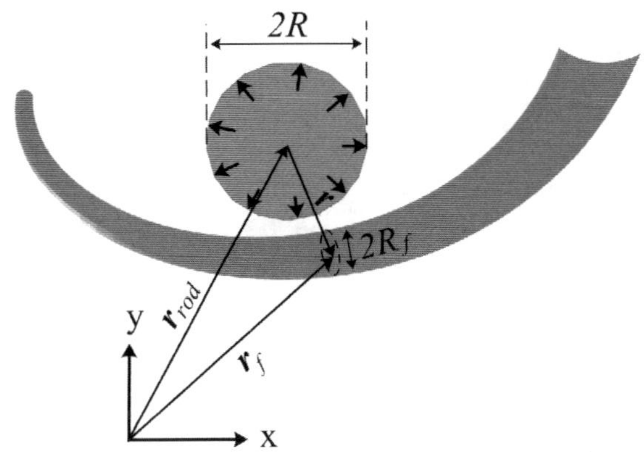

Figure 4. Schematics of the contact force modelling approach. The rod is represented by a repulsive radial force field which is proportional to the penetration of the flagellar modules (or robot body) into it.

y and z axes. In particular, we considered a linear strain distribution. In total, each module is characterized by 7 DoF (1 for the revolute joint and 6 for the soft filament). To summarize, the strain base used for the body is $\boldsymbol{I}(6)$, for the revolute joint it is $[1, 0, 0, 0, 0, 0]'$, and for soft filament it is given by:

$$\boldsymbol{B}_{filament} = \begin{bmatrix} 1 & X & 0 & 0 & 0 & 0 \\ 0 & 0 & 1 & X & 0 & 0 \\ 0 & 0 & 0 & 0 & 1 & X \\ 0 & 0 & 0 & 0 & 0 & 0 \\ 0 & 0 & 0 & 0 & 0 & 0 \\ 0 & 0 & 0 & 0 & 0 & 0 \end{bmatrix}, \quad (2)$$

where X is a curvilinear abscissa in the longitudinal direction of the filament. The total degrees of freedom of the model is $6 + 4 \times 7 = 34$. We used 15 integration points (Gauss quadrature points) to compute the actuation and the external forces on each soft *filament*. The added mass and the drag and lift matrices, representing the fluid interaction on the body, are then saved as properties of SoRoSim linkage class. In total, there are eight actuators on the robot: four angle-controlled revolute joints representing the motors used to rotate the soft *filaments* and four soft actuators (two of which are used later in the simulation).

3.2. Contact Force Modelling

With regards to the modelling of the actuation and the external forces, including the fluid-structure interaction, the reader is referenced to the previous work by the authors [1]. Here, we address the modelling of the contact interaction between the flagellar modules and the targeted rod, Fig. 2a and b. Let us consider a cross-section of the rod and the filament at an arbitrary value of the global coordinate z (corresponding to the location of an integration point on the soft link), Fig. 4. The contact force is then modelled as a repulsive force field acting in the radial directions of the rod.

Let $\boldsymbol{r}_{rod}$ and $\boldsymbol{r}_f$ be the position vectors towards the center of the rod and the center of a cross-section of the soft filament, which is closest to the rod center, at an arbitrary z coordinate. Thus, the position of the cross-section of the soft filament with respect to the rod is given by $\boldsymbol{r} = \boldsymbol{r}_f - \boldsymbol{r}_{rod}$. It should be noted that the z component of $\boldsymbol{r}$ is 0. The magnitude of the force is proportional to the penetration of the soft filament into the rod. Finally, the contact force is given by:

$$\boldsymbol{F}_C = \begin{cases} K_C(R + R_f - ||\boldsymbol{r}||)\hat{\boldsymbol{r}}, & \text{if } R + R_f > ||\boldsymbol{r}|| \\ 0, & \text{otherwise} \end{cases}, \quad (3)$$

where K_C is a stiffness constant, R is the radius of the rod, R_f is the radius of the cross-section of the soft filament, $||\boldsymbol{r}||$ is the magnitude of $\boldsymbol{r}$, and $\hat{\boldsymbol{r}}$ is the unit vector in the $\boldsymbol{r}$ direction. The force is computed at every integration points on the soft link, transformed into generalized force

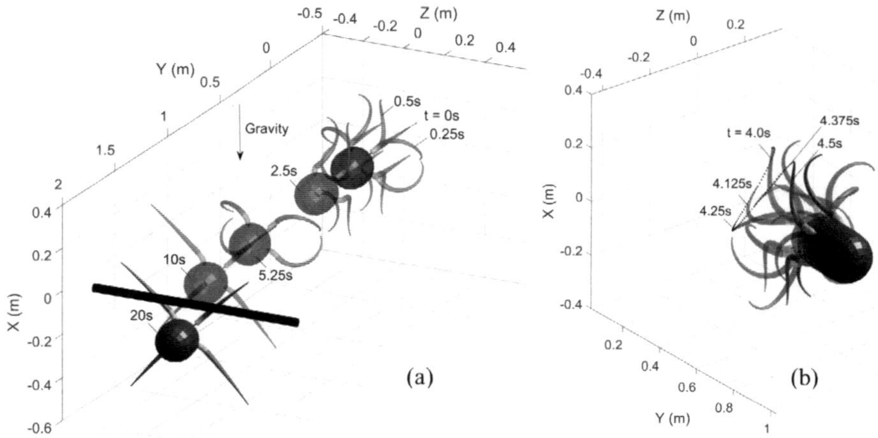

Figure 5. Locomotion of the flagellate robot: (a) Superimposed robot images from the start to the final position. All four motors are rotated at 2 rps and they are stopped after 5.25s. (b) Superimposed images of the robot for one rotation of the motors. The dashed line shows the tip trajectory of a module.

vectors using appropriate Jacobians, and added to the generalized force vector ($\boldsymbol{F}$) in equation (1). Similarly, we also model the contact force on the body of robot. The value of K_C used for the simulation is 500 N/m.

4. Simulation Results

The envisaged scenario consists of three phases: the robot navigates towards the target (the rod), it brakes to approach it, and finally the cables are pulled to grasp the rod and anchor the soft modules to it. With regards to the navigation part, the four motors are rotated in alternate directions but with the same speed, to obtain the desired forward motion [1]. In particular, the motors rotate at 2rps for a total of 5.25 seconds. Once the robot approaches the target, the motors are stopped and the soft filaments assume a spread-out configuration, offering the maximum drag required to slow the robot to a stop. This second part lasts for 20 seconds, when the vehicle reaches the target position and stops below it. The snapshots of these first two parts of the simulation are shown in Fig. 5. The interaction forces due to the fluid environment on the soft filaments lead to its helical shape, which provides a positive thrust for the propulsion. The drag force exerted by the fluid on the spread-out configuration of the robot would lead to the slowing down of the robot. Hence, the propulsion and breaking are accomplished by environment interaction rather than explicit design.

The final anchoring phase starts once the top two modules are aligned below the rod. The tension in each cable is applied as a ramp input $0 - 8$ N for 2 seconds and is kept constant afterwards. As a result of the actuation and contact force, the soft filaments wrap around the rod and pull the robot up. The results of the anchoring process are shown in Fig. 11. After the contact of the shell with the rod at $t = 22.2s$, the robot is pushed down. The firm spiral grip of the soft filament Fig. 6b is an emergent behavior due to the resultant actuation and contact force.

5. Discussion and Conclusion

In this paper, we presented a multi-functional soft modular robot that provides a new versatile tool for underwater intervention, overcoming some of the most stringent limitations of the

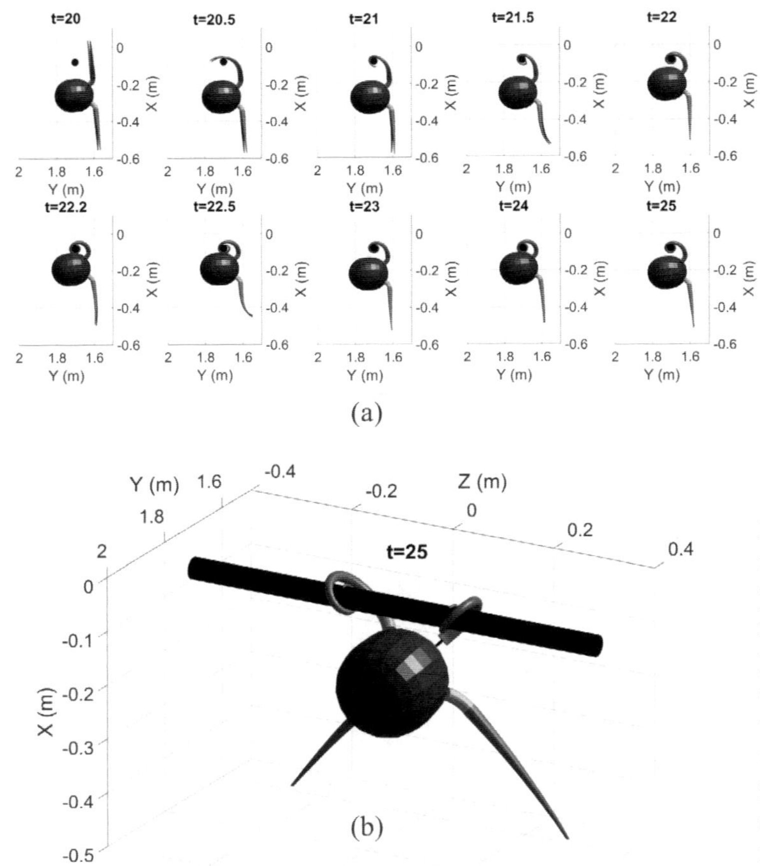

Figure 6. Simulation of the manipulation task: (a) Superimposed images of the robot at different times during the gripping phase. The two top modules contact the rod at around 20.5s, while the body touches the rod at 22.2s. (b) 3D view of the configuration at 25s. The soft filaments spirals around the rod to form a firm grip.

current state-of-the-art. We investigated the locomotion and manipulation capabilities of soft filaments that can bend and twist subjected to internal actuation and fluid interaction forces. This combination leads to the emergence of intelligent deformations and motions that aid the locomotion and gripping with the need of a few actuators.

We demonstrated the feasibility of our idea by simulating the dynamics of the system using the GVS approach to provide a multi-functional device which, through some simple control inputs (the motor's rotation and the cable tension) is capable of propelling the robot and actuating it to safely interact with the surroundings. This result is an implicit consequence of the compliant nature of the soft materials that are employed for the proposed design, which enriches the system of an embodied intelligence that would not be possible with a traditional rigid technology. Exploiting its natural redundancy, the proposed design can perform manipulation tasks by caging the object or the structure of interest. Caging provides a way to control and move a body without immobilizing it, gaining tolerance towards uncertainty and easing manipulation tasks in unstructured environments. In this way, the proposed flagellum-inspired system shapes a new class of soft robots especially suited for intervention and inspection in fluid-filled conduits of diverse size and fluid composition. Similar techniques could be applied to underwater ROVs with soft or underactuated components.

Needless to say, the next task involves the fabrication of the proposed design. This is definitely a challenging task, which requires, first of all, an in-depth analysis of the proposed actuation mechanism to provide the grasping capabilities to the soft modules. In this scenario, the final design should be characterized by the perfect co-existence of the two actuators (for the locomotion and for the manipulation), which are supposed to work without limiting each other. This will definitely be the the most complicated part, but at the same time it represents the main novelty of our approach. Moreover, being composed of soft modules mechanically and electronically connected, the proposed system provides a unique opportunity for developing innovative control algorithms based on parallel-distributed-processing where each module can be considered as an agent of a connected network contributing to the global goal.

6. Acknowledgments

This work was supported by the US Office of Naval Research under GRANT: N62909-21-1-2033 and in part by the Khalifa University of Science and Technology under GRANTS: CIRA-2020-074 and RC1-2018-KUCARS.

References

[1] Armanini C, Farman M, Calisti M, Giorgio-Serchi F, Stefanini C, Renda F. Flagellate Underwater Robotics at Macroscale: Design, Modeling, and Characterization. IEEE Transactions on Robotics. 2021:1-17.

[2] Aracri S, Giorgio-Serchi F, Suaria G, Sayed ME, Nemitz MP, Mahon S, et al. Soft Robots for Ocean Exploration and Offshore Operations: A Perspective. Soft Robotics. 0;0(0):null. PMID: 33450174. https://doi.org/10.1089/soro.2020.0011.

[3] Giorgio-Serchi F, Arienti A, Corucci F, Giorelli M, Laschi C. Hybrid parameter identification of a multi-modal underwater soft robot. Bioinspiration & Biomimetics. 2017 Feb;12(2):025007. https://doi.org/10.1088/1748-3190/aa5ccc.

[4] Park SJ, Gazzola M, Park KS, Park S, Santo VD, Blevins EL, et al. Phototactic guidance of a tissue-engineered soft-robotic ray. Science. 2016 Jul;353(6295):158-62. https://doi.org/10.1126/science.aaf4292.

[5] Marchese AD, Onal CD, Rus D. Autonomous Soft Robotic Fish Capable of Escape Maneuvers Using Fluidic Elastomer Actuators. Soft Robotics. 2014;1(1):75-87. PMID: 27625912. https://doi.org/10.1089/soro.2013.0009.

[6] Stefanini C, Orofino S, Manfredi L, Mintchev S, Marrazza S, Assaf T, et al. A novel autonomous, bioinspired swimming robot developed by neuroscientists and bioengineers. Bioinspiration & Biomimetics. 2012 may;7(2):025001. https://doi.org/10.1088/1748-3182/7/2/025001.

[7] Li G, Chen X, Zhou F, Liang Y, Xiao Y, Cao X, et al. Self-powered soft robot in the Mariana Trench. Nature. 2021 Mar;591(7848):66-71. https://doi.org/10.1038/s41586-020-03153-z.

[8] Giorgio Serchi F, Arienti A, Laschi C. Biomimetic Vortex Propulsion: Toward the New Paradigm of Soft Unmanned Underwater Vehicles. IEEE/ASME Transactions on Mechatronics. 2013;18(2):484-93. https:/doi.org/10.1109/TMECH.2012.2220978.

[9] Sfakiotakis M, Kazakidi A, Chatzidaki A, Evdaimon T, Tsakiris DP. Multi-arm robotic swimming with octopus-inspired compliant web. In: 2014 IEEE/RSJ International Conference on Intelligent Robots and Systems; 2014. p. 302-8. https://doi.org/10.1109/IROS.2014.6942576.

[10] Arienti A, Calisti M, Giorgio-Serchi F, Laschi C. PoseiDRONE: Design of a soft-bodied ROV with crawling, swimming and manipulation ability. In: 2013 OCEANS - San Diego; 2013. p. 1-7. https://doi.org/10.23919/OCEANS.2013.6741155.

[11] Katzschmann RK, DelPreto J, MacCurdy R, Rus D. Exploration of underwater life with an acoustically controlled soft robotic fish. Science Robotics. 2018;3(16):eaar3449. https://www.science.org/doi/abs/10.1126/scirobotics.aar3449.

[12] Laschi C, Mazzolai B, Cianchetti M. Soft robotics: Technologies and systems pushing the boundaries of robot abilities. Science Robotics. 2016;1(1). https:/doi.org/10.1126/scirobotics.aah3690.

[13] Galloway KC, Becker KP, Phillips B, Kirby J, Licht S, Tchernov D, et al. Soft Robotic Grippers for Biological Sampling on Deep Reefs. Soft Robotics. 2016;3(1):23-33. PMID: 27625917. https://doi.org/10.1089/soro.2015.0019.

[14] Whitesides GM. Soft Robotics. Angewandte Chemie International Edition. 2018;57(16):4258-73. https://onlinelibrary.wiley.com/doi/abs/10.1002/anie.201800907.

[15] Alici G. Softer is Harder: What Differentiates Soft Robotics from Hard Robotics? MRS Advances. 2018;3(28):1557–1568. https://doi.org/10.1557/adv.2018.159.

[16] Calisti M, Giorgio-Serchi F, Stefanini C, Farman M, Hussain I, Armanini C, et al. Design, Modeling and Testing of a Flagellum-inspired Soft Underwater Propeller Exploiting Passive Elasticity. In: 2019 IEEE/RSJ International Conference on Intelligent Robots and Systems (IROS); 2019. p. 3328-34. https://doi.org/10.1109/IROS40897.2019.8967700.

[17] Hughes J, Culha U, Giardina F, Guenther F, Rosendo A, Iida F. Soft Manipulators and Grippers: A Review. Frontiers in Robotics and AI. 2016;3:69. https://www.frontiersin.org/article/10.3389/frobt.2016.00069.

[18] Rus D, Tolley MT. Design, fabrication and control of soft robots. Nature. 2015 May;521(7553):467-75. https://doi.org/10.1038/nature14543.

[19] Renda F, Armanini C, Lebastard V, Candelier F, Boyer F. A Geometric Variable-Strain Approach for Static Modeling of Soft Manipulators With Tendon and Fluidic Actuation. IEEE Robotics and Automation Letters. 2020;5(3):4006-13. https://doi.org/10.1109/LRA.2020.2985620.

[20] Boyer F, Lebastard V, Candelier F, Renda F. Dynamics of Continuum and Soft Robots: A Strain Parameterization Based Approach. IEEE Transactions on Robotics. 2020:1-17. https://doi.org/10.1109/TRO.2020.3036618.

[21] Renda F, Boyer F, Dias J, Seneviratne L. Discrete Cosserat Approach for Multisection Soft Manipulator Dynamics. IEEE Transactions on Robotics. 2018 Dec;34(6):1518-33. https://doi.org/10.1109/TRO.2018.2868815.

[22] Mathew AT, Hmida IB, Armanini C, Boyer F, Renda F. SoRoSim: a MATLAB Toolbox for Soft Robotics Based on the Geometric Variable-strain Approach. CoRR. 2021;abs/2107.05494. https://arxiv.org/abs/2107.05494.

[23] SoRoSim: An open-chain robotics simulation toolbox;. Version 2.32, Updated 11 Jul 2021. https://github.com/Ikhlas-Ben-Hmida/SoRoSim and https://www.mathworks.com/matlabcentral/fileexchange/83038-sorosim.

IOP Conf. Series: Materials Science and Engineering 1261 (2022) 012009 doi:10.1088/1757-899X/1261/1/012009

Leveraging Embodied Intelligence for Dexterous Robotic Manipulators Through Iterative Co-design

Kai Junge & Josie Hughes

CREATE Lab - Computational Robot Design & Fabrication Lab, EPFL, 1015 Lausanne, Switzerland

E-mail: kai.junge@epfl.ch, josie.hughes@epfl.ch

Abstract. The role of embodied intelligence (EI) has the potential to overcome current limitations in the fabrication, control, and resulting behavior to create robust and effective dexterous robotic manipulators. To develop hands that truly exploit EI, we must design hands by considering the entire system: the physical body, sensory systems, and the brain (the controller). However, we lack clear approaches and methods that enable this system level design for hands. We introduce an iterative approach for co-design which seeks to utilize simulation and real world evaluation to maximize the performance by distributing EI across the different elements of the system. To achieve this vision we require hands that can be rapidly fabricated with variability in the design space. Thus, to further the development of robotic hands that utilize EI we need streamlined fabrication pipelines which incorporate spatially distributed sensors, complex geometries and materials, and control distributed at the sensory-motor and high task planning domains.

1. Introduction

Robotic manipulation has long been an active research field involving interdisciplinary researchers from domains including neuroscience, engineering, computer science, and materials science [30, 33, 1]. A key focus and open challenge is achieving more human like dexterity [24, 28]. Dexterity describes the ability to perform complex interactions with objects and the environment including in-hand manipulation, the ability to translate and rotate objects [31], and the ability to exploit environmental constraints, for example sliding a coin over a table to grasp it off the edge. Dexterous manipulation is particularly challenging as it inherently relies on complex interactions between the robot, object, and the environment [8, 23].

In recent years, there have been some notable examples of robot hands that showcase examples of this dexterous behavior [32, 7, 39, 34]. However, the capabilities of these hands are still limited when compared to our human counterparts. For example, they have limitations for error recovery, robustness to different environments, or behavioral diversity [11]. Extending the capabilities of robot hands will unlock many application ranging from service robots, prosthesis, medical or even agricultural robotics.

Many different approaches have been explored to move towards achieving dexterous manipulation [31, 40, 2, 26, 3]. This includes developing methods for fabricating complex bio-inspired hands and joints [38, 16], and developments in materials science and soft robotics to allow the creation of more compliant hands and manipulators [20, 27]. Deep learning, highly successful at many computational and learning tasks, has also been introduced to the field of manipulation [29, 32]. However, when compared with humans, AI based machines still struggle to exploit their physicality and situation with the environment, while introducing large complexities into the system [7]. In order to achieve a significant change in current dexterous manipulation capabilities, this "conventional-AI" must be augmented or combined with so called "embodied AI" which allows the computational power to be combined with exploitation of the physicality of the system. To move towards this approach we must understand how to design the "body" of robotic hands in conjunction with the algorithmic intelligence of the "brain" of these hands.

Considering human hands and fingers as a blueprint for highly dexterous manipulation, their "co-design" of the highly complex physical components (skin, sensors, bones, ligaments) and control system (nervous system and brain) has developed over millennia of biological evolution. Artificially developing hands and methods of design that rival evolution continues to be a significant challenge for engineers. Whilst complex multi-fingered hands have been developed numerously in the past, there is limited use of the design optimization, modeling, and controller development. In addition, due to complexity of the design and fabrication of sensory-motor systems, the exploitation and design of tactile sensors is also often not incorporated or considered as part of the design process for dexterous hands. This problem of negotiating complexity calls for a new approach towards designing and developing dexterous manipulation systems.

In this chapter, we introduce and expand upon the idea of iterative co-design: the simultaneous design iteration of the hardware and controller for multi-fingered hands. This concept aims to exploit embodied intelligence (EI) to provide behavioral advantages by designing the brain (controller) and body (mechatronic design) simultaneously. Importantly this extends to sensory systems at both the physical (sensory design and properties) and computational level (sensory-motor interactions). We hypothesize that a design methodology which simultaneously co-designs the EI and the computational intelligence of a system will be able to achieve behaviors that are otherwise not possible, or would require significantly more complexity or cost when designed separately. In the remainder of this chapter, we first address the challenges of design of robot hands, followed by a brief review of the related works with a discussion on what research directions are missing. A framework for iterative co-design of dexterous hands is then presented, before we conclude with a discussion on the challenges and future work.

2. Problem definition

A dexterous multi-fingered system is a highly complex system with multiple degrees of freedom (dof), intricate structures, actuators, and levels of control. Naturally, the process of designing and realizing such a system is therefore complex. Fig. 1 represents the progression of the development of a robot hand from the design stages to the use case. The brain and body of the hand is designed, fabricated, and the interactions planned. Through interactions between this hand system and the environment the behavior of the hand emerges. As labeled in the diagram, certain stages of the development incur "costs" C while the output behavior can be considered a "reward" R. Examples of how these quantities could relate with a robotic hand are shown in Table 1.

The objective of the design process is to maximize $R_{behavior}$. However, simply aiming for the maximum behavioral reward is likely to result in a overly complex system which is challenging/resource-intensive to optimize at best and unable to optimize at worse. To combat the problem with complexity, we must take into consideration of the costs of development and devise a method to minimize those costs while maximizing the behavioral reward. This can be done at two levels within the robot development.

In other words, we must consider an objective function $\mathcal{Q}$ which simultaneously maximize the behavioral reward and minimize the costs constructed by some function $q(\cdot)$. We can consider the total cost to be the combination of various costs that arise throughout the full robot development stage and care about, given by $c(\cdot)$. The exact form of the function $q(\cdot)$ and $c(\cdot)$, and how the

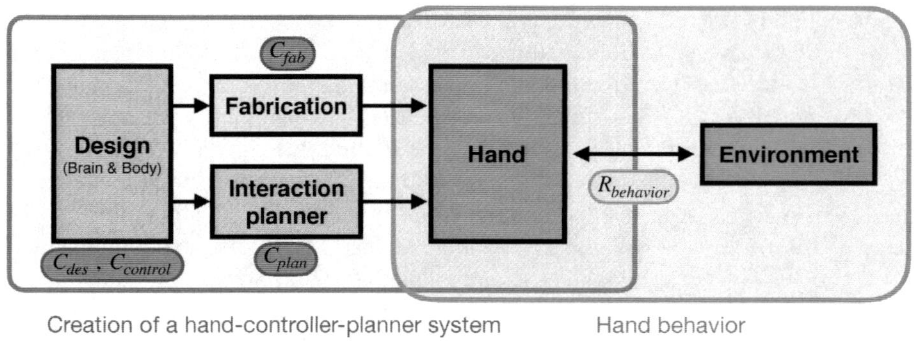

Figure 1. Simplified diagrammatic representation of the development and use of a robotic hand. The costs and rewards arising throughout the process is indicated by C and R.

Table 1. Cost and rewards throughout the robot hand development and usage

Quantity	Description	Example
C_{des}	Cost of mechatronic design	Number of actuators Time required to modify an existing design
$C_{control}$	Cost of controller complexity	Number of gain parameters to tune Number of layers in the neural network
C_{plan}	Cost of planning and running the controller	Computational resource for planning or training
C_{fab}	Cost of fabrication	Time taken to fabricate a hand Monetary cost of purchasing a hand
$R_{behavior}$	Reward of dexterous behavior	Reachable work space by the fingertip Success rate for a task

specific rewards and costs defined, is highly dependent on the situation.

$$\mathcal{Q} = q\left(R_{behavior}, C_{total}\right)$$
$$C_{total} = c\left(C_{des} + C_{control} + C_{plan} + C_{fab}\right) \tag{1}$$

Using this analysis of the "cost" of hands we can analyze the role that EI can play in both the design, creation and use of hands. Typically, concepts EI is used to either maximize the behavioral output with minimal gains in cost (e.g. using soft materials to provide robustness) or, to minimize the costs whilst maintaining behavioral output (e.g. muscle synergies). However, if we can develop methods that allow us design systems where EI can be used to affect both the costs and rewards, we may be able to better optimize the use of EI across the entire system. Hence develop hands with both better performance and lower cost.

2.1. Approaches towards robot hand development

In this section we discuss two high level approaches to the development of robotic hands: sequential and iterative co-design. We aim to highlight how the costs C_{total} is closely linked to the structure of the approach itself, and therefore claim that the cost can be minimized by proposing an alternative robotic development approach: iterative co-design.

When designing any robotic system, and in particular a robotic hand, the standard/traditional approach has been a sequential process shown by the top flow chart of Fig. 2. In this approach, the robotic hand is developed step by step starting from the mechatronic design, to the fabrication, sensor addition, and finally a controller to operate the hand.

Iterative co-design presents an alternative robotic system development approach and is shown by the bottom flow chart of Fig. 2. Firstly, this approach is considered a *co-design* since the

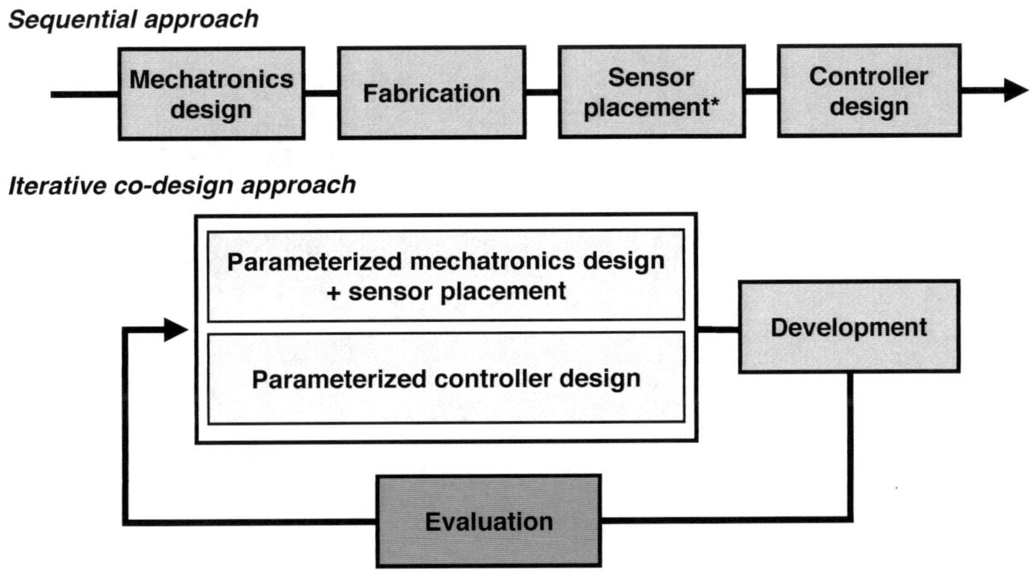

Figure 2. Block diagram of two methods of developing a robotic hand: Sequential approach and Iterative co-design approach. *Depending on the system, sensor placement can be included in the mechatronics design process

controller and the hardware is designed at once. Secondly it is *iterative* because after each development cycle, the robotic system is evaluated which can be used to update the parameters of the system. The development step refers to different ways a robot hand can be realized (e.g.: physical fabrication, simulated hand model, etc). Between the two approaches, there are trade-offs in costs incurred throughout the process. However, we claim that there are intrinsic properties within the iterative co-design approach that maximize $\mathcal{Q}$ and hence the preferred approach for the future of robotic hand design.

The sequential approach being the standard/traditional approach has a number of key benefits which should not be overlooked. The key advantage is to exploit specialization. Robotics being intrinsically multi-disciplinary, it is not scalable for a single individual or institutions to perform all aspects of the development. By distributing the development to different individuals/institutions with different tailored skills, one can argue the robot development is more efficient. Furthermore, the development process does not rely on multiple iteration, which further speeds up the process. However, with highly complex systems such as robot hands, there are clear disadvantages of this process. When a robot is developed sequentially, at any given step, design decision made earlier in the process affect the design decisions later in the process which has the tendency for the C_{total} to increase significantly.

For robotic hands, commonly the mechanical system is designed to allow for maximum flexibility for the user since the use case is unknown. For example, by maximizing the number of joints and actuators, the designer of the controller is less constrained by the possible motion of the hand. However, such design decisions generate complexity which may not contribute to increasing the performance. In this same example, as the number of dof and actuators in the hand increase, the controller becomes inherently complex and difficult to adapt to new scenarios. Furthermore, certain design decisions made earlier in the process can be a hindrance unknowingly for later processes - an unintentional cost incurred due to the structure of the sequential approach. Especially if the hardware is purchased, the inability to modify the design in addition to the monetary cost of the platform all contribute to increase C_{total}.

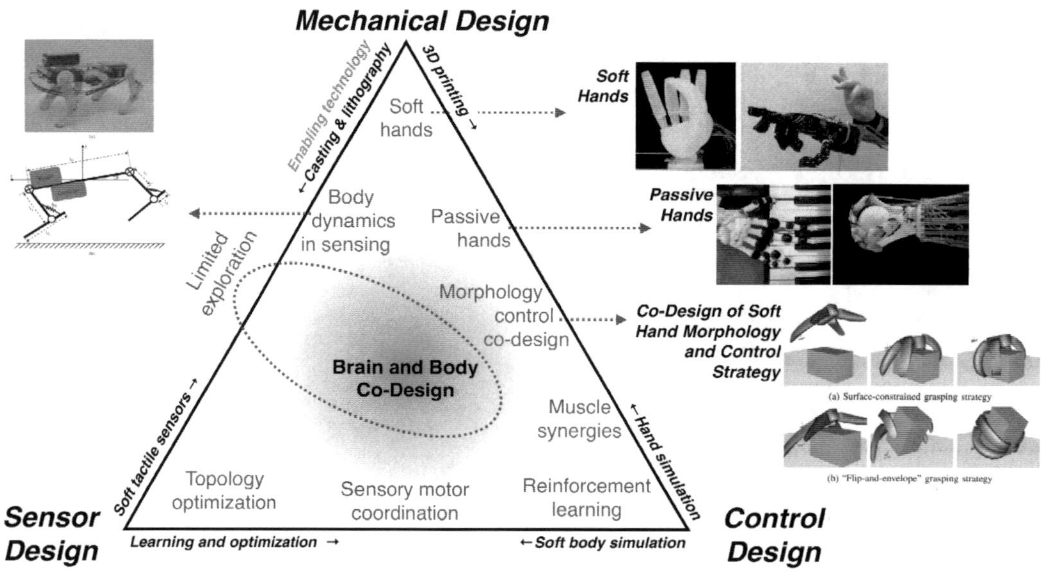

Figure 3. Organization of related work on co-design with a focus on robotic hands utilizing EI.

Iterative co-design offers an alternative approach to benefit from EI. Robotic systems and in particular robot hands have a strong coupling between the brain and body. The behavior of a robotic hand is hugely dependent on the mechatronics design (body): the number of fingers, how they bend, what sensors are placed and how they are placed, etc. Equally, the controller (brain) defines the interaction with the world and thus the behavior. We hypothesize that through co-design, the interaction between the brain and body can be captured and optimized, leveraging the advantages of EI. This would result in a system with a lower complexity/cost compared to a system developed through the sequential approach (e.g.: simpler controller, lower dof) but can achieve similar or improved behavior. In other words, by leveraging the EI of hands through the design method, we are able to systematically reduce C_{total}. Furthermore, the iterative nature of the approach allows for a optimization process to be defined, leading to increasing $\mathcal{Q}$.

3. Existing approaches to co-design

To understand the existing methods of co-design in robotic hand development, we consider the diagram in Fig. 3. This figure represents a three dimensional design space between mechanical, control, and sensor design where co-design sit at the center of this design space. This analysis allows us to identify key works that utilize co-design as a method of exploiting EI. Please note this is intended as a means to highlight notable works as opposed to a full review.

Soft hands are one significant area of research where EI is exploited through mechanical design to enable compliance or adaptive grasping approaches with minimal control [20]. Enabled by casting and lithography techniques made possible by 3D printing, key examples include the RBO hand [12, 36], or the anthropomorphic compliant DLR hand [10]. At the intersection of control and mechanical design, there has been progress made in recent years. This includes simulation work that enables the co-design of morphology and grasping control for soft robots [13]. This, and similar approaches [6, 37], are increasingly becoming feasible due to the development of simulators that can be accurately represent soft structures and complex contact with the environment. An alternative approach which focuses on fully exploiting EI is that of the passive

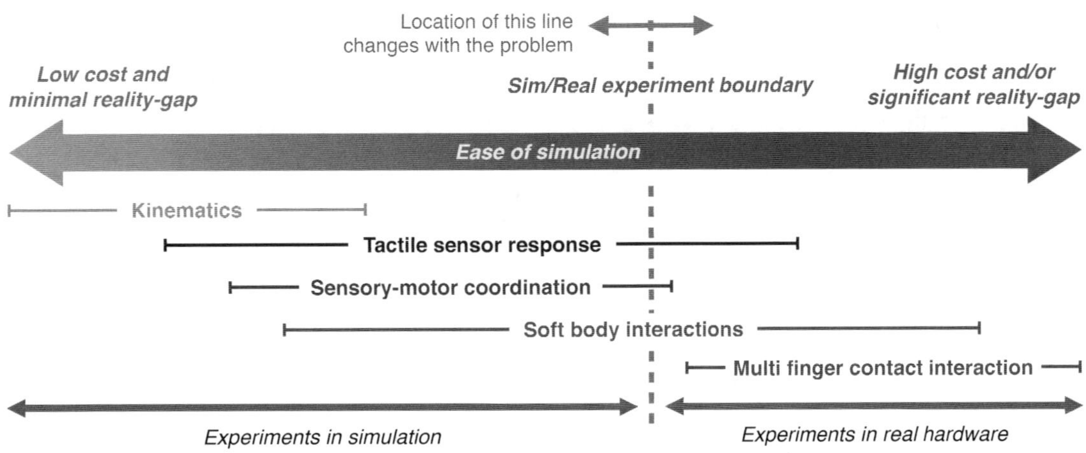

Figure 4. Qualitative placement of various robotics tasks on a spectrum based on the ease of simulating that task. Tasks on the spectrum can be split into two sections by the green dotted line: one where it is cheaper to experiment in simulation, another where it is cheaper to experiment using a physical setup.

hand – where only the wrist is actuated and the design or mechanical properties of the hand are designed or tuned to optimize the dexterity and range of motion [19, 16]. Control approaches leveraging this have been shown to enable piano playing and grasping of objects.

Moving towards the control domain, muscle synergies which focus on utilizing a low-dimensional representation or control input to achieve a high-dimensional rich behavioral/actuated output have been successfully demonstrated for manipulation [15, 17, 9]. This bio-inspired concept focuses more on exploiting the EI of control within the hands, but requires mechanical design that have sufficient complexity to allow such control methods to be exploited[14, 5].

Sensory-motor co-ordination sits at the intersection of sensor and controller design, and is necessary for many of the increasingly capable in-hand manipulation tasks being demonstrated [25]. Much of the research in this area focuses on developing controllers (including learning approaches) that leverage sensor design. There is limited exploration of how the sensory design affects the controller and the resultant performance. Whilst there has been some exploration of algorithms for topology optimization or morphology optimization of soft sensors [21, 35], this largely focuses on optimizing to improve the reconstruction of tactile information opposed to optimizing for control purposes.

As highlighted in Fig. 3, despite the developments in the area of co-design there appears to be limited work in the intersection of incorporation of sensor design into the design process and hands. Existing work in sensory topology has highlighted the gains that can be made solely through topology optimization, thus, there is an opportunity to combine and accentuate this other EI contributions. Equally works such as [22] show the importance to consider the body mechanics (in this case body dynamics) to assist sensing capabilities. We argue that to fully leverage and exploit EI within hands, we need to move to an approach for co-design that considers mechanical, control, and sensors, and that, to date, this design space has not been explored. We hypothesized that this is largely due to three current constraints or limitations which we discuss in detail in the following section.

IOP Conf. Series: Materials Science and Engineering 1261 (2022) 012009 doi:10.1088/1757-899X/1261/1/012009

3.1. The challenges in co-design

Three key challenges can be identified when implementing the iterative co-design process for robotic hands. The first challenge is the limitation of cheap simulation tools with a minimal reality gap. While simulation capabilities have increasingly improved over the decade, for robotics research involving physical hardware, accurate and general purpose simulation tools are not widely available. The second challenge is the lack of a cheap, fast, and reliable fabrication method to iterate upon the hardware. While customizable hardware can be developed, it tends to incur a high cost in C_{fab} (in the form of monetary cost and time) preventing multiple hardware iterations and exploration from being practical. The third challenge is the challenges in parameterization of the design. There are multiple ways of parameterizing both the hardware and the controller, and an effective method of doing so while maintaining real world constraints (such as the tolerances in manufacturability) is unsolved.

To better understand these challenges, consider Fig. 4. This figure aims to illustrate a spectrum of robotics tasks arranged by their general trend of qualitative "ease of simulation". Tasks which lie on the left side of the spectrum can be simulated cheaply and has a minimal reality gap. Such tasks only have a few well defined dof and modeling techniques are well-explored. Towards the right side, experimentation in simulation becomes challenging. This is because a) the simulation is computationally expensive, and/or b) the reality gap is so difficult to close such that the simulation is not meaningful. The full simulation of hands: complex multi dof mechanical interactions involving softness and friction with modeled sensors that match reality, is a highly challenging task which lies on the right side of this figure.

One way to combat the lack of simulation tools, is to look towards real world experiments. The green dotted line in Fig. 4 is a slider which divides robotics tasks on the "ease of simulation" spectrum between those which are suitable to execute in simulation or instead on a physical setup. The position of the slider is qualitative and depends on the task and resources available. Using real experiments, we are able to negotiate the limitation of simulation tools by offloading certain tasks to be evaluated in real life. Conceptually, this reduces costs of setting up and executing a complex simulation reducing C_{total}. This calls for a further need to tackle the second key challenge: the lack of cheap, reliable hardware.

Even if we are able to achieve a good experimental setup utilizing cheap simulation and cheap hardware, to leverage the best of both worlds, the third challenge remains. An effective method of identifying parameters of a complex system does not exist, and is usually left to the designer. Since the co-design process iterates on the parameters, their definition governs the final performance of the system. Furthermore, the parameter set must be tractable in both simulation and fabrication. If simulation and fabrication capabilities are improved, the constraints on parameters being tractable are loosened, allowing for more diverse exploration of the design space.

From this discussion, we see the three challenges in iterative co-design are loosely linked. If simulation capabilities are improved, more complex experiments can be conducted in simulation, reducing the need for hardware experiments. Conversely, if a low cost and reliable hardware can be fabricated, real world experiments can substitute the need for complex simulations. Improvements in simulation and fabrication techniques also allow for more diverse sets of parameters to be used. Although addressing one challenge would contribute towards solving the other challenge, we claim the development of a low cost and reliable hardware is most critical in the context of robotic hands. This is because regardless of the simulation capabilities or the definition of parameters the hardware must be fabricated at some point. With current simulation capabilities, even with the most computational intensive machines, there will always be some reality gap which requires calibration with a real system to close the gap.

Therefore, we believe the necessary steps to address the three key challenges are: a) accelerating the development of a low cost and reliable robotic hardware; b) construct an iterative

IOP Conf. Series: Materials Science and Engineering 1261 (2022) 012009 doi:10.1088/1757-899X/1261/1/012009

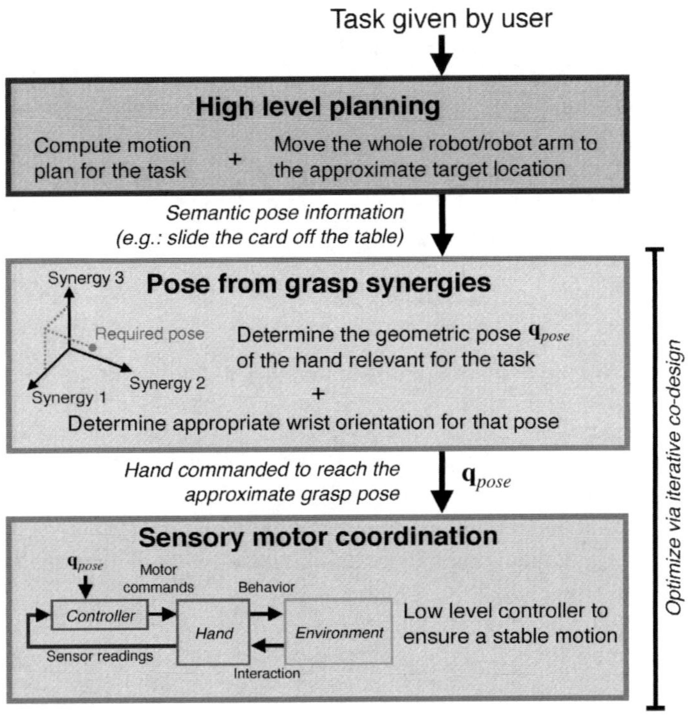

Figure 5. Three layers of abstraction to control a robotic hand system.

co-design process which utilizes simulation and real-life experiments and accounts for the cost of hardware development; and c) explore different methods of defining a suitable parameter set, while expanding/diversifying with improvements in simulation and fabrication.

4. A framework for iterative co-design

In the remainder of this chapter, a framework for implementing iterative co-design for robotic hands will be discussed. We first focus on how hand control and planning can be abstracted and divided into three layers of control. This is followed by an outline and discussion of a possible iterative co-design process.

4.1. Control abstraction of robotic hands

4.1.1. Overview of the abstraction layers To begin the discussion of co-design, we must consider how the hand will be controlled. Fig. 5 presents three level of abstractions for controlling a robotic hand. This abstraction structure is directly inspired from how humans interact with objects using our hands. Take the example of opening a door. At the highest level of abstraction, we move our body, arm, and hand towards the object (the door handle). In the middle level of abstraction we make an approximate pose of our hand (power grasp of the handle). Finally, at the lowest level, we use our tactile and proprioceptive sensors to achieve a robust action (readjusting our grip, location of fingers, force applied to the door handle, etc).

In the first two layers, the motion is planned and adjusted primarily with visual feedback (e.g. adjusting your hand position with respect to the door handle). In the final layer however, the motion is more reactive and is purely based on the low level controller.

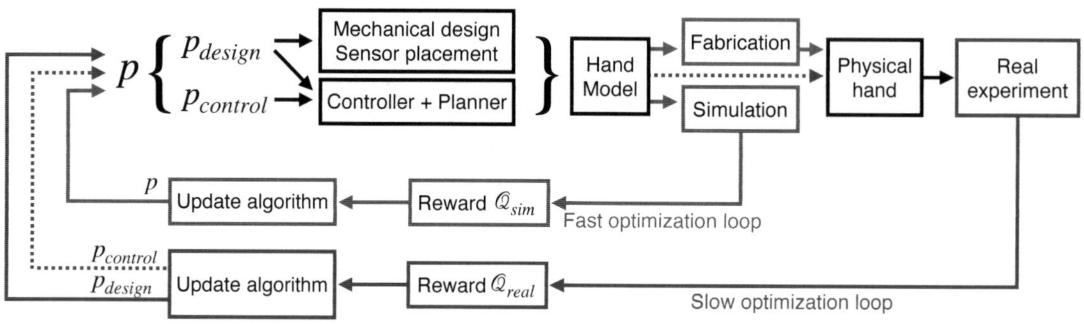

Figure 6. Block diagram of a possible implementation framework of iterative co-design.

4.1.2. Abstraction layers in detail and its relationship to iterative co-design For iterative co-design, the high level planner can be neglected (and will be neglected for the remaining discussion), as this planner is concerned with semantic understanding of the task and positioning of the hand, but not with the direct interaction of the hand with the environment. The final two layers depends largely on the morphology, sensing, and the sensory-motor control, and hence will be the focus of iterative co-design.

The middle layer of abstraction: "Pose from grasp synergies" are generated based on the semantic pose information. Grasp synergies are a collection of low dimensional representations of key *poses* of the hand, where a large collection of poses can be derived and chosen by the mid-level controller by "interpolating" between such synergies. In this way, the set of all poses becomes a continuous quantity - allowing possibilities of gradient based optimization to be incorporated in this co-design process. However, potential discontinuities and unreachable spaces due to the physical configuration of joints and actuators must be determined and optimized for the task.

The lowest layer: "Sensory motor coordination" focuses on how the sensors in the hands are used to make the approximate pose determined in the layer above into a reliable action which interacts with the environment. The structure and parameters of the low level controller is the quantity to be optimized through the iterative co-design process.

4.2. Iterative co-design framework

Fig. 6 illustrates a high level framework for the implementation of iterative co-design. We assume to have access to a low cost and reliable hardware platform as discussed in section 3.1 where its design is parameterized. The core of this framework is p, the set of all parameters of the hand. p can be divided into p_{design} and $p_{control}$, the parameters for the hardware design and for the controller respectively. The aim is to maximize $\mathcal{Q}$ by tuning p.

The *iterative* part of the block diagram takes the hand system based on p_i (the set of parameters at some iteration i), evaluates the objective function (Q_{sim} or Q_{real} based on if the experimentation was in simulation or in real life), and then updates the parameters $p_{i+1} \rightarrow p_i$. The red "inner" loop updates p by running simulated experiments while the blue "outer" updates p by running real experiments. Through the use of a low cost hardware to run real world experiments, we aim to optimally exploit the relative advantages of simulation and real world experimentation as discussed in section 3.1.

In both loops the reward must be calculated. Defining a reward function mathematically is one of the most challenging tasks of the implementation since a universal performance metric for dexterous manipulation does not exist and the details are dependent on the task type and environment. Furthermore, matching the high level intention of the framework designer and form of the reward metric is challenging as we know from related challenges in artificial intelligence and

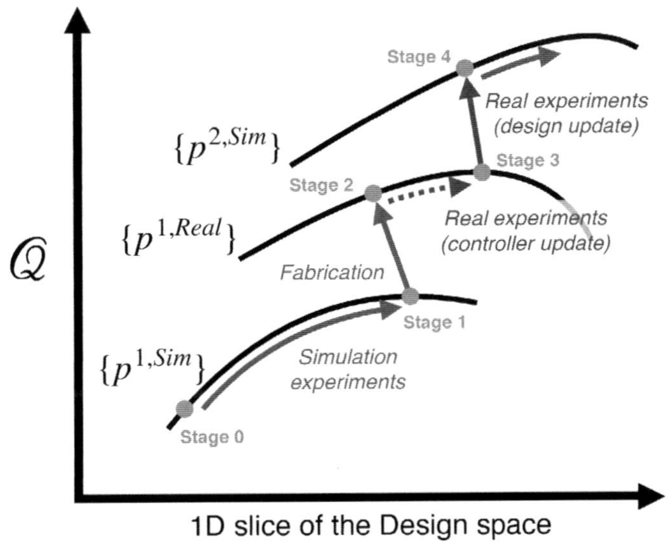

Figure 7. Conceptual plot to visualize how the iterative co-design process is expected to increase Q.

reinforcement learning such as reward hacking [4] and misalignment [18]. The implementation of the updating algorithm is also a challenging yet key component of the framework. This updating algorithm can be completely model free such as Bayesian optimization, or include some information about the hand model to speed up the optimization process.

4.3. Iteration process

The iteration and optimization process can be divided into three stages.

(i) Update p through simulation
(ii) Update $p_{control}$ through real experiments after fabrication
(iii) Update p_{design} through real experiments

Fig. 7 illustrates how we expect the reward to increase as we repeat the iteration process. The colors and formatting of the arrows match the that in Fig. 6. We start the process at **Stage 0** with some initial design parameter. We first use simulation to optimize the parameters (red solid arrow). We expect to use a low cost simulation, and hence this is the *fast optimization loop* which can run arbitrarily many times. A key point is that during this optimization process we only have "access" to a set of parameters of the design space the simulation can explore denoted by $\{p^{1,Sim}\}$ (the superscript number reflects the iteration count while *Sim* or *Real* reflects the optimization setting). The whole design space cannot be covered due to limitations of the simulation, computational costs, and how designs may get stuck in a local minima.

Once **Stage 1** is reached (the optimal point achievable via simulation experiments) the hand will be fabricated to conduct real world experiments (green solid arrow). Since the real world experiments allow further exploration of the design space, fabrication of the hand "unlocks access" to a new set of parameters $\{p^{1,Real}\}$ which can now be explored and reaching **Stage 2**.

Once the hardware is fabricated, multiple iterations to update $p_{control}$ can be performed (blue dotted arrow). We limit this step to $p_{control}$ because updating a controller is expected to be significantly faster than updating (re-fabricating) the hardware. Once **Stage 3** is reached we

IOP Conf. Series: Materials Science and Engineering 1261 (2022) 012009 doi:10.1088/1757-899X/1261/1/012009

have exhausted the optimization possibilities using the current hardware setup. Finally, p_{design} can be updated to reach **Stage 4** (solid blue arrow) - restarting the iteration process.

There are of course clear potential roadblocks with this iteration method. For example, the choice of updating algorithm is critical to produce meaningful improvements in Q. Equally, as we expand p, exploration of the full search space becomes difficult. The trade-off between complexity and reward must always be considered.

5. Conclusion

Exploiting EI is one means by which the capabilities of current robot hands could be extended to improve their performance in increasingly complex and unstructured environments. In addition to core technologies that provide physical intelligence (fabrication methods, sensors, materials), we also require methods to optimize the design of hands to incorporate EI. In order to fully exploit the possibility of EI, we believe that co-design of the brain and body of the robotic system is critical. In this chapter we propose a method for this: iterative co-design, which leverages both the simulated and real world as a new method for hand design.

To implement this iterative co-design approach, one fundamental challenge is the need of a physical platform with complex structures and embedded sensors which can be fabricated rapidly and repeatably. The fabrication and design should be achievable on the timescale of hours. Such a hardware platform allows the parameterized design to be rapidly iterated and experiments conducted that can not be performed in simulation. We also require the hardware to have a significant design space that spans different materials, sensor, and geometries.

One aspect of this approach which we have only touched upon is the task definition and the evaluation or bench-marking metrics. The iterative optimization process relies on evaluating the hand performance and updating the design parameters. Likewise the aim of simultaneously developing the brain and body of hands, is to find a design solution which can perform increasingly complex, broad, and potentially even unseen tasks. Hence, evaluation metrics must be formulated that captures the performance of a range of tasks effectively and accurately.

References

[1] Robotics and neuroscience. *Current Biology*, 24(18):R910–R920, 2014.

[2] Dexterous grasping under shape uncertainty. *Robotics and Autonomous Systems*, 75:352–364, 2016.

[3] John Amend and Hod Lipson. The jamhand: Dexterous manipulation with minimal actuation. *Soft Robotics*, 4(1):70–80, 2017. PMID: 29182098.

[4] Dario Amodei, Chris Olah, Jacob Steinhardt, Paul Christiano, John Schulman, and Dan Mané. Concrete problems in ai safety, 2016.

[5] Giuseppe Averta, Edoardo Battaglia, Cosimo Della Santina, Manuel G Catalano, and Matteo Bianchi. A synergistic behavior underpins human hand grasping force control during environmental constraint exploitation. In *International Conference on NeuroRehabilitation*, pages 67–71. Springer, 2018.

[6] Dominik Bauer. *Automated design of tendon-driven soft foam hands using Markov-Chain-Monte-Carlo optimization methods.* PhD thesis, Master's thesis, Karlsruhe Institute of Technology, 2018. 10, 32, 39, 40, 2018.

[7] Aditya Bhatt, Adrian Sieler, Steffen Puhlmann, and Oliver Brock. Surprisingly robust in-hand manipulation: An empirical study. *Robotics: Science and Systems XVII*, 2021.

[8] Aude Billard and Danica Kragic. Trends and challenges in robot manipulation. *Science*, 364(6446):eaat8414, 2019.

[9] Christopher Y. Brown and H. Harry Asada. Inter-finger coordination and postural synergies in robot hands via mechanical implementation of principal components analysis. In *2007 IEEE/RSJ International Conference on Intelligent Robots and Systems*, pages 2877–2882, 2007.

[10] Jörg Butterfaß, Markus Grebenstein, Hong Liu, and Gerd Hirzinger. Dlr-hand ii: Next generation of a dextrous robot hand. In *Proceedings 2001 ICRA. IEEE International Conference on Robotics and Automation (Cat. No. 01CH37164)*, volume 1, pages 109–114. IEEE, 2001.

[11] Ryan Coulson, Chao Li, Carmel Majidi, and Nancy S Pollard. The elliott and connolly benchmark: A test for evaluating the in-hand dexterity of robot hands. In *2020 IEEE-RAS 20th International Conference on Humanoid Robots (Humanoids)*, pages 238–245. IEEE, 2021.

[12] Raphael Deimel and Oliver Brock. Soft hands for reliable grasping strategies. In *Soft Robotics*, pages 211–221. Springer, 2015.

[13] Raphael Deimel, Patrick Irmisch, Vincent Wall, and Oliver Brock. Automated co-design of soft hand morphology and control strategy for grasping. In *2017 IEEE/RSJ International Conference on Intelligent Robots and Systems (IROS)*, pages 1213–1218. IEEE, 2017.

[14] Cosimo Della Santina, Matteo Bianchi, Giuseppe Averta, Simone Ciotti, Visar Arapi, Simone Fani, Edoardo Battaglia, Manuel Giuseppe Catalano, Marco Santello, and Antonio Bicchi. Postural hand synergies during environmental constraint exploitation. *Frontiers in neurorobotics*, 11:41, 2017.

[15] Tao Geng, Mark Lee, and Martin Hülse. Transferring human grasping synergies to a robot. *Mechatronics*, 21(1):272–284, 2011.

[16] Kieran Gilday, Josie Hughes, and Fumiya Iida. Wrist-driven passive grasping: interaction-based trajectory adaption with a compliant anthropomorphic hand. *Bioinspiration & Biomimetics*, 16(2):026024, 2021.

[17] Giorgio Grioli, Manuel Catalano, Emanuele Silvestro, Simone Tono, and Antonio Bicchi. Adaptive synergies: An approach to the design of under-actuated robotic hands. In *2012 IEEE/RSJ International Conference on Intelligent Robots and Systems*, pages 1251–1256, 2012.

[18] Evan Hubinger, Chris van Merwijk, Vladimir Mikulik, Joar Skalse, and Scott Garrabrant. Risks from learned optimization in advanced machine learning systems, 2021.

[19] JAE Hughes, P Maiolino, and Fumiya Iida. An anthropomorphic soft skeleton hand exploiting conditional models for piano playing. *Science Robotics*, 3(25), 2018.

[20] Josie Hughes, Utku Culha, Fabio Giardina, Fabian Guenther, Andre Rosendo, and Fumiya Iida. Soft manipulators and grippers: a review. *Frontiers in Robotics and AI*, 3:69, 2016.

[21] Josie Hughes, Luca Scimeca, Perla Maiolino, and Fumiya Iida. Online morphological adaptation for tactile sensing augmentation. *Frontiers in Robotics and AI*, 8, 2021.

[22] F. Iida and R. Pfeifer. Sensing through body dynamics. *Robotics and Autonomous Systems*, 54(8):631–640, 2006. Morphology, Control and Passive Dynamics.

[23] Charles C. Kemp, Aaron Edsinger, and Eduardo Torres-Jara. Challenges for robot manipulation in human environments [grand challenges of robotics]. *IEEE Robotics Automation Magazine*, 14(1):20–29, 2007.

[24] Fouad F Khalil and Pierre Payeur. *Dexterous robotic manipulation of deformable objects with multi-sensory feedback-a review*. IntechOpen, 2010.

[25] Cecilia Laschi, Gioel Asuni, Eugenio Guglielmelli, Giancarlo Teti, Roland Johansson, Hitoshi Konosu, Zbigniew Wasik, Maria Chiara Carrozza, and Paolo Dario. A bio-inspired predictive sensory-motor coordination scheme for robot reaching and preshaping. *Autonomous Robots*, 25(1):85–101, 2008.

[26] Qiujie Lu, Nicholas Baron, Angus B. Clark, and Nicolas Rojas. Systematic object-invariant in-hand manipulation via reconfigurable underactuation: Introducing the ruth gripper. *The International Journal of Robotics Research*, 40(12-14):1402–1418, 2021.

[27] Qiujie Lu, Liang He, Thrishantha Nanayakkara, and Nicolas Rojas. Precise in-hand manipulation of soft objects using soft fingertips with tactile sensing and active deformation. In *2020 3rd IEEE International Conference on Soft Robotics (RoboSoft)*, pages 52–57, 2020.

[28] Raymond R. Ma and Aaron M. Dollar. On dexterity and dexterous manipulation. In *2011 15th International Conference on Advanced Robotics (ICAR)*, pages 1–7, 2011.

[29] Lucas Manuelli, Wei Gao, Peter Florence, and Russ Tedrake. kpam: Keypoint affordances for category-level robotic manipulation, 2019.

[30] Matthew T. Mason. Toward robotic manipulation. *Annual Review of Control, Robotics, and Autonomous Systems*, 1(1):1–28, 2018.

[31] A.M. Okamura, N. Smaby, and M.R. Cutkosky. An overview of dexterous manipulation. In *Proceedings 2000 ICRA. Millennium Conference. IEEE International Conference on Robotics and Automation. Symposia Proceedings (Cat. No.00CH37065)*, volume 1, pages 255–262 vol.1, 2000.

[32] OpenAI, Marcin Andrychowicz, Bowen Baker, Maciek Chociej, Rafal Jozefowicz, Bob McGrew, Jakub Pachocki, Arthur Petron, Matthias Plappert, Glenn Powell, Alex Ray, Jonas Schneider, Szymon Sidor, Josh Tobin, Peter Welinder, Lilian Weng, and Wojciech Zaremba. Learning dexterous in-hand manipulation, 2019.

[33] Ryuta Ozawa and Kenji Tahara. Grasp and dexterous manipulation of multi-fingered robotic hands: a review from a control view point. *Advanced Robotics*, 31(19-20):1030–1050, 2017.

[34] Yu She, Shaoxiong Wang, Siyuan Dong, Neha Sunil, Alberto Rodriguez, and Edward Adelson. Cable manipulation with a tactile-reactive gripper, 2020.

[35] Thomas George Thuruthel, Josie Hughes, and Fumiya Iida. Joint entropy-based morphology optimization of soft strain sensor networks for functional robustness. *IEEE Sensors Journal*, 20(18):10801–10810, 2020.

[36] Vincent Wall, Gabriel Zöller, and Oliver Brock. A method for sensorizing soft actuators and its application to the rbo hand 2. In *2017 IEEE International Conference on Robotics and Automation (ICRA)*, pages

IOP Conf. Series: Materials Science and Engineering 1261 (2022) 012009 doi:10.1088/1757-899X/1261/1/012009

4965–4970. IEEE, 2017.

[37] Jie Xu, Tao Chen, Lara Zlokapa, Michael Foshey, Wojciech Matusik, Shinjiro Sueda, and Pulkit Agrawal. An end-to-end differentiable framework for contact-aware robot design. *arXiv preprint arXiv:2107.07501*, 2021.

[38] Zhe Xu and Emanuel Todorov. Design of a highly biomimetic anthropomorphic robotic hand towards artificial limb regeneration. In *2016 IEEE International Conference on Robotics and Automation (ICRA)*, pages 3485–3492, 2016.

[39] Shenli Yuan, Lin Shao, Connor L. Yako, Alex Gruebele, and J. Kenneth Salisbury. Design and control of roller grasper v2 for in-hand manipulation, 2020.

[40] Henry Zhu, Abhishek Gupta, Aravind Rajeswaran, Sergey Levine, and Vikash Kumar. Dexterous manipulation with deep reinforcement learning: Efficient, general, and low-cost, 2018.

Intelligent Soft Hands and Benchmarking towards General-Purpose Robotic Manipulation

Kieran Gilday and Fumiya Iida

The Bio-Inspired Robotics Lab, Department of Engineering, University of Cambridge, UK

E-mail: kg398@cam.ac.uk, fi224@cam.ac.uk

Abstract. In order to progress the development of intelligent soft hands for general-purpose use in humanoid robots, social assistive robots, adaptive manufacturing, prosthetics and more, we need to rethink our approach to benchmarking. Where previously, hands are compared by their performance in a limited set of tasks, resulting in performance optimisations in the subjective, most common tasks. Instead, we must focus on increasing the hand's potential at the lowest level, by improving the underlying passive behaviours, in terms of increased behavioural diversity and cheap control.

1. Introduction

It is often noted how the apparent capabilities of our artificial manipulators are severely lacking when compared to the capabilities of our digital systems [1]. Why can we make computers able to exceed human performance by many orders of magnitude in certain tasks, but struggle to develop manipulators with similar capabilities to a child? This is a false comparison; the real challenge is developing any general-purpose systems. We can design a hand to perform the same task over-and-over with great success, but just as we struggle to develop artificial *general* intelligence we struggle to develop hands which can achieve human-like performance in general unstructured environments. We use benchmarking to drive development and solve the most pressing challenges, though benchmarking for general-purpose use is not straightforward [2].

A general-purpose robotic hand is not only useful in the everyday common tasks, but also able to adapt to a huge variety of novel situations. Toda [3] stated: 'man's efficiency in coping with his environment depends not on how well he performs isolated tasks, but on how well he can co-ordinate several different functions in order to solve the problems of daily life.' One perspective of this is decomposing a complex control system. Though, we believe the underlying passive hand behaviours provide critical functionality, where behaviours emerge from interactions between the parallel, loosely-coupled hand sub-components and the environment [4]. The hand needs to have enough diversity in its behaviours to be able to function and interact with the environment in these novel situations [5]. In addition, the complexity of the hand should not overburden the control system in the common use cases [6]. A hand with these properties can be said to have *embodied intelligence*, and only this, paired with an adequate control system, will lead to general-purpose performance. To drive development, the question is then how to benchmark this intelligence? And is it even possible/desirable?

Existing benchmarks in the field of robotic hands are fundamentally flawed for driving intelligent hand design. Existing benchmarks can be broken down into two main categories:

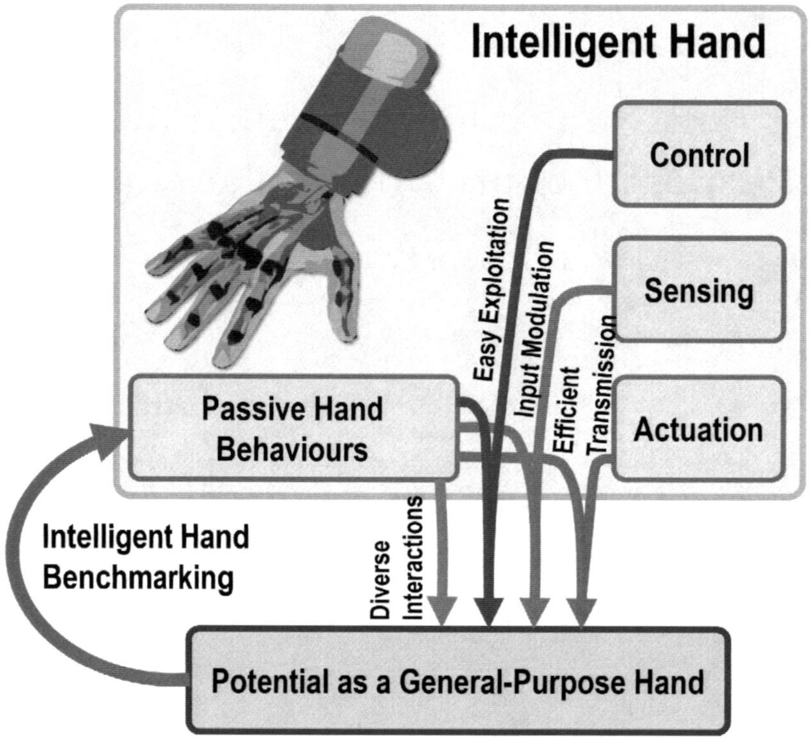

Figure 1. Passive hand behaviours are critical in the performance of hands for general-purpose use. In order to develop more intelligent hands, we must benchmark the potential of the hand and use that to inform development of the underlying passive behaviours.

those scoring based on performance in a selected set of real-world tasks [7]–[9], and those based on achieving some higher-level behaviours [10]–[12]. In the case of task-based benchmarking, there are two main issues. The first issue is the subjective bounds of the selected tasks. If a limited set of tasks is chosen, little indication is given to the performance in niche tasks, which is the hallmark of general-purpose hands. If a much broader set of tasks is chosen, the benchmark becomes unwieldy and impractical to run. The second issue is the combined evaluation of the hand's passive behaviours and the control system; these benchmarks are most effective at driving controller development. Since the benchmarking scores are so dependent the controller, it is impossible to effectively compare the underlying physical hands. The case of the higher-level behaviour-based benchmarking is a step in the right direction, where there is reward for higher range of behaviours up to the bounds of the benchmark, such as diversity of grasping postures. Though, many benchmarks in this category also rely on the controller, in addition to being limited in the behaviours that are evaluated (posture from internal actuation, and shape adaptation being the most common), others, such as passive motion, forces and damping, are overlooked.

If current benchmarks are unsuitable for intelligent hand development, there are two options: forgoing benchmarking entirely or creating a new benchmark. There is a strong argument for not benchmarking, since benchmarking is most useful at optimisation and comparing designs objectively when the differences are too small for more subjective evaluations. While robotic hands have had significant development over decades, development of passive and soft hands is relatively immature [13], the existence of benchmarks could potentially stifle creativity.

We believe it *is* possible to create a new benchmark which drives forward the potential of intelligent hands. Since a hand's potential in general-purpose use so heavily relies on the underlying passive behaviours, the benchmark needs to be able to evaluate this potential in order to be fed back into the passive design. Fig. 1 outlines this concept. The passive hand behaviours are fundamental, upon which control, actuation and sensing can be developed, leading to hands with higher real performance [14].

In the remainder of this article, we review existing benchmarking options and what we can learn from them, either to adopt or avoid. Then we introduce our concept for general-purpose hand benchmarking, based on promoting behavioural diversity and cheap control.

2. Benchmarking Options

2.1. Existing Benchmarks: Task-based

Task-based benchmarks or competitions have proven extremely useful for driving development towards well-defined goals [15]. The Amazon Robotics Challenge focused on pick and place in a factory setting, with the main challenges in perception and grasping in somewhat unstructured environments [16]. RoboCup strongly promotes the development of artificial intelligence and robot integration, in addition to coordination between agents [17].

Rlbench is designed to accelerate progress in a number of manipulation areas, including reinforcement learning and few-shot learning [8]. This benchmark has scalability in mind, as true task diversity is required to prevent overfitting, with at least 100 unique tasks and readily expandable with user submitted tasks.

The YCB object and model set intends to provide a complete grasping and manipulation benchmark with a carefully selected set of widely used objects and tasks to represent everyday use, as well as being practical for researchers to obtain, use and store [7]. This promotes real-world experiments, with the main challenge in dexterous capabilities from the control systems in tasks such as assembly or folding.

Mnyusiwalla et al. [9] propose a bin picking benchmark for whole system benchmarking for the most common robot operations in industry, pick and place. This task is reproduce-able and promotes development towards one of the largest markets for industrial robots, in logistics warehouses. A strong case is made for whole system evaluation, from recognition and pre-grasping to transporting and placing, with high value placed on performance which will limit development outside of a subset of tasks.

In all the above benchmarking tasks, integration is key. The hands, actuators and controllers must be evaluated together. Since different bodies necessitate different controllers, these tasks are unable to effectively compare the capabilities of the underlying physical systems.

2.2. Existing Benchmarks: 'Behaviour' Based

There is increasing interest in benchmarks with less focus on specific tasks. Some benchmarks are able to evaluate select lower-level hand behaviours, others are based-off the recognition of 'potential dexterity', and others are concerned with how we can discern the intelligence in the robotic system.

The Kapandji thumb test (Fig. 2) is a common example, originally designed for rehabilitation of human thumb range of motion but appropriated to evaluate the critical opposable thumb range of motion in anthropomorphic robotic hands [10]. The score based on range of motion gives good indication of numerous capabilities, from diversity in grasping postures to possible manipulations.

Grasping taxonomies are often used to demonstrate the range of useful postures, similar to the Kapandji thumb test, but with a more practical demonstration. The Feix 'Grasp' taxonomy comprehensively covers the most common human grasps on primitive objects [11]. Generally, the more grasp types a hand can form, the more potential it has to succeed in arbitrary grasping tasks, the shortcoming of this is it gives little indication whether the grasps can be achieved in a

IOP Conf. Series: Materials Science and Engineering 1261 (2022) 012010 doi:10.1088/1757-899X/1261/1/012010

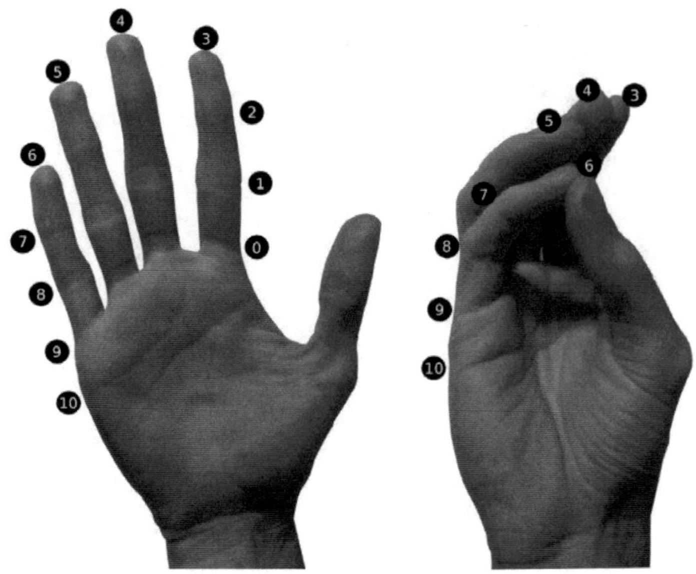

Figure 2. Kapandji thumb opposition scores. A score of 0 indicates no opposition, a score of 10 indicates maximal opposition [18].

real, unstructured environment, where the object isn't always placed favourably and preliminary interactions are required. Manipulation behaviours can be classified in a similar way, these are much more focused on hand-environment interactions [19]. Though, to be evaluated, require a suitable controller.

The FFP Index (form-features-performance) moves emphasis away from just performance benchmarking by evaluating how close the form of the hand is to the human hand and the quality of the features present, primarily actuation and types of sensing [12]. This provides a more holistic view of the hand and gives a broad, but coarse, metric with which to compare a spectrum of hands.

Recognising the potential of hands is critical. A complex hand may prove less capable than a much simpler system if it has inadequate control, this is evidenced in the notable gap in 'potential dexterity' and 'real dexterity' [20], where dexterity alludes to the complexity of tasks that can be performed. Some rough quantifications of potential dexterity have been proposed with the Dexterity Index, where features of a hand's mechanical structure, such as degrees of freedom or contact surfaces, are used to consider high-level capabilities of fingertip/whole-hand grasping/manipulation, in addition to the types of sensors present, where sensors have been identified as critical to in-hand manipulation performance [21]. Once again, this index provides quite a coarse evaluation, though recognition of sensing capabilities as an indication of potential dexterity is, in our opinion, a critical future step.

Finally, the Turing-Like Handshake Test [22] provides an interesting take on benchmarking, where a measure of intelligence can be given by a system's ability to fool an interrogator into believing it is human. This currently exists in a haptic system, where a human interrogator attempts a handshake with a system with force control provided by either a human or a computer. This evaluates the ability of the controller to generate human-like handshakes, surprisingly challenging with precise force, timing and reactions required [23]. There is potentially a way of extending this concept to evaluate a hand's passive dynamics, if the obvious discriminators (temperature, oiliness...) between human and artificial hands can be concealed.

All of these benchmarking tasks either have little, or no, consideration of hand-environment

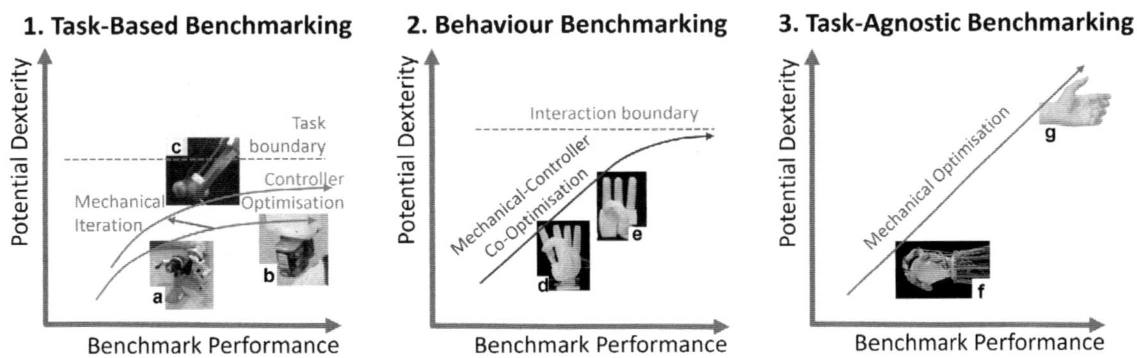

Figure 3. Development trajectories under different benchmarks. 'Potential dexterity' of a hand is a potential, quantifiable indicator of real-world performance in general-purpose tasks, including metrics such as number of kinematics, mechanical complexity and level of system integration. Benchmark performance is defined by each benchmark. **1**. Task-based benchmarks encourage controller optimisation and are bounded by task descriptions. (a) HERB planning optimisation [7]. (b) Deep neural networks can improve manipulation benchmarking performance of simple grippers [24]. (c) [9] inspired caging gripper [25], updating mechanical design often requires restarting controller development (short-term loss of benchmark performance). **2**. Some behaviour based benchmarks encourage co-optimisation, though are limited by not testing hand-environment interactions, e.g., Kapanji thumb test optimisation, (d) RBO 2 [26] compared to (e) BCL-13 [27], has an upper bound. **3**. New approach to benchmarking sets potential dexterity/embodied intelligence as its goal. Encouraging cheap control and environmental exploitation as controller optimisation is considered independently, (f) interaction-based design [14], (g) and target performance equivalent to a human.

interactions, or require a sufficiently advanced controller to properly evaluate. Therefore, for the purpose of evaluating hand design, these benchmarks are not sufficient.

2.3. No Benchmarking

A discussion of benchmarking options would not be complete without the option of no universal benchmark at all. Many robotic hands have been developed with an internal benchmark or specification, or no benchmark altogether, with varying degrees of success. There are many recognised challenges in developing hands, we can only focus on so much at one time and there is disagreement on the priorities. This leads to divergence in hand development. While it is clear this complicates comparison, divergence is not always a negative and can lead to greater novelty in solutions or even whole new research directions.

When there is agreement on the primary development challenges, getting everyone to agree on a common benchmark is much easier. The passive hand is the fundamental system that actuation, sensing and control are built on top of. Mechanical dexterity of the human hand was a major factor in the development of human intelligence [20]. Therefore, we believe development of the passive hand and its behaviours to be the priority and a benchmark to unify and accelerate development in this area is critical.

2.4. Future Benchmarking Directions

Benchmarking promotes development in a particular direction. Task-based benchmarks encourage controller optimisations, especially as mechanical iterations often decrease performance in the short-term while the controller catches up, Fig. 3.1 shows this typical development trajectory, and examples of developments stemming from the YCB set [7] and Bin-picking benchmark [9]. Some behaviour-based benchmarks encourage mechanical optimisation towards embodied intelligence, though with additional emphasis on controllers, e.g., the improved Kapanji thumb test performance of the BCL-13 hand [27] over the RBO Hand 2 [26]. Reliance on controllers and limited scope for evaluating interactions bounds the development here (Fig. 3.2). With no benchmarking there is wider exploration of the space and optimisations tend to be towards an internal specification or task-set, which will likely succumb into the pitfalls of existing benchmarking methods. Finally, Fig. 3.3 shows the target benchmark towards achieving human-like performance. This should have embodied intelligence, 'potential dexterity' or similar quantifiable concept, as its performance metric, inherently providing corresponding improvement when increasing benchmarking score.

3. Task-Agnostic Benchmarking towards General-Purpose Robotic Manipulation

There is a motivation behind every benchmarking task. There exist benchmarks for many different aspects of robotic hand research, from driving development of sensors and actuators to controllers. We see a clear void in benchmarking options for the underlying passive behaviours. In order to benchmark these, we need a clear goal. Our benchmark should reward the potential capabilities of the hand in general-purpose, unstructured environments. In order to have a higher potential, the hand must exhibit high behavioural diversity, to interact with the environment in novel situations, in addition to being easily exploitable by a control system. Intelligence within the body should simplify the control, especially in relatively simple activities.

While it is better to incentivise behavioural diversity rather than a subjective set of ideal behaviours, if our benchmark only incentivised behavioural diversity, it would likely just encourage more-and-more complex and unmanageable systems. While this may generate hands with more potential, there are still limitations in the actuation, control and sensing, all of which are critical to build on top of the hand for more complex capabilities. We not only need behavioural diversity, but also the 'correct' behaviours. Where 'correct' behaviours are not those that are most 'useful' to complete common or bounded tasks, instead they are behaviours which allow cheap control. Therefore, instead of thinking about generating the 'correct' behaviours, we should instead think about generating the 'correct' interactivity. Where we incentivise interactions, both within and outside the body, which simplify actuation, sensing and control. In this way, the body can be more easily exploited, translating to better 'real' potential with the same controller.

We do not claim to know the best way to benchmark for the intelligent hand. Though we can offer some guidelines:

- **Evolution/Developmental dimension**: the benchmarking goals themselves will likely evolve over time, though clarity is critical to unify and accelerate development. Our objective is to embed intelligence into the hand. This can be split into two parts, firstly to promote behavioural diversity, and secondly to ensure the emerging passive behaviours are easily exploitable. This must be reflected in the benchmarking score, or 'fitness function,' with consideration that this approach should be open-ended and not artificially bound development.
- **Situatedness**: since the hand behaviours emerge from interactions, any objective measure of behavioural diversity must be based on abstracted interaction tasks. For example, a benchmarking set can score based-on: range of motion when acted on externally, range of passive forces from static and dynamic interactions, and redundancy in posture with passive shape adaptation to primitive objects.
- **Embodiment focus**: evaluation of passive and emergent behaviours should be the focus, rather than on the control system. The function and coordination of every passive

component within the hand is as critical as the active components. To isolate their effect, the controller must be minimal [20]; with the theory that if we successfully embed intelligence within the physical hand, then control in the complex tasks should also be cheaper [6]. We don't want to evaluate the control system itself at this stage, except to decompose it to its' simplest form which is still adequate to perform the benchmark.

- **Measure of intelligence**: for this benchmark to be widely adopted, it needs to be convincing to those carrying out benchmarking tests. Having an analogy to measuring 'intelligence' makes benchmarking much more intuitive and persuasive. This can be an alternative test of potential, rather than trying to definitively prove this benchmark is a good predictor of general performance, which has the same pitfalls as traditional benchmarking. Exploiting the powerful, but subjective, human brain to compare real and artificial systems can give strong evidence either way. A 'Turing-like' test [22], say an interrogator who differentiates between the passive behaviours of gloved human and robot hands, would give trusted results when trying to achieve 'life-like' hand behaviours.
- **Reproducibility of test results**: as with all benchmarks, for systematic/collaborative research, it should be easy for anyone to run. The YCB object and model set was carefully selected with affordability, storage, transport, etc. [7]. Therefore, tests should be conducted with off-the-shelf and affordable measuring instruments like force sensitive resistors, basic data loggers and generic setups that can be easily replicated across different robot platforms with common techniques like 3D printing.

Now we have these guidelines, which fundamentally encourage the emergence of embodied intelligence, we can begin to see what a task-agnostic benchmark would look like. The benchmark could consider different mechanical levels and provide performance metrics of how closely they mimic human behaviours, e.g., at the joint-, finger- and hand-level, compliant degrees of freedom, interaction forces and passive shape adaptation respectively. The benchmark could be based-on a unified set of behavioural diversity benchmarks, both existing and new, e.g., the Kapanji thumb test, new benchmarks which measure *ranges* of passive forces when interacting the hand with the environment (**situatedness**) and/or perform Turing-like tests (**measure of intelligence**). These can be performed with extremely limited in-hand actuation. By setting the hand into a passive, or almost passive, state and driving interactions through control of the wrist [14], focus is taken away from control, allowing **reproducibility of results** with comparison over different platforms, and towards **embodiment** by directly evaluating passive hand behaviours and feeding back results into the design (Fig. 1). For the **evolution/developmental dimension**, the metrics can be weighted and summed. Additionally, other factors can be included, such as a negative score from the level of control required to complete the benchmark; whether the hand could complete each task with a single actuator or significant control effort was required to coordinate finger/hand motions. In this way, runaway complexity is inhibited and cheap control is encouraged.

There are additional major challenges to overcome in general-purpose hand design, from rapid fabrication of complex systems, design parameterisation, actuation and sensing technologies to intelligence and learning in the controller. Critically, rapid fabrication and design parameterisation are interconnected with benchmarking. While other challenges are important, we reiterate the need to realign development towards embodied intelligence in the hand.

4. Conclusions

The future of general-purpose robot hands is in distributed intelligence in the sensing, actuation, brain *and* body. Intelligence within robotic hands is lacking. One of the primary reasons for this is current views on benchmarking steering development more towards complex control systems. Benchmarking provides a unifying goal to accelerate development, out of the numerous benchmarks for robotic hands, very few encourage development of the underlying passive system.

In order to drive forward development of intelligent, general-purpose hands, an alternative approach to the current benchmarks is essential. Perhaps benchmarking at this stage should be scrapped altogether, and research should be allowed to diverge until consensus can be found again and benchmarking can be reintroduced for optimisation. Though, we believe the correct benchmarks can be a useful tool in driving development in this area.

We haven't only criticised existing benchmarks in this article, we have also proposed the outline of a new comprehensive, open-ended, task-agnostic approach to benchmarking. This focuses on encouraging behavioural diversity in the hand, and rather than defining 'useful' behaviours as those used in the most common grasping and manipulation tasks, instead, 'useful' behaviours should be the emerging behaviours that are readily exploitable with simple controllers.

ACKNOWLEDGEMENTS

References

[1] E. Mattar, "A survey of bio-inspired robotics hands implementation: New directions in dexterous manipulation," *Robotics and Autonomous Systems*, vol. 61, no. 5, pp. 517–544, 2013.

[2] T. Watanabe, K. Yamazaki, and Y. Yokokohji, "Survey of robotic manipulation studies intending practical applications in real environments-object recognition, soft robot hand, and challenge program and benchmarking," *Advanced Robotics*, vol. 31, no. 19-20, pp. 1114–1132, 2017.

[3] M. Toda, "The design of a fungus-eater: A model of human behavior in an unsophisticated environment," *Behavioral Science*, vol. 7, no. 2, pp. 164–183, 1962.

[4] R. Pfeifer and C. Scheier, *Understanding intelligence*. MIT press, 2001.

[5] J.-B. Mouret and S. Doncieux, "Encouraging behavioral diversity in evolutionary robotics: An empirical study," *Evolutionary computation*, vol. 20, no. 1, pp. 91–133, 2012.

[6] G. Montúfar, K. Ghazi-Zahedi, and N. Ay, "A theory of cheap control in embodied systems," *PLoS computational biology*, vol. 11, no. 9, e1004427, 2015.

[7] B. Calli, A. Singh, A. Walsman, S. Srinivasa, P. Abbeel, and A. M. Dollar, "The ycb object and model set: Towards common benchmarks for manipulation research," in *2015 international conference on advanced robotics (ICAR)*, IEEE, 2015, pp. 510–517.

[8] S. James, Z. Ma, D. R. Arrojo, and A. J. Davison, "Rlbench: The robot learning benchmark & learning environment," *IEEE Robotics and Automation Letters*, vol. 5, no. 2, pp. 3019–3026, 2020.

[9] H. Mnyusiwalla, P. Triantafyllou, P. Sotiropoulos, *et al.*, "A bin-picking benchmark for systematic evaluation of robotic pick-and-place systems," *IEEE Robotics and Automation Letters*, vol. 5, no. 2, pp. 1389–1396, 2020.

[10] A. Kapandji, "Clinical test of apposition and counter-apposition of the thumb," *Annales de chirurgie de la main: organe officiel des societes de chirurgie de la main*, vol. 5, no. 1, p. 67, 1986.

[11] T. Feix, J. Romero, H.-B. Schmiedmayer, A. M. Dollar, and D. Kragic, "The grasp taxonomy of human grasp types," *IEEE Transactions on human-machine systems*, vol. 46, no. 1, pp. 66–77, 2015.

[12] A. Vazhapilli Sureshbabu, G. Metta, and A. Parmiggiani, "A systematic approach to evaluating and benchmarking robotic hands—the ffp index," *Robotics*, vol. 8, no. 1, p. 7, 2019.

[13] J. Hughes, U. Culha, F. Giardina, F. Guenther, A. Rosendo, and F. Iida, "Soft manipulators and grippers: A review," *Frontiers in Robotics and AI*, vol. 3, p. 69, 2016.

[14] K. Gilday, J. Hughes, and F. Iida, "Wrist-driven passive grasping: Interaction-based trajectory adaption with a compliant anthropomorphic hand," *Bioinspiration & Biomimetics*, vol. 16, no. 2, p. 026 024, 2021.

[15] J. Hughes, K. Gilday, L. Scimeca, S. Garg, and F. Iida, "Flexible, adaptive industrial assembly: Driving innovation through competition," *Intelligent Service Robotics*, vol. 13, no. 1, pp. 169–178, 2020.

[16] N. Correll, K. E. Bekris, D. Berenson, *et al.*, "Analysis and observations from the first amazon picking challenge," *IEEE Transactions on Automation Science and Engineering*, vol. 15, no. 1, pp. 172–188, 2016.

[17] M. Asada, H. Kitano, I. Noda, and M. Veloso, "Robocup: Today and tomorrow—what we have learned," *Artificial Intelligence*, vol. 110, no. 2, pp. 193–214, 1999.

[18] D. J. Leamy, J. Kocijan, K. Domijan, *et al.*, "An exploration of eeg features during recovery following stroke–implications for bci-mediated neurorehabilitation therapy," *Journal of neuroengineering and rehabilitation*, vol. 11, no. 1, pp. 1–16, 2014.

[19] I. M. Bullock and A. M. Dollar, "Classifying human manipulation behavior," in *2011 IEEE International Conference on Rehabilitation Robotics*, IEEE, 2011, pp. 1–6.

[20] A. Bicchi, "Hands for dexterous manipulation and robust grasping: A difficult road toward simplicity," *IEEE Transactions on robotics and automation*, vol. 16, no. 6, pp. 652–662, 2000.

[21] L. Biagiotti, F. Lotti, C. Melchiorri, and G. Vassura, "How far is the human hand? a review on anthropomorphic robotic end-effectors," 2004.

[22] I. Nisky, G. Avraham, and A. Karniel, "Three alternatives to measure the human-likeness of a handshake model in a turing-like test," *Presence*, vol. 21, no. 2, pp. 156–182, 2012.

[23] G. Avraham, I. Nisky, H. L. Fernandes, *et al.*, "Toward perceiving robots as humans: Three handshake models face the turing-like handshake test," *IEEE Transactions on Haptics*, vol. 5, no. 3, pp. 196–207, 2012.

[24] J. Tobin, R. Fong, A. Ray, J. Schneider, W. Zaremba, and P. Abbeel, "Domain randomization for transferring deep neural networks from simulation to the real world," in *2017 IEEE/RSJ international conference on intelligent robots and systems (IROS)*, IEEE, 2017, pp. 23–30.

[25] L. Beddow, H. Wurdemann, and D. Kanoulas, "A caging inspired gripper using flexible fingers and a movable palm," in *2021 IEEE/RSJ International Conference on Intelligent Robots and Systems (IROS)*, IEEE, 2021, pp. 7195–7200.

[26] R. Deimel and O. Brock, "A novel type of compliant and underactuated robotic hand for dexterous grasping," *The International Journal of Robotics Research*, vol. 35, no. 1-3, pp. 161–185, 2016.

[27] J. Zhou, J. Yi, X. Chen, Z. Liu, and Z. Wang, "Bcl-13: A 13-dof soft robotic hand for dexterous grasping and in-hand manipulation," *IEEE Robotics and Automation Letters*, vol. 3, no. 4, pp. 3379–3386, 2018.

IOP Conf. Series: Materials Science and Engineering 1261 (2022) 012011 doi:10.1088/1757-899X/1261/1/012011

Morphological Computation and Control Complexity

Thomas George Thuruthel, Fumiya Iida

The Bio-Inspired Robotics Lab, Department of Engineering, University of Cambridge, UK.

E-mail: tg444@cam.ac.uk

Abstract.
Morphological computation proposes the idea that in a physical system, certain computational processes can be off-loaded to the body. However, the concept has still eluded serious theoretical quantification attempts, unlike traditional computational theory. This perspective examines the notion of morphological computation from the well established theories of traditional computation and computational complexity, drawing parallels between the two, to understand the differences and similarities. Further, we look at the quantification efforts of morphological computation and attempt to link it to the unexplored field of control complexity. We argue that the development of complexity theory for control problems is necessary to study and utilize the concept of morphological computation, if it is possible.

1. Introduction

The notion of *morphological computation* is becoming ever prevalent in the field of robotics [1], especially since the rise of soft robotic technologies [2]. This perspective article examines the notion from the well established theories of traditional computation and computational complexity, drawing parallels between the two, to provide answers to some utilitarian questions; *How does morphological computation theory differ from traditional computational theory? Can morphological computation be quantified? Can physical control problems be classified based on their complexity?*. The organization of the paper is as follows. The first section introduces morphological computation from a pancomputational viewpoint, discusses the properties of traditional computation hardware and field of computational complexity. We conclude with a comparison of morphological computation through the lens of traditional computation. The second section looks into works that attempt to quantify morphological computation for control tasks and their merits. We further propose viable routes to develop the field of control complexity and conclude with a brief discussion.

1.1. What is Computation?

The most general way to look at a computational process is to describe it from a dynamical systems' perspective [3]. To quote [3] directly, "A dynamical system is described in terms of its abstract state space, the system's current state within its state space, and a rule that determines its motion through its state space. In a classical computational system, that rule is given explicitly by the computer program; in a physical system, that rule is the underlying physical law governing the behavior of the system". The pancomputational paradigm (everything is a computing system) is hence adopted by the authors. This implies that any computable function can be represented by a carefully designed dynamical system with an appropriate input-output encoding. Morphological computation, hence, refers to the computation that the body (including the structure, actuators, and sensors) performs while executing a physical task. Note that this is very similar to mechanical computation [4], only differing in their application domain; mechanical computers are designed for solving abstract computational problems whereas morphological computation is associated with solving physical control problems.

1.2. Computational Hardware's and their Properties

The specific dynamical system (hardware) we use for computing depends on the application, requirements, and constraints. The digital electronic computer is the most common amongst all the available artificial hardware systems. They are characterized by transmission of information by discrete and binary electrical signals. This makes them less susceptible to noise, good data transmission bandwidth at low-energy consumption's, efficient data storage and retrieval options, a high cascadability (output of one logic gate can be easily connected to another input) and fan-out (multiple input gates can be connected from a single output gate) due to ease of logic-level restoration and input-output isolation. Note that most of the input-output devices (monitor, for example) are still analog and hence digital computers have to convert information from the digital-to-analog form or vice versa, which is easily doable with electronic circuits.

A close counterpart of the digital electronic computer would be the analog electronic computer, where information is transmitted in continuous electric signals. This would provide them higher bandwidth and faster calculation speeds, but they are easily affected by environmental noise, parasitic effects and a lack of a reliable variable information storage mechanism. Mechanical computers and mechanical analog computers are characterized by information transmission through physical mechanisms (For a detailed review of the field, please refer to [5]). Hence, their computing bandwidth is very low (information travels at the speed of sound when compared to the much faster drift velocity of electrons), highly susceptible to noise and energy inefficient. As expected from a pancomputational viewpoint, there are numerous other computing mechanism like optical computing, quantum computing, DNA computing, etc. Although digital electronic computers seem the best choice for general computing applications, there are scenarios where each type of computing hardware can be better. For example, mechanical computers have been useful in extreme environments [6]; faster and energy efficient computing is potentially achievable in analog electronic computers and optical computers [7, 8, 9]; massive parallelization and large storage is possible in DNA computers [10, 11], and so on. The natural equivalent of a computing hardware, the nervous system, is actually characterized by a poor information bandwidth (the speed of information transfer through a combination of electrical and chemical signals happens at a speed of 70-120 m/s, comparable to mechanical computers [12]). However, these systems have good energy efficiency, massive asynchronous parallelization abilities, compactness and incredible robustness to damage (due to their redundant architecture).

1.3. Task/Problem Complexity

In computational theory, the complexity of a problem is usually measured by the amount of resources required to solve the problem by the application of a particular algorithm. Typically, the resource of concern is time and memory. This formalization allows us to classify problems and provides an estimate of the time and memory requirements imposed on the computational hardware. The Big 'O' notation is the most common metric for calculating how the time complexity of a problem scales with the size of the input. Note that the time complexity of each problem is always relative to the input encoding, however the actual resource utilization will remain the same as the encoding transformation causes a change in input size corresponding to the change in algorithm complexity. In other words, a computational problem cannot be simplified by just changing the input encoding or the computational hardware. Later we will see that this is not the case for control tasks.

1.4. Characteristics of Morphological computation

Having looked at the digital electronic computation, we can now compare the computational characteristics done through a mechanical system. We assume that physical information travels through the mechanical system purely by mechanical waves, without any signal transduction in between. Although, the notion of morphological computation was introduced to perform non-traditional computation, it is still insightful to predict how it would fare in traditional computational tasks. Natural mechanical systems have smooth and continuous dynamics, thereby making it difficult to isolate their logic networks (or its equivalent). Mechanical systems are also plagued with energy dissipative elements like friction and damping. Both these effects reduces the cascadability and fan-out capacity of mechanical computers (This is not necessarily a disadvantage when it comes to physical tasks, as we will investigate later). The information storage capacity and bandwidth of a mechanical system is more difficult to be analyzed. Even though mechanical waves travel at the speed of sound, which is comparable to the speed of information transfer in the brain (still much slower than the drift velocity of electrons in digital computers), the large loss of data due to damping, friction and poor input-output isolation, would reduce the information processing bandwidth of these systems. Hence, we see that all mechanical systems for computing (traditional) purposes are designed to be discrete by using bistable structures or origami architectures [13, 14, 15, 16]. There have been examples of using continuous mechanical systems for traditional computing, however, the fidelity of the system is unproven and they are still 'programmed' using digital computing hardware's [17].

2. Quantifying Physical Complexity

The field of computational complexity solely focuses on abstract numerical problems. Blum complexity axioms specify desirable properties of complexity measures on the set of computable functions [18]. There are no equivalent axioms for control tasks. Although, there are numerical problems to be solved while performing a physical task, the notion of complexity associated with physical tasks is non-existent to the best of our knowledge. If the complexity of physical tasks can be quantified and classified, the same metrics can be used to evaluate the complexity of the physical body and controller(brain). There have been attempts to quantify morphological complexity and controller complexity that we will analyze and attempt to connect.

A change in dynamical properties of a robot(agent) would affect its control architecture to perform the *same task*. This idea has sometimes been without theoretical evidence been used to put forth the notion of a morphology-controller trade-off [19]. Typically, we tend to associate the complexity of a physical system to its observable degrees of freedom (a piece of metal and silicone can otherwise exhibit the same *levels* of passive deformation) and nonlinear dynamical properties, especially since the rapid growth of interest in soft robotics. The controller complexity is typically linked to the number of active elements, the bandwidth of the sensors and the computational complexity of the mathematical processing involved. A commonly used example that is used to illustrate the 'morphology-controller trade-off' is the universal robotic gripper [20]. Here, the specific design of the robot reduces the number of actuators/sensors required, and the mathematical processing involved in control when compared to the traditional rigid designs for grasping an object. Although, on the surface, this example seems to indicate a reduced *complexity* of the controller, we have to be careful to conclude the same. First, we do not have a good measure to quantify the complexity of the controller. Second, we also need the complexity of the task to be quantifiable and comparable. Although, the universal gripper is good at generating stable grasp of an object, the final objective is still to vary the state of the object in a desirable manner (for example, placing it in a known location in a certain pose) and this would require more sensing and control elements even with the universal gripper. Another typically omitted detail is that the universal gripper has to be mounted on a robotic manipulator to be able to perform the task.

There have been few works on developing metrics for quantifying morphological computation, and none on control complexity (to the best of our knowledge). A lot of them have focused on the use of physical body to solve traditional computational problems. Our objective is to look into metrics for solving physical tasks. Hence, these works have been omitted. The most promising work on quantifying morphological computation was presented by Keyan Zahedi et. al. [21]. They introduce two concepts for measuring morphological computation using information theory. They state that morphological computation is the effect of the current world state (X) on the next world state (X'), which is not assignable to the controller (C). The world state also includes the internal states of the agent in their formulation. In other words, the world state X contains the internal state of the agent and the states of the task objective (if they are independent of the agent). Based on this, two metrics can be calculated; one that measures the effect of the action on the next world state; the second measures the effect of the current world state on the next world state. At maximal morphological computation, we would expect the first measure to be at its minima and the second to be at its maxima. Information theory measures are then used to quantify these metrics.

The same idea was investigated from a control system perspective in [22]. Here, it was proved that the optimal open-loop controller obeys the relation $I(X';C) = 0$ (not a sufficient condition). This means that the mutual information between the random variables X' and C is zero or that they are independent. An optimal open loop controller would lead to the maximum entropy reduction from the initial state to the desired state ($H(X) - H(X'|C)$). This is the same as the first concept introduced in [21] for quantifying morphological computation. Note that the choice of controller is relevant to infer any information about the contribution of the open-loop response and hence the morphology itself, even if the controller is open loop. For a closed loop controller, it was also proven that the amount of entropy that can be extracted from a system with initial state X is:

$$\Delta H_{closed} \leq \Delta H_{open}^{max} + I(X;C) \tag{1}$$

Where, $I(X;C) = H(X) - H(X|C)$.

Quantification of morphological complexity is pointless until we can quantify the complexity of a mechanical task itself. From the above works, the closest candidate for the task complexity would be given by:

$$\Delta H_{task} = H(X) - H(X')_{des} \tag{2}$$

Where $H(X')_{des}$ is dependent on the number of independent variables involved and the bounds on the uncertainties that the user specifies. This formulation however does include the states of the agent in it, meaning that that task complexity cannot be quantified without considering the morphology. This is unlike traditional computational problems, whose complexity is independent of the hardware. The complexity of the morphology can be linked to $H(X)$ and the complexity of the controller to $H(C)$ (similar to [23]). In which case, we can then expand the closed-loop controller formulation given in equation 1 to:

$$\Delta H_{closed} \leq \Delta H_{open}^{max} + H(X) + H(C) - H(X,C) \tag{3}$$

From this we can conclude that one way to reduce the complexity of the controller is by increasing $\Delta H_{open}^{max} + H(X)$, which can be done by changing the morphological properties (without changing the cardinality of X), a conclusion which was also found in [23]. In [23], the complexity of the controller was defined as the variance of the control signal (the premise being that a linear controller is the simplest), which for a one-dimensional input is closely related to the entropy. If the cardinality of the state is changed (for example by having additional degrees of freedom), the complexity of the controller is arbitrarily changed as the complexity of the task also changes in the process.

3. Conclusion

Unlike traditional computational problems, it seems that the complexity of control tasks cannot be studied in isolation from the agent(morphology) itself. Hence, hard problems in control are contingent on the robot. For a given task and morphology, the minimum controller complexity (this problem was addressed in [24]) and the optimal open-loop response can be estimated, the latter being the best estimate of morphological computation. Quantification of morphological complexity can be useful to check the *efficiency* of a design, i.e a design whose $(\Delta H^{max}_{open}/\Delta H^{max}_{closed})$ is closer to one would require the least sensory information to perform a task. They would be more robust to sensory noise, however, more sensitive to control output noise [25]. Note that this reduction in control complexity due to the appropriate design of the robot cannot be likened to a morphological complexity-control trade-off. It is possible that for a given task, a simple design with rigid components might provide the optimal open-loop efficiency. Estimation of entropy measures rely on numerical techniques. This makes the process of evaluating a design with its associated controller difficult. It is an open challenge to investigate other metrics for control complexity measures that can be expressed in a closed form, at least with certain simplifying assumptions. A future direction of interest is to look at well known robotic designs and control tasks, evaluating their corresponding task complexity measures, morphological computation metrics and controller complexity to better understand the role of morphology in control and its relation to traditional computational theory.

3.1. Acknowledgments

This work was supported by the SHERO project, a Future and Emerging Technologies (FET) program of the European Commission (grant agreement ID 828818).

References

[1] Pfeifer R, Iida F and Gómez G 2006 *International Congress Series* vol 1291 (Elsevier) pp 22–29
[2] Laschi C and Cianchetti M 2014 *Frontiers in bioengineering and biotechnology* **2** 3
[3] Stepney S 2012 *Handbook of natural computing* **2**
[4] Swade D D 1993 *Scientific American* **268** 86–91
[5] Yasuda H, Buskohl P R, Gillman A, Murphey T D, Stepney S, Vaia R A and Raney J R 2021 *Nature* **598** 39–48
[6] Sauder J, Hilgemann E, Johnson M, Parness A, Hall J, Kawata J and Stack K 2017
[7] Guo N, Huang Y, Mai T, Patil S, Cao C, Seok M, Sethumadhavan S and Tsividis Y 2016 *IEEE Journal of Solid-State Circuits* **51** 1514–1524
[8] Almeida V R, Barrios C A, Panepucci R R and Lipson M 2004 *Nature* **431** 1081–1084
[9] Wang C Y, Chen C W, Jau H C, Li C C, Cheng C Y, Wang C T, Leng S E, Khoo I C and Lin T H 2016 *Scientific reports* **6** 1–7
[10] Church G M, Gao Y and Kosuri S 2012 *Science* **337** 1628–1628
[11] Adleman L M 1994 *Science* **266** 1021–1024
[12] Lovinger D M 2008 *Alcohol research & health*
[13] Ion A, Wall L, Kovacs R and Baudisch P 2017 *Proceedings of the 2017 CHI Conference on Human Factors in Computing Systems* pp 977–988
[14] Song Y, Panas R M, Chizari S, Shaw L A, Jackson J A, Hopkins J B and Pascall A J 2019 *Nature communications* **10** 1–6
[15] Yasuda H, Tachi T, Lee M and Yang J 2017 *Nature communications* **8** 1–7
[16] Halg B 1990 *IEEE Transactions on Electron Devices* **37** 2230–2236
[17] Nakajima K, Hauser H, Li T and Pfeifer R 2015 *Scientific reports* **5** 1–11
[18] Blum M 1967 *Journal of the ACM (JACM)* **14** 322–336
[19] Pfeifer R and Scheier C 2001 *Understanding intelligence* (MIT press)
[20] Brown E, Rodenberg N, Amend J, Mozeika A, Steltz E, Zakin M R, Lipson H and Jaeger H M 2010 *Proceedings of the National Academy of Sciences* **107** 18809–18814

[21] Zahedi K and Ay N 2013 *Entropy* **15** 1887–1915
[22] Touchette H and Lloyd S 2004 *Physica A: Statistical Mechanics and its Applications* **331** 140–172
[23] Rückert E A and Neumann G 2013 *Artificial Life* **19** 115–131
[24] Montúfar G, Ghazi-Zahedi K and Ay N 2015 *PLoS computational biology* **11** e1004427
[25] Åström K J 2002 *Department of Mechanical and Environmental Engineering University of California Santa Barbara* **333**

IOP Conf. Series: Materials Science and Engineering 1261 (2022) 012012 doi:10.1088/1757-899X/1261/1/012012

Physical Intelligence in Biomechanics

Shiqiang Wang[1], Yongqi Shi[1] and Li Wen[1,2]

[1] School of Mechanical Engineering and Automation, Beihang University, Beijing, 100191, People's Republic of China

[2] Author to whom any correspondence should be addressed

E-mail: liwen@buaa.edu.cn

Abstract. The intelligence of the biological agents is enabled by their neural intelligence in their brains, in the meantime, their physical intelligence encoded in the bodies plays a nonnegligible role. The development of intelligent bio-inspired robots depends on an in-depth understanding of physical intelligence, particularly in biomechanics. Here, we briefly review the physical intelligence of biological organisms from three aspects: material, structure, and morphology. We envision that bio-inspired physical intelligence would boost the development of future intelligent lifelike machines.

1. Introduction

Robots can perform complex tasks in structured environments (such as factories, warehouses, etc.). However, in unstructured environments with uncertainties, the capabilities of robots are severely impaired. Biological organisms can easily thrive in a variety of complex and unstructured environments. This is partly thanks to the physical intelligence encoded in the body [1]. By understanding mechanisms of physical intelligence in biomechanics, we can forge a path for more intelligent robots. The body of an organism is composed of three elements: material, structure, and morphology. Naturally, the physical intelligence encoded in the body also depends on these three elements to implement. We will illustrate their essential role in physical intelligence-based biological functions through the following typical examples.

IOP Conf. Series: Materials Science and Engineering 1261 (2022) 012012 doi:10.1088/1757-899X/1261/1/012012

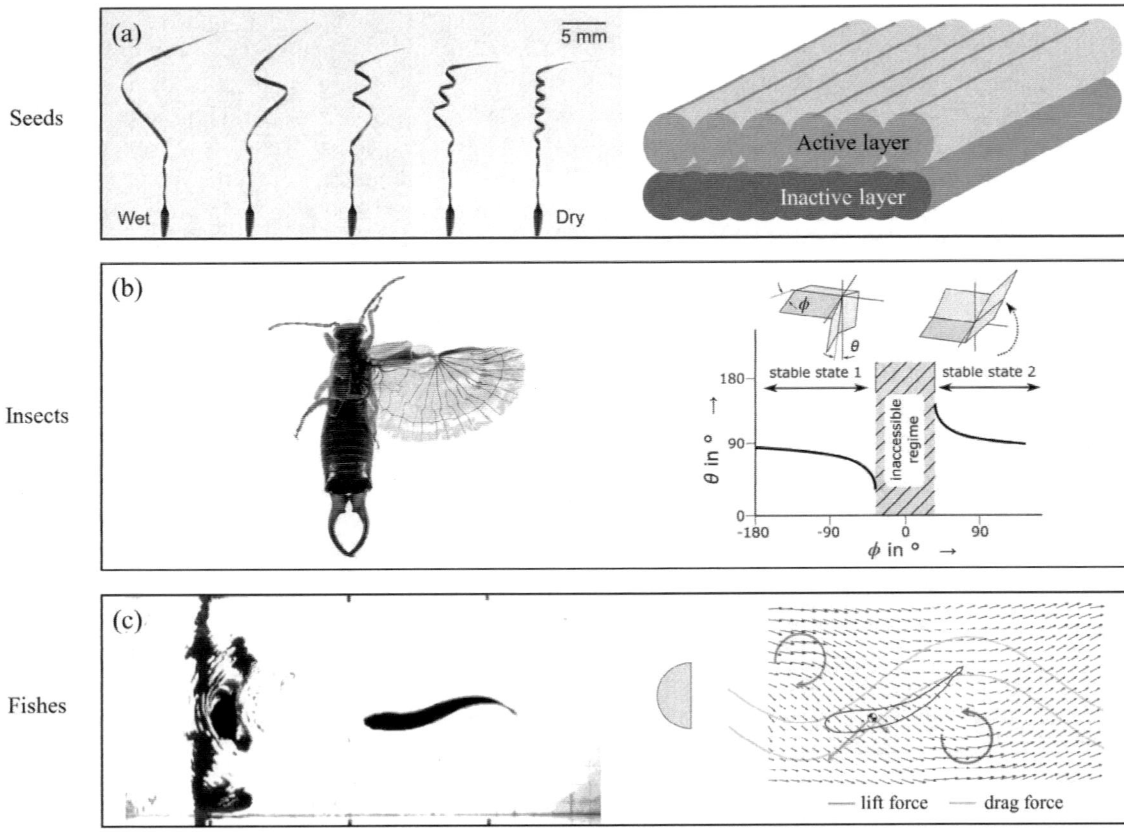

Figure 1. Representative physical intelligence in biomechanics. (a) The awns of Pelargonium seeds can respond to changes in environmental humidity, which enable the self-burial of the seeds into the soil by helically coiling [2,3]. (b) The earwig wing has self-folding and self-locking functions realized by origami and bistable structures [4,5]. (c) A dead fish swims upstream because of the interaction between its body morphology and the vortices [6,7].

2. Material

Regarding the materials of physical intelligence, there are many lessons that can be learned from nature. For example, the self-burial behavior of some plant seeds can increase their chances of germination and survival, which is achieved by awns that respond to changes in environmental humidity. For example, the awns of Pelargonium seeds uncoil to stretch when wet but coil when dry, resulting in coiling motion to propel the seeds into the soil [8,9] , as shown in figure 1(a). Similarly, the dispersal unit of wild wheat bears a pair of awns, which bend during the day because the air is dry and straighten at night due to increased humidity. The daily humidity cycle provides motility to the seeds. Silicified hairs covering the awns make the seeds only move downwards and eventually into the soil [2,3]. Ice plants growing in arid environments have a unique seed dispersal mechanism [10]. In the dry state, five protective valves on the seed capsule cover the seed compartments to prevent premature dispersion of seeds. When sufficiently hydrated with water, each valve unfolds outwards and backwards over an angle of about 150° within a few minutes. It increases the possibility of seeds dispersed under conditions conducive to germination. The tissues and organs of leptocephali (larvae of eels) are mainly composed of active transparent hydrogels, which can achieve optical camouflage in water to reduce the risk of being found by natural enemies [11,12].

3. Structure

The leaves of the Venus flytrap can be described as bistable structures. The rapid closure of the Venus flytrap is made by harnessing mechanical instability, which is one of the fastest movements ($\approx$100 ms) in the plant kingdom [13]. Origami structures can be found in the leaves of plants and the wings of insects. Leaves of hornbeam and beech have a corrugated Miura-Ori pattern that enables them to unfold during budding [14]. The greater the angle between the lateral and central veins, the more compactly the leaf can be folded within the bud, but the longer it takes to unfold. The earwig wing is a typical origami structure, with the highest folding ratio (1:10 to 1:18) in the animal kingdom, and can achieve rapid self-folding without the use of muscles [4,5]. The mid-wing mechanisms act as a bistable locking structure to keep the unfolded wing in its open state during flight, as shown in figure 1(b). Tensegrity structures have been found in biological organisms at different length scales. For example, the cytoskeletons are similar to tensegrity structures. The microtubules and microfilaments correspond to compression and tension members, respectively [15,16]. In vertebrates, stiff bones and tensile muscles are connected tightly to ensure that their structure is both firm and flexible [17,18].

4. Morphology

Morphology also plays an important role in the behavior and movement strategies of biological organisms. Tumbleweed has a spherical skeleton. When dry autumn comes, the plant detaches from its roots and harnesses the force of the wind to roll and bounce on flat land to disperse seeds [19]. Maple seeds autorotate as they descend because of their geometry and the location of their center of gravity. By generating a stable leading-edge vortex, the autorotating seeds of maples attain high lift, enhancing the descent time and dispersal performance [20]. A dead fish in Kármán vortex street can swim upstream, looking the same as a live fish [6,7], as shown in figure 1(c). This is due to the passive thrust generated by the interaction between its body morphology and the vortices. Adaptive morphology enables biological organisms to overcome the limitations of fixed morphology. Because different environments require specific morphologies, many animals have evolved deformable bodies to expand their operational range. For example, octopuses can squeeze through holes much smaller than their bodies. A bird's wing can be quickly morphed for different flight modes to obtain optimal flight performance. Extended wings are more suitable for slow glides and turns, and swept wings are more suitable for fast glides and turns [21]. Birds also use folded wings to fly in cluttered environments and traverse gaps slightly larger than their bodies [22]. Adaptive morphology can also help animals quickly evade predators. For example, when attacked by its nemesis – predatory wasps, the golden wheel spider turns its body into a wheel shape and rolls down dunes after a short run [23]. When disturbed, the web-toed salamander curls itself into a loop and rolls down the slope of the rock surface, much faster than running on it [24].

5. Conclusion

Biological organisms can survive and operate autonomously in unstructured environments because of the neural intelligence in their brains and the physical intelligence encoded in their bodies. Decomposing the body into three key elements: material, structure, and morphology, contributes to a better understanding of the mechanisms of physical intelligence in biomechanics. Hygro-responsive materials that widely exist in plant seeds play a critical role in the dispersal and germination of seeds. Bistable, origami, and tensegrity structures are also found in biological organisms, enabling them to achieve flexible and efficient movement. Fixed morphology formed through long-term evolution can be coordinated with adaptive morphology, enabling biological organisms to perform various tasks in different environments. As a new paradigm, physical intelligence would progress the future robots beyond those that focus only on computational intelligence. Bio-inspired design is a critical method for developing physical intelligence for robots, Biological systems have formed advanced physical intelligence by evolution, and researchers can get inspiration from them. For example, stimuli-responsive functional materials are capable of changing their properties (e.g., shape, color, stiffness) in response to an external stimulus (e.g., light, electrical current, force), including dielectric elastomers, hydrogels, shape memory polymers, liquid crystal elastomers, magnetic materials and so on. With the addition of intelligent materials, the robot's body can take over actuation and the task of sensing and processing information from the environment. Furthermore, smart structures have unconventional and programmable mechanical properties, responding to environmental changes in morphologically and functionally adaptive ways. Bistable, origami, tensegrity structures, and other smart structures are essential choices to realize the physical intelligence of robots. Finally, adaptive morphology provides additional degrees of freedom, increasing the robot's adaptability to the changes in environment and tasks. So that morphologically adaptive robots do not have to be redesigned and rebuilt when facing unexpected environments. Bio-inspired intelligent materials, structures, and morphologies can be integrated into the robot's body through bio-inspired design to introduce biological agents' physical intelligence into the robot.

Acknowledgments

This work was supported by the National Science Foundation support projects, China (Grant No. 91848105), in part by National Key R&D Program of China (Grant No. 18YFB1304600), and National Science Foundation support projects 61822303, 61633004, 91848206.

References

[1] Sitti M 2021 Physical intelligence as a new paradigm *Extreme Mech Lett.* **46** 101340
[2] Jung W, Kim W and Kim H Y 2014 Self-burial mechanics of hygroscopically responsive awns *Integr. Comp. Biol.* **54** 1034-42
[3] Ha J, Choi S M, Shin B, Lee M, Jung W and Kim H 2020 Hygroresponsive coiling of seed awns and soft actuators *Extreme Mech Lett.* **38** 100746
[4] Haas F, Gorb S and Wootton R J 2000 Elastic joints in dermapteran hind wings: materials and wing folding *Arthropod Struct. Dev.* **29** 137-46
[5] Faber J A, Arrieta A F and Studart A R 2018 Bioinspired spring origami *Science* **359** 1386-91
[6] Beal D N, Hover F S, Triantafyllou M S, Liao J C and Lauder G V 2006 Passive propulsion in vortex wakes *J. Fluid Mech.* **549** 385
[7] Liao J C 2007 A review of fish swimming mechanics and behaviour in altered flows *Philos. Trans. R. Soc. B-Biol. Sci.* **362** 1973-93
[8] Elbaum R, Zaltzman L, Burgert I and Fratzl P 2007 The role of wheat awns in the seed dispersal unit *Science* **316** 884-6
[9] Elbaum R, Gorb S and Fratzl P 2008 Structures in the cell wall that enable hygroscopic movement of wheat awns *J. Struct. Biol.* **164** 101-7
[10] Harrington M J, Razghandi K, Ditsch F, Guiducci L, Rueggeberg M, Dunlop J W C, Fratzl P, Neinhuis C and Burgert I 2011 Origami-like unfolding of hydro-actuated ice plant seed capsules *Nat. Commun.* **2**
[11] Johnsen S 2001 Hidden in plain sight: the ecology and physiology of organismal transparency *Biol. Bull.* **201** 301-18
[12] Miller M J, Rutgers R, Haythorne B, Yavuzdoğan T, Obata S, Wu T, Rutgers H, Powell J and Tsukamoto K 2013 Observations of large muraenid leptocephali in coastal Indonesia: locations of sightings and behaviour of the larvae *Mar. Biodivers.* **6**
[13] Forterre Y, Skotheim J M, Dumais J and Mahadevan L 2005 How the Venus flytrap snaps *Nature* **433** 421-5
[14] Kobayashi H, Kresling B and Vincent J F V 1998 The geometry of unfolding tree leaves *Proc. R. Soc. B-Biol. Sci.* **265** 147-54
[15] Wang N, Butler J P and Ingber D E 1993 Mechanotransduction across the cell surface and through the cytoskeleton *Science* **260** 1124-7

[16] Wang N, Naruse K, Stamenović D, Fredberg J J, Mijailovich S M, Tolić-Nørrelykke I M, Polte T, Mannix R and Ingber D E 2001 Mechanical behavior in living cells consistent with the tensegrity model *Proc. Natl. Acad. Sci. U. S. A.* **98** 7765-70
[17] Levin S M 2002 The tensegrity-truss as a model for spine mechanics: biotensegrity *J. Mech. Med. Biol.* **2** 375-88
[18] Scarr G 2014 *Biotensegrity* (United Kingdom: Handspring Publishing)
[19] Armour R H and Vincent J F V 2006 Rolling in nature and robotics：a review *J. Bionic Eng.* **3** 195-208
[20] Lentink D, Dickson W B, van Leeuwen J L and Dickinson M H 2009 Leading-edge vortices elevate lift of autorotating plant seeds *Science* **324** 1438-40
[21] Lentink D, Müller U K, Stamhuis E J, de Kat R, van Gestel W, Veldhuis L L M, Henningsson P, Hedenström A, Videler J J and van Leeuwen J L 2007 How swifts control their glide performance with morphing wings *Nature* **446** 1082-5
[22] Williams C D and Biewener A A 2015 Pigeons trade efficiency for stability in response to level of challenge during confined flight *Proc. Natl. Acad. Sci. U. S. A.* **112** 3392-6
[23] Garcia-Paris M and Deban S M 1995 A novel antipredator mechanism in salamanders: rolling escape in hydromantes platycephalu*s* *J. Herpetol.* **29** 149-51
[24] Henschel J R 1990 Spiders wheel to escape *S. Afr. J. Sci.* **86** 151-152

IOP Conf. Series: Materials Science and Engineering 1261 (2022) 012013 doi:10.1088/1757-899X/1261/1/012013

Mechanical Sensing in Embodied Agents

Matteo Lo Preti[1,2*], Thomas George Thuruthel[3*], Kieran Gilday[3], Lucia Beccai[1], Fumiya Iida[3]

[1] Soft BioRobotics Perception Lab, Istituto Italiano di Tecnologia, Genova, IT
[2] The Biorobotics Institute, Sant'Anna School of Advanced Studies, Pisa, IT
[3] The Bio-Inspired Robotics Lab, Department of Engineering, University of Cambridge, UK
[*] These authors contributed equally to this work.

E-mail: matteo.lopreti@iit.it

Abstract.
Sensors enable autonomous systems to obtain information about their internal states and the environment for guiding their actions. It is as essential for these sensors to reject disturbances as to gather the correct information. There are numerous trade-offs and considerations in designing these sensory systems. For instance, natural agents evolved a vast diversity of highly optimized sensory organs to perform their tasks. This work focuses on how these sensory systems estimate mechanical stimuli. We look at some of the strategies and design principles found in nature to understand fundamental trade-offs and design considerations when acquiring and processing mechanical information.

1. Introduction

Embodied intelligence advances the argument that a behavior cannot manifest without considering the body, brain, sensors, and environment [1, 2, 3]. This argument expanded the field of inquiry into the sensori-motor system and its interaction with the environment, attributing the role of the mind to the ultimate goal of guiding actions [1]. In the pursuit of differentiating the computational function of the body from the brain, researchers have recently introduced the idea of *morphological computation* [4]. Due to the seamless interdependent nature of the sensori-motor system, it becomes difficult to study each component per se. Nonetheless, in this work, we look at the role of embodiment in the sensory system, particularly that involved with estimating mechanical cues. Hence, we first need to define what mechanical sensing is.

Definition. *Mechanical sensing is the ability of a system to encode mechanical stimuli, e.g., displacement and force, owing to its mechanical properties (intrinsic or structural), and convert them into an electrical signal.*

The role of the sensory system is to gather information about the world and the body in a form that the agent can use to guide its actions. A sensor is then a device that selects and transforms this information into a usable, i.e., electrical, form. Indeed, a signal that undergoes processing can be considered "usable". Data selection (or reduction) is necessary to ease the processing burden on the agent, both in terms of available memory and computational time. Keeping irrelevant data out of the controller is as a critical function of sensory systems as providing relevant information [5].

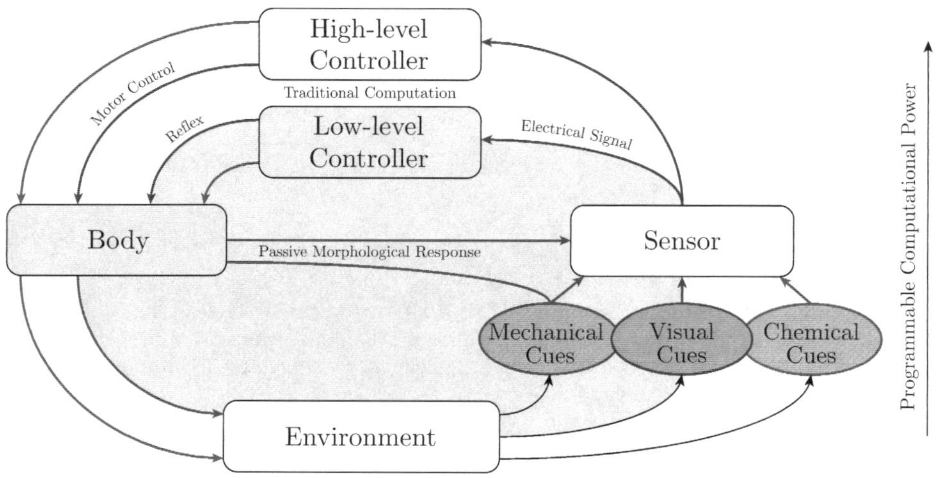

Figure 1: Sensing is a time-intensive process that involves the fusion and processing of different cues perceived at different speeds. The sensori-motor system is characterized by the body and sensing morphology and by a trade-off mainly between response time and energy efficiency. The computational effort is highlighted by the loop area of the High-level Controller, Low-level Controller, and Body, reflecting the processing time.

As Fig. 1 shows, the design architecture of the sensori-motor system is primarily governed by the environment that imposes physical constraints, the task at hand, and energy considerations. For example, visual sensors are not well developed in deep underwater animals and tiny organisms without a nervous system (as visual processing is computationally heavy). The speed of information varies on the type of the physical cue. For instance, the speed at which information travels by mechanical waves is limited to the speed of sound in that medium. In contrast, the speed of information is much faster in electric circuits, and it is only limited by the electron drift velocity.

Transduction of information is necessary for the data to be processed by computing hardware. However, transducing an energy form into another has costs in terms of time, energy, and noise addition. This aspect is highlighted in Fig. 1, where the area of the different processing loops is proportional to the processing time. The red loop refers to a purely open-loop mechanical response, providing the fastest response if optimally designed, since this information transformation loop does not involve signal transduction. On the other hand, the loop will be limited in the kind of computational processing it can perform [6]. Hence, it becomes challenging to consider all these factors in designing artificial agents.

The blue and yellow loops in Fig. 1 refer to the transduction and flow of information with a low- and a high-level controller, respectively. These loops can be seen in analogy with vertebrates' peripheral and central nervous systems to clarify this concept. The former needs the signal from external cues to be transduced into an electrical source. Then, the signal is elaborated to produce a motor response. On the other hand, a high-level controller involves more sophisticated cognitive processes, like perception. In this case, the brain organizes, identifies, and interprets transformed sensory information.

Observing natural phenomena and mechanisms can lead to a better understanding of design rules for developing new robots. We look at the sensori-motor system of several natural and artificial agents to find their common threads and analyze the overall role of morphology in sensing and the fundamental trade-offs imposed by delays in the sensory system.

2. Sensing for Control

The brain integrates inputs across multiple sensory modalities to obtain a rich and coherent understanding of the environment [7, 8]. All sensory modalities have their characteristic ranges, bandwidth, and processing difficulties. These differences largely determine the employment of one modality over the other. For example, in humans, auditory signals have a frequency range from 20-20.000 Hz, tactile sensing can go up to 1000 Hz, while visual data have the lowest frequency range. In terms of range, visual signals can travel over large distances compared to the poor range of mechanical signals. As mentioned before, the speed of information varies depending on the type of physical cue. Note that mechanical information can pass through the sensors and the body itself, unlike other physical cues, as shown by the passive morphological response loop in figure 1. This perspective focuses primarily on the sensing of mechanical stimuli. An applied force is a mechanical cue that very diverse organisms (such as microbes, roundworms, insects, and mammals) detect and respond to [9]. It is caused by the actions or by the passive adaptation of the body to the environment, and it induces deformation in the body and sensors. Hence, the placement of the sensors, the interface between the sensor and the environment, and its co-dependency with the morphological properties of the enclosing body play an essential role in what a living organism can sense [10]. This close association also means that every closed-loop sensorimotor response initiated by mechanical sensors can, in theory, be replaced by an open-loop mechanical system [11]. In other words, the controller that performs the computation can be replaced by a mechanical system (with an energy source) to perform the same computation [6]. A closed-loop sensorimotor system would be less sensitive to uncertainties in real-world scenarios under unmodelled disturbances. In contrast, an open-loop mechanical system would be insensitive to sensor noise [11].

2.1. Active vs. passive mechanical response

Every biological agent has a certain degree of passive and active components in responding to environmental instances. Full-state feedback control would provide the best performance measures in robotic design without considering cost, weight, fabrication, noise, and processing delays. However, once we consider real-world scenarios where we have uncertainties in state estimation, partial observability, and delays incurred due to sensory transduction (time constant of the sensor), we inevitably introduce fundamental limitations to the performance of the system [12]. Hence, there is always a trade-off between passive and active responses in an agent deployed in real-world situations.

For instance, a delay in an open-loop system merely causes a delay in its output, while delays in a closed-loop system cause instabilities. There are multiple ways to reduce instabilities. All of them involve the development of a predictive model for delay compensation [13]. When the predictive models become complex, passive (open-loop) or closed-loop responses at low gain become the only viable option. The fundamental limit of passive systems is their inability to inject energy into the system/environment controllably. Next, we look at how some of these trade-offs are handled in nature, the role of sensor morphology in simplifying the transduction process, and, finally, how the body morphology regulates the information structuring in these sensors.

3. Case Studies

3.1. Handling Sensory Delays in the Control Loop

Embodiment plays a major role in short-cutting the control loop in various tasks. As case studies, we report two examples of totally different organisms (humans and fruit flies) that nature created to minimize latencies.

IOP Conf. Series: Materials Science and Engineering 1261 (2022) 012013 doi:10.1088/1757-899X/1261/1/012013

3.1.1. Human Hand Touch plays an essential role in our perception of the environment, our ability to manipulate objects [14], and human social interaction [15]. Tactile information is perceived by receptors in the skin, especially in the hands [16, 17], which can encode various modalities, such as texture, shape, temperature, force, pressure, and pain. Such sensory information is perceived and processed with different coding strategies through the afferent nervous system [18, 19]. However, the processing of tactile information in the brain is time-intensive. Yet, techniques have been developed to handle this delay by using predictive models as previously discussed [13].

The pick&lift task (see Fig. 2) is a prototypical basic manipulation task deeply studied in neuroscience. It consists of different phases of interaction with an object and environment, as indicated in Fig.3. Here, the tactile afferents' responses generated at contact events are known as neural signatures, and they represent critical sensorimotor control points [20]. These control points are used to generate anticipatory control actions to reduce latency. Embodiment plays a primary role since the mechanical deformations, in turn, modify the sensori-motor loop to achieve the overall anticipatory motion [14]. Indeed, the morphology of the fingertip (shown in Fig. 2) plays a role since its deformation produces a complex distribution of stresses that, in turn, affects the mechanoreceptors and the tactile response.

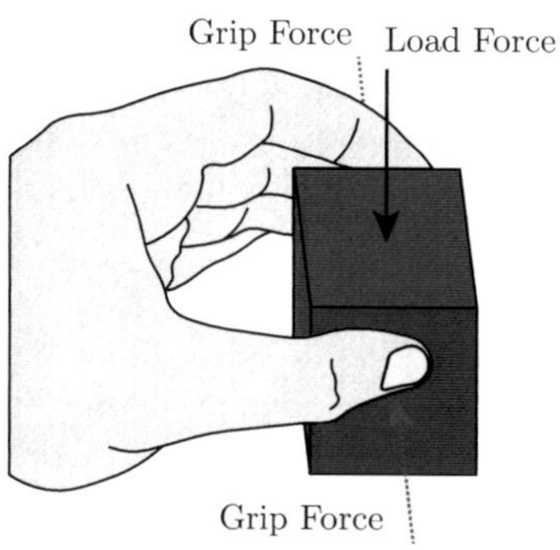

Figure 2: Scheme of the pick&lift task. The thumb and the index fingers contact with the test object and exert a grip force that must overcome the load force (normal to the object surface) to lift the object itself.

In this context, skin wrinkles play a crucial role in the control loop when the hand operates in a wet environment. Notably, the action of wrinkles is reflected in the modulation of grip and load forces throughout a basic manipulation task. This aspect has been studied by performing experiments in a wet environment, where it is known that wrinkles are enhanced. In this context, Keraklas et al. showed that individuals with wrinkles could perform pick&lift better than ones without them. [21]. According to Changizi et al. [22], wrinkles act like channels that draw water away when the fingertip is pressed upon a surface. Therefore, objects are grasped more readily. Indeed, the force required to grip wet objects decreases by 20%, with a value in line with that needed for dry objects [23]. In general, the morphological changes due to wrinkling induce a variation in the mechanical interaction at the fingertip/object interface, and thus in the sensitivity of the inner receptors [24].

3.1.2. Fruit Flies Flying insects have strict control requirements due to the need for energy efficiency and the disastrous implications of control failures. These organisms must minimize sensory delays [25], which are a source of instability in closed-loop control systems. Flight stabilization can be achieved through both active and passive mechanisms. Even though passive designs are the easiest design-wise, active ones are the most recurrent [26].

IOP Conf. Series: Materials Science and Engineering 1261 (2022) 012013 doi:10.1088/1757-899X/1261/1/012013

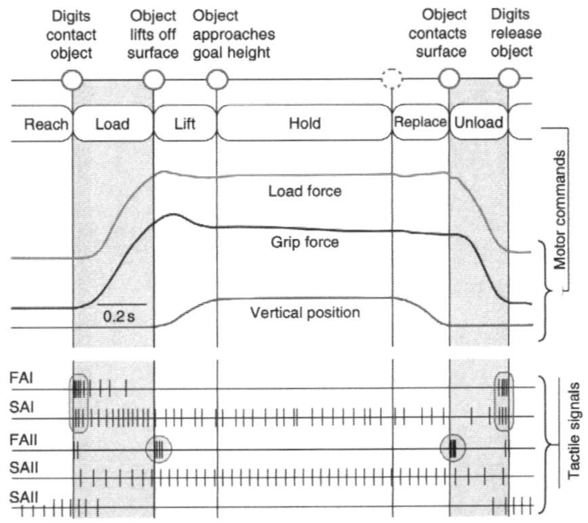

Figure 3: The information used in a basic manipulation task is encoded in the temporal pattern of spikes in ensembles of tactile afferents. Static contact responses are encoded by the fast-adapting type I (FAI) afferents, whereas slow-adapting type I (SAI) and fast-adapting type II (FAII) tactile afferents encode dynamic events. At subgoal events, tactile afferents burst distinctly when digits make/break contact with an object, while FAII respond when an object is lifted off/released from/on the surface. Adapted from [14].

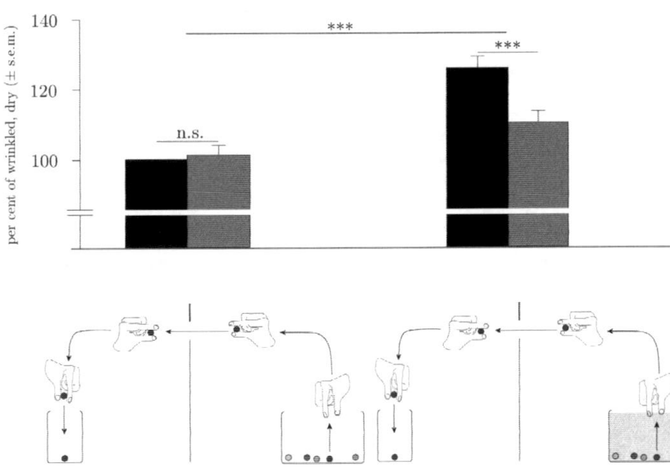

Figure 4: Transfer time (standardized to the time taken to transfer dry objects with unwrinkled fingers) is shortest for dry objects, independent of wrinkling, but faster for submerged objects with wrinkled (red bar) fingers than with unwrinkled (black bar) fingers ($p < 0.001$). Items are picked up with the thumb and the index finger of the right hand, passed through a hole to the left hand, and put into a box with a hole in the lid. Adapted from [21].

Flying insects have highly-specialized body rotation sensors to reduce the detrimental effects of delays. These sensors can provide electrical inputs directly to motor neurons [27] similar to the reflex pathways (Figure 5). However, this is only possible because they have specialized sensors to measure the relevant state variables and the control mapping from the estimated state to the motor action is relatively simple. The sensory information comes from mechanoreceptors at the

IOP Conf. Series: Materials Science and Engineering 1261 (2022) 012013 doi:10.1088/1757-899X/1261/1/012013

base of the halteres, a dumbbell-shaped organ that actively beats back and forth antiphasic to the motion of the wings. Since a moving object in a rotating reference frame experiences inertial (Coriolis) forces, the mechanoreceptors can measure the forces generated by the halteres that are linearly proportional to the angular velocity of the body.

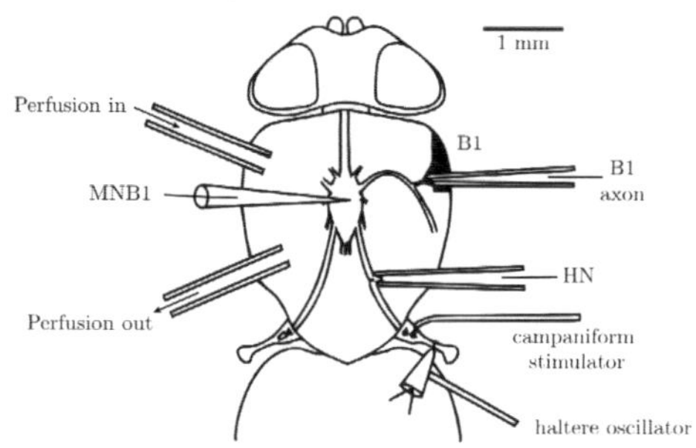

Figure 5: Fruit flies can respond to sensory information quickly due to their optimized sensor-neuron-motor pathway. Adapted from [27].

3.2. Extended embodiment for sensing
Both the body and the environment are perceived from an egocentric point of view [28]. In this regard, the concept of body is not fixed, and both the dynamics of its changes with respect to the agent and that of items used by an agent can be learned [29]. Concerning these aspects, we describe two cases in which the sensory system can use passive instruments to sense the surroundings and adapt to their changes.

3.2.1. Spiders Tactile sensors found in spider species are specialized for discerning particular mechanical cues [5]. Almost all spider species have poor visual capabilities, probably due to their relatively limited nervous system [30]. Spiders use two very similar sensors for detecting medium flow and contact detection (Figure 6). The medium flow sensors or Trichobothria are among the most sensitive biological sensors observed [31, 10]. They have low base stiffness and behave like a rigid rod that effortlessly moves with a flowing airfield. The trichobothria can easily modify their mechanical frequency response by varying their length. Tactile hairs, a modified morphology of the trichobothria, are specialized to detect contact. They are characterized by a tapered morphology and a much higher base stiffness. These characteristics lead to overall lower deflection at the base of the hair, even stress distribution along the structure, and a nonlinear stimuli response which the spider utilizes to be more sensitive at low forces [32]. Spider slit sensilla is another structure developed for strain sensing and vibrations detection. The slit geometry coupled with the stiffness gradient that arises from a rigid exoskeleton and the soft viscoelastic pad enables ultrasensitive displacement detection [33]. Additionally, spiders also show one of the most pronounced examples of modifying the environment for information structuring [34].

Web-building spiders are a fascinating example of embodiment within the ambient to extend cognition. These animals show complex behaviors like route planning, tactics learning, and adjustment to altered conditions [35]. A part of the acquisition, processing, storage, and use of information is outsourced from the central nervous system [36]. This outsourcing leads to less demanding representations of the external world [37, 38, 39].

The mutual manipulability criterion has been used to assess whether web threads and configurations can be seen as integral parts of the cognitive systems [40]. This criterion is widely used to test the constitutive relevance, i.e., to what extent an entity is relevant to explain

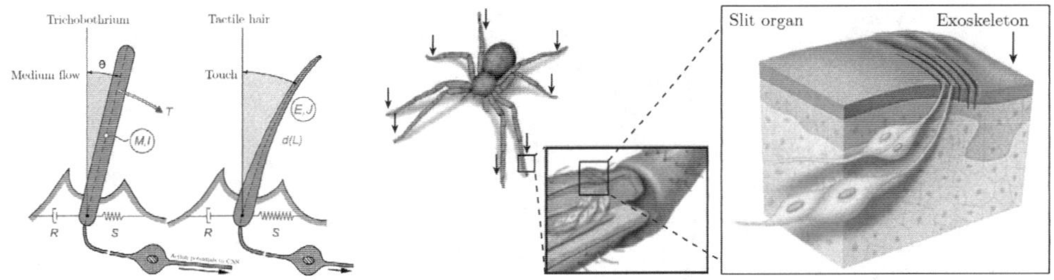

Figure 6: Spider example of how small mechanical design changes affect the sensing principle of the mechanoreceptor. Adapted from [5, 33].

the behavior of a larger system. In the case of spiders, it was investigated whether sensations arising from external devices (the web) are relevant to the brain. If changes in the web lead to changes in the cognitive system and vice versa, then the entity is said to be part of the cognitive system. Nakata et al. found out that spiders can focus on a web portion by varying the pulling force exerted upon a part of the threads. This behavior was shown to enhance capture success [41]. Moreover, this attention can be induced. Whenever prays were exclusively put in the same area of the web, spiders learned to increase the pulling force exerted on those threads and, accordingly, to respond faster to stimuli in that direction [42]. These findings allowed us to conclude that the web threads and configurations are integral parts of the cognitive systems [43].

3.2.2. Whiskers Whiskers are highly sensitive, long, and stiff tactile hairs growing in bunches. Animals use whiskers, or vibrissae, to detect information like the precise location, size, and texture of an object and changes in fluid flow to follow underwater prays or navigate the environment. In living organisms, an applied force does not directly affect sensory receptors. Instead, they are activated by the induced deformations transmitted by mediating tissues. In the vibrissal system, tactile information is conveyed by a receptorless and not innervated whisker hair to follicle mechanoreceptors, which then provide input to the brain [44]. The mechanoreceptors are located at the base of the whiskers, in the follicle-sinus complex [45]. The follicle-whisker junction is rigid, so the whisker acts as a mechanical transducer between the objects being touched and the mechanoreceptors of the follicle-sinus complex.

Interactions between the whisker, its follicle, and the associated intrinsic muscle do not occur at a single point. Instead, it occurs along a few millimeters of the follicle length, which contains mechanoreceptors of several types and spatial orientations. The density of a whisker decreases going from the proximal towards the distal region. Hartmann et al. [46] reported how isolated whiskers exhibit natural resonances and strong damping depending on their length, playing an important role in detecting object boundaries.

Active touch comprises several components, such as motor, physical, mechanical, and neuronal transformations. There is also evidence that pre-neuronal morphological transformations are one of these components. For example, rats push their whiskers against objects during localization to induce meaningful morphological coding through the bending moment at the whiskers. This technique is used to obtain a three-dimensional spatial representation of the environment. Indeed, object position can be efficiently determined from whisker mechanics alone, suggesting a role for pre-neuronal morphological computation in active vibrissal touch [47].

Whiskers work by altering their shape to adapt to a particular situation. In case of damage, the sensitivity of the damaged whisker will be reduced by shrinking the barrel, while the sensitivity of the neighboring parts appears to increase. Different levels of elasticity can be

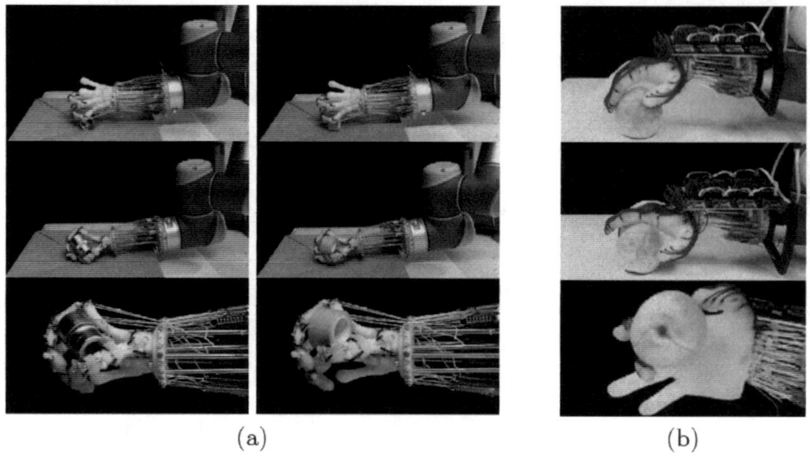

Figure 7: Fully passive hand with mechanical finger synergies trained for sphere 3-finger grasping on an Hex Socket (⌀= 40mm) and on Tape (⌀= 35mm) [55] (a) and sensorized soft hand for passive grasping (b). Different hand-environment interactions generate different finger behaviours, hence different sensing responses. Adaptive grasping behaviours allow for repeatable and predictable physical and sensory response.

produced by reshaping the medulla layer. The characteristics of this flexible body are modulated to outsource this computational function to the body. Kaneko et al. [48] were the first to investigate the correlation between the curvature of a flexible wire and tactile information. Nguyen et al. [49] proposed an artificial whisker that could actively adjust its sensitivity when the body length was either trimmed or broken. The design proposed by these researchers is at the basis of the compensation strategy needed to develop an active robotic whisker whose malfunctioning can be compensated by varying their sensitivity and that of their neighborhood. Two current limitations concern the lack of algorithms to evaluate the breakage condition and more evidence that allocating the compensatory action to the body is more practical than involving the central processing units.

3.3. Information self-structuring through the morphology of the body

The kinematic design of the body and its mechanical properties also play an important role in funneling and filtering tactile information. Such adaptations become more relevant for complex systems that rely on dense mechanical information processing for control, as shown in two examples.

3.3.1. "Mechanical" Synergies: Synergy control is a popular method to solve the dimensionality problem without sacrificing flexibility for controlling high Degrees-of-Freedom (DoF) hands [50]. In human hands, synergies map low dimensional inputs to the control of a much larger amount of digits. This enables rapid action and adaptation in changing environments without overburdening the central nervous system.

The neural basis for hand synergies is well understood [50, 51], where signals from the Central Nervous System activate characteristic patterns of muscles. Kinematic synergies are the tendencies of different digits to move together [52], the origin of which is rooted in the physical design of the hand and the interconnected nature of tendons driving different joints [53]. Kinematic synergies have been applied to great success in the control of robotic hands, allowing highly adaptive grasping with two [54], one, and even zero [55] degrees of actuation (Fig. 7).

Kinematic synergies act in tandem with reflexes and higher-level hand control, providing a rapid-acting response to small uncertainties in the environment by dynamically balancing hand-environment interactions over multiple joints. This bypasses the central processing system for incredibly cheap and fast control [55], and conditions the hand morphology to activate mechanoreceptors predictably. One example is the passive redistribution of prehensile grasping forces in compliant hands [56]. Other open-loop passive hands, such as Hughes' piano-playing hand, show the importance of morphology in conditioning hand-environment interactions [57]. In another example from Sanfilippo et al. [58], a modular, sensorized, synergy-driven, compliant hand shows the importance of sensing feedback for dexterity. It can achieve this with relatively simple control and cheap sensors, in part, due to the hand's morphology.

3.3.2. Pre-neuronal Compression: Further than shaping and modulating the response at the receptor, the sensor morphology and its behaviors within a passive dynamical body can contribute to compressing information. The skin of the human hand contains $\approx$ 17.000 mechanoreceptors to provide detailed and localized tactile feedback [59]. Physical interactions stimulate these receptors through the complex, nonlinear, viscoelastic skin [60]. Dynamic touch elicits mechanical waves in the tactile frequency range that spread throughout the whole hand [61]. The sheer quantity of raw tactile information would quickly overwhelm our cognitive system. Therefore, this information needs to be compressed. There is evidence to suggest that some compression occurs in the dynamic mechanical response of the skin [62]. Shao et al. showed that dynamic touch elicits mechanical waves that efficiently encode tactile information [62]. Optimal encoding of thousands of natural whole-hand stimuli results in a compact representation with primitive spatiotemporal patterns. This suggests the interplay of body dynamics and the sensory system provides a critical modulation of dynamic tactile information such that it can be encoded with extreme efficiency. Sensing redundancy is essential for more general-purpose systems, such as the human hand, and highly specialized receptors are not relied upon. Dense areas of mechanoreceptors can gather vast quantities of tactile data and preneuronal compression is essential for efficient perception of this tactile information.

4. Discussions

The notion of morphological computation and its related field of soft robotics emphasize the importance of passive adaptation of the body. Although passive adaptation provides an avenue for "cheap" control, several fundamental limits and trade-offs need to be considered. Primarily, passive systems are characterized by their inability to control their energy output, whereas delays in state estimation constrain active systems. Hence, passive responses are more observed in static control tasks like grasping, whereas active control is required in more dynamic tasks (as they involve sharper energy changes). Simple systems that require active show a highly specialized sensory system. This is seen in the case of flying insects and, to a lesser extent, in the mechanoreceptors of spiders. In this case, the sensori-motor loop is easily streamlined to perform the desired task while reducing sensory lag and avoiding instability in the controller. Complex systems rely on passive and active responses with predictive models to overcome the performance limitations due to delayed state feedback. Passive responses reduce the burden on the sensory system and the controller, while the active component maintains the power requirements. This can be observed in the case of the human hand, where passive skin adaptation is used along with closed-loop tactile feedback for manipulation.

Unlike audiovisual sensing, mechanical cues are encoded by highly distributed and dense sensory structures (or receptors). Hence, there is a higher need to pre-process the information before it reaches the nervous system. In nature, we see multiple strategies to achieve this. Primarily, mechanical structures are optimized to yield a highly sensitive response to the desired mechanical cues, measured by the sensory neurons at the base. The tactile sensors found in spiders and whiskers of mammals are good examples of this. In natural organisms, the topology of the receptors evolved to avoid an unneeded information overload. Moreover, the co-dependency between the sensor and the enclosing body aims at increasing both the robustness to damage and input specialization. Indeed, the sensor and the body cannot be considered two separate entities.

The combination of body and brain will allow artificial systems to cope with real environments [63]. Indeed, the biomechanical principles of many biological sensory organs can be identified and classified according to the functionality and the environment in which they operate [10]. On the artificial side, there are still some bottlenecks related to the manufacturing complexity of the materials[64] and in tuning three-dimensional fabrication technologies to obtain specific robust morphologies of sensors in the body. In any case, the major challenge resides in the absence of a well-structured framework that allows full co-design and co-evolving of materials, structure, and control to achieve pre-determined functionalities.

Acknowledgment

This work was supported by the PROBOSCIS project, a FET program of the European Commission(grant agreement ID 863212), and by the SHERO project, a FET program of the European Commission (grant agreement ID 828818).

References

[1] Wilson M 2002 *Psychonomic bulletin & review* **9** 625–636 ISSN 1531-5320
[2] Cangelosi A, Bongard J, Fischer M H and Nolfi S 2015 Embodied intelligence *Springer Handbook of Computational Intelligence* (Springer) pp 697–714
[3] Pfeifer R and Bongard J 2006 *How the body shapes the way we think: a new view of intelligence* (MIT press) ISBN 0262288524
[4] Pfeifer R and Gómez G 2009 Morphological computation–connecting brain, body, and environment *Creating brain-like intelligence* (Springer) pp 66–83
[5] Barth F G 2004 *Current opinion in neurobiology* **14** 415–422 ISSN 0959-4388
[6] Yasuda H, Buskohl P R, Gillman A, Murphey T D, Stepney S, Vaia R A and Raney J R 2021 *Nature* **598** 39–48
[7] Stein B E and Meredith M A 1993 *The merging of the senses.* (The MIT Press) ISBN 0262193310
[8] Ernst M O and Banks M S 2002 *Nature* **415** 429–433 ISSN 1476-4687
[9] Katta S, Krieg M and Goodman M B 2015 *Annual review of cell and developmental biology* **31** 347–371
[10] Blandin A A, Bernardeschi I and Beccai L 2018 Biomechanics in soft mechanical sensing: From natural case studies to the artificial world
[11] Åström K J 2002 *Department of Mechanical and Environmental Engineering University of California Santa Barbara* **333**
[12] Chen J, Fang S and Ishii H 2019 *Annual Reviews in Control* **47** 155–177
[13] Middleton R H 1991 *Automatica* **27** 281–292
[14] Johansson R S and Flanagan J R 2009 *Encyclopedia of Neuroscience* **8** 583–594
[15] Cascio C J, Moore D and McGlone F 2019 *Developmental cognitive neuroscience* **35** 5–11 ISSN 1878-9293
[16] Johansson R S and Vallbo A B 1979 *The Journal of physiology* **286** 283–300 ISSN 0022-3751
[17] Johansson R S, Landstro U and Lundstro R 1982 *Brain research* **244** 17–25 ISSN 0006-8993
[18] Wu S, Amari S i and Nakahara H 2002 *Neural Computation* **14** 999–1026 ISSN 0899-7667
[19] Binder M D, Hirokawa N and Windhorst U 2009 *Encyclopedia of neuroscience* vol 3166 (Springer Berlin, Germany)
[20] Franzén O and Westman J 2016 *Information processing in the somatosensory system* vol 57 (Macmillan International Higher Education) ISBN 1349115975
[21] Kareklas K, Nettle D and Smulders T V 2013 *Biology Letters* **9** ISSN 1744957X
[22] Changizi M, Weber R, Kotecha R and Palazzo J 2011 *Brain, behavior and evolution* **77** 286–290 ISSN 0006-8977
[23] Davis N J 2021 *PLoS ONE* **16** 1–10 ISSN 19326203 URL `http://dx.doi.org/10.1371/journal.pone.0253185`
[24] Yamashita H, Wang Z, Hirai S and Shibuya K 2017 *IEEE Transactions on Industrial Informatics* **13** 2496–2506 ISSN 1551-3203

[25] Elzinga M J, Dickson W B and Dickinson M H 2012 *Journal of The Royal Society Interface* **9** 1685–1696
[26] Ristroph L, Ristroph G, Morozova S, Bergou A J, Chang S, Guckenheimer J, Wang Z J and Cohen I 2013 *Journal of The Royal Society Interface* **10** 20130237
[27] Fayyazuddin A and Dickinson M H 1996 *Journal of Neuroscience* **16** 5225–5232
[28] Starzyk J A 2008 *Motivation in embodied intelligence* (Citeseer) ISBN 9537619176
[29] Holmes N P, Calvert G A and Spence C 2004 *Neuroscience letters* **372** 62–67 ISSN 0304-3940
[30] Barth F G 1982 *Journal of comparative physiology* **148** 175–185 ISSN 0340-7594
[31] Humphrey J A C, Barth F G, Reed M and Spak A 2003 The physics of arthropod medium-flow sensitive hairs: biological models for artificial sensors *Sensors and sensing in biology and engineering* (Springer) pp 129–144
[32] Barth F G and Dechant H E 2003 Arthropod cuticular hairs: tactile sensors and the refinement of stimulus transformation *Sensors and sensing in biology and engineering* (Springer) pp 159–171
[33] Kang D, Pikhitsa P V, Choi Y W, Lee C, Shin S S, Piao L, Park B, Suh K Y, Kim T i and Choi M 2014 *Nature* **516** 222–226 ISSN 1476-4687
[34] Klärner D and Barth F G 1982 *Journal of comparative physiology* **148** 445–455 ISSN 0340-7594
[35] Peckmezian T and Taylor P W 2015 *Behavioural processes* **113** 99–104 ISSN 0376-6357
[36] Shettleworth S J 2009 *Cognition, evolution, and behavior* (Oxford university press) ISBN 0199886385
[37] Shapiro L 2019 *Embodied cognition* (Routledge) ISBN 1315180383
[38] Pfeifer R, Iida F and Lungarella M 2014 *Trends in cognitive sciences* **18** 404–413 ISSN 1364-6613
[39] Cappuccio M L 2017 *Phenomenology and the Cognitive Sciences* **16** 425–448 ISSN 1572-8676
[40] Craver C F 2007 *Explaining the brain: Mechanisms and the mosaic unity of neuroscience* (Oxford University Press) ISBN 0199299315
[41] Nakata K 2010 *Proceedings of the Royal Society B: Biological Sciences* **277** 29–33 ISSN 0962-8452
[42] Nakata K 2013 *Biology letters* **9** 20130052 ISSN 1744-9561
[43] Japyassú H F and Laland K N 2017 *Animal Cognition* **20** 375–395 ISSN 14359448
[44] Huet L A, Rudnicki J W and Hartmann M J 2017 *Soft Robotics* **4** 88–102 ISSN 21695180
[45] Ebara S, Kumamoto K, Matsuura T, Mazurkiewicz J E and Rice F L 2002 *Journal of Comparative Neurology* **449** 103–119 ISSN 0021-9967
[46] Hartmann M J, Johnson N J, Towal R B and Assad C 2003 *Journal of Neuroscience* **23** 6510–6519 ISSN 02706474
[47] Bagdasarian K, Szwed M, Knutsen P M, Deutsch D, Derdikman D, Pietr M, Simony E and Ahissar E 2013 *Nature Neuroscience* **16** 622–631 ISSN 10976256 URL `http://dx.doi.org/10.1038/nn.3378`
[48] Kaneko M, Kanayama N and Tsuji T 1998 *IEEE Transactions on robotics and automation* **14** 278–291 ISSN 1042-296X
[49] Nguyen N H and Ho V A 2021 *Soft Robotics* ISSN 2169-5172
[50] Santello M, Baud-Bovy G and Jörntell H 2013 *Frontiers in computational neuroscience* **7** 23
[51] Santello M 2014 Synergistic control of hand muscles through common neural input *the human hand as an inspiration for robot hand development* (Springer) pp 23–48
[52] Tagliabue M, Ciancio A L, Brochier T, Eskiizmirliler S and Maier M A 2015 *Frontiers in human neuroscience* **9** 165
[53] Schieber M H and Santello M 2004 *Journal of applied physiology* **96** 2293–2300
[54] Della Santina C, Piazza C, Grioli G, Catalano M G and Bicchi A 2018 *IEEE Transactions on Robotics* **34** 1141–1156
[55] Gilday K, Hughes J and Iida F 2021 *Bioinspiration and Biomimetics* **16** ISSN 17483190
[56] Pozzi M, Malvezzi M and Prattichizzo D 2016 *IEEE Robotics and Automation Letters* **2** 329–336
[57] Hughes J, Maiolino P and Iida F 2018 *Science Robotics* **3**
[58] Sanfilippo F, Zhang H and Pettersen K Y 2015 The new architecture of modgrasp for mind-controlled low-cost sensorised modular hands *2015 IEEE International Conference on Industrial Technology (ICIT)* (IEEE) pp 524–529
[59] Dargahi J and Najarian S 2004 *The international journal of medical robotics and computer assisted surgery* **1** 23–35
[60] Wang Q and Hayward V 2007 *Journal of biomechanics* **40** 851–860
[61] Shao Y, Hayward V and Visell Y 2016 *Proceedings of the National Academy of Sciences* **113** 4188–4193
[62] Shao Y, Hayward V and Visell Y 2020 *Science advances* **6** eaaz1158
[63] Miriyev A and Kovač M 2020 *Nature Machine Intelligence* **2** 658–660 ISSN 25225839 URL `http://dx.doi.org/10.1038/s42256-020-00258-y`
[64] Howard D, Eiben A E, Kennedy D F, Mouret J B, Valencia P and Winkler D 2019 *Nature Machine Intelligence* **1** 12–19 ISSN 25225839 (*Preprint* `1901.05704`) URL `http://dx.doi.org/10.1038/s42256-018-0009-9`

IOP Conf. Series: Materials Science and Engineering 1261 (2022) 012014 doi:10.1088/1757-899X/1261/1/012014

Coevolution of internal representations in physical human-robot orchestration – models of the surgeon and the robot in robotic surgery

Ilana Nisky[1,2,3], **Leone Costi**[3], **and Fumiya Iida**[3]

[1] Department of Biomedical Engineering, Ben-Gurion University of the Negev, Beer-Sheva, 84105, Israel

[2] Zlotowski Center for Neuroscience, Ben-Gurion University of the Negev, Beer-Sheva, 84105, Israel

[3] Department of Engineering, University of Cambridge, Cambridge, UK

E-mail: `nisky@bgu.ac.il, lc830@cam.ac.uk, fi224@cam.ac.uk`

Abstract. In teleoperated Robot-Assisted Minimally-Invasive Surgery (RAMIS), a surgeon controls the movements of instruments inside the patient's body via a pair of robotic joysticks. RAMIS has transformed many surgical disciplines, but its full potential is still to be realized. In this chapter we propose a pathway towards overcoming several bottlenecks that are related to transparency and stability of the teleoperation channels that mediate RAMIS. We describe the traditional system centered and the more recent human-centred approaches to teleoperation, and the special considerations for RAMIS as an application of teleoperation. However, the human-centered approach is still one sided view focusing on the surgeon but neglecting the learning capabilities of robotic systems. Hence, we consider a more general idea of physical human-robot orchestration with coevolution of mutual internal representations – of the human and the robot, and discuss it in comparison to human-human collaboration over teleoperated channels.

1. Introduction

In teleoperated Robot-Assisted Minimally-Invasive Surgery (RAMIS), a surgeon manipulates a pair of robotic joysticks that control the movements of instruments and an endoscopic camera inside the patient's body (Fig. 1(a)). For the patient, the benefits of RAMIS, compared to open surgery, are low infection rates, reduced blood loss, and a shorter recovery time [1]. For the surgeon, the benefits, compared to traditional minimally invasive surgery, are improved vision, dexterity, and precision [2, 3]. RAMIS has thus transformed many surgical disciplines [4], and in just one decade, the yearly number of robotic procedures performed with the da Vinci (Intuitive Surgical Inc.) increased form 136,000 in 2008 [5] to over a million in 2018 [6]. Even wider application is expected as new systems are developed and enter the market [7, 8, 9]. However, the full potential of RAMIS is yet to be realized, and debate is still ongoing regarding its safety, efficacy, and cost effectiveness [5].

To improve outcomes in RAMIS, combining robotic intelligence [10] with models of the surgeon [11] is needed. An example of a challenge that can benefit from such approach is the absence of force feedback in teleoperated RAMIS [12] that limits the quality and safety of

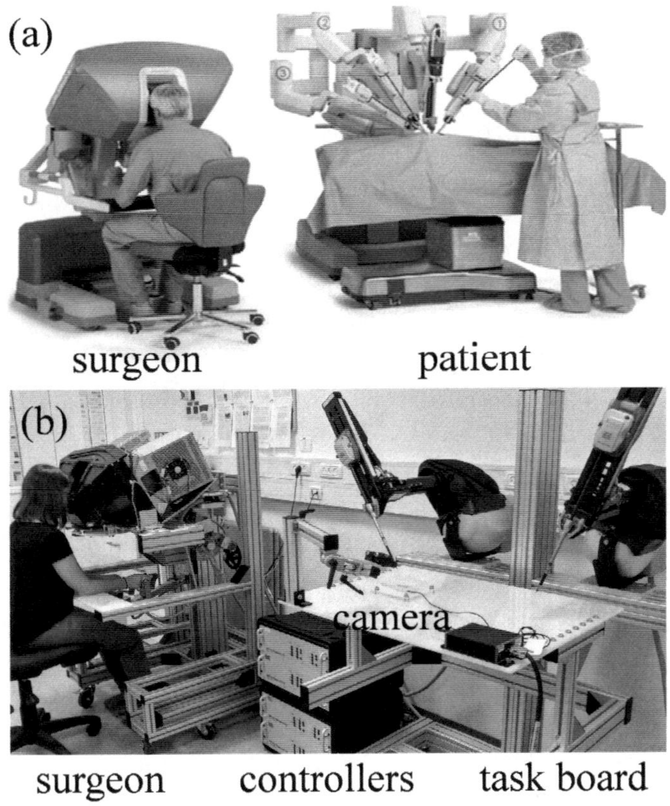

Figure 1. Examples of Robot-assisted Minimally Invasive Surgical systems: (a) a clinical da Vinci Si, Intuitive Surgical Inc. (image from www.intuitivesurgical.com) (b) a da Vinci Research Kit, a former da Vinci Standard system [260], at the Biomedical Robotics Lab of Ben-Gurion University of the Negev.

procedures [13]. Despite progress in the design of systems with force feedback [14, 15], their benefit to surgery is still debated [16, 17]. Instead, the efforts of many groups have focused on using tactile stimulation instead of direct force feedback [18, 19, 20, 21, 22, 23, 24, 25, 26], or on the automation of surgery [27, 28, 29]. However, the judgement of surgeons coupled with their fine manipulation abilities are crucial for the success of surgery. Furthermore, motor learning affects the handover between automated and human operation [30], and hence, even in autonomous surgery the surgeon must have an accurate representation of the controlled system for effective handover. Hence, it is critical to keep the surgeon physically in the control loop, and integrate feedback with teleoperation controllers that use models of the surgeons' motor control. That is, it would be beneficial if intelligent robotic systems, that are inherently equipped with learning capabilities, build models of the actions of the users, especially in complex surgical maneuvers [31, 17, 32, 33].

Even more than in the case of force feedback, assistance from artificial intelligence in RAMIS [34] can be made more effective with such models. Virtual fixtures [35, 36, 37, 38, 39], also known as active constraints [40, 41], provide the shared control framework for robotic assistance to the control of the instruments. They can guide the surgeon to a desired path (for example around delicate anatomical structures) in teleoperated and collaborative manipulation based on pre- and intra-operative information. In teleoperation, these assistive forces are presented in addition to feedback from the environment, and as a result inherently sacrifice transparency to gain easy and accurate performance of a task [35]. Recent studies showed that robotic assistance is most effective when the controllers obey fundamental laws of human movements [42, 43] and are predictable [44, 45]. Based on these results, we posit that surgeons may better accept the assistance of virtual fixtures if (1) their control will converge to natural human-like trajectories, (2) they can be tailored to each surgeon, and (3) they can be adapted as the movement and internal representations of the surgeons and the robotic systems evolve.

We recently developed a human-centered approach to teleoperation that, unlike classical studies that focused on the robotic system in isolation [46] puts the human user in the center [47, 48, 49]. However, focusing solely on the human still provides a one-sided view. Many physical human-robot interaction applications can benefit from mutual adaptation between the human and the robot. In this chapter, we review the human-centered design approach and its application to RAMIS in Section 2, and consider a more general idea of physical human-robot orchestration with coevolution of mutual internal representations in Section 3.

2. Human-centered transparency, stability, and internal representations

2.1. Human-centered transparency

Transparency, a measure of the fidelity of a teleoperation system, has been a focal objective in teleoperation since the birth of the first teleoperated systems. Due to complexity of modeling the human operator, the concept has traditionally been restricted to the communication channel and the leader-follower system, not considering the user (see Fig. 2(a)). According to this approach transparency requires matching the displacement (or velocity) and force between the local manipulator to the ones on the remote manipulator, entirely ignoring the physical and cognitive processing by the human. We believe that a human-centered approach to define transparency is more suited to achieve a true transparency during teleoperation. In [47], a new multidimensional transparency definition was proposed that is composed of 3 different components: *perceptual transparency*, making the human operator not able to distinguish between the system and an identity channel, *local motor transparency* making the human motion and forces identical while operating over the teleoperated system or via an identity channel, and *remote motor transparency*, making the remote's agent motion and forces identical while operating over the teleoperated system or an identity channel. Such a framework places the human operator at the center of the system and is used to engineer the most natural possible experience during teleoperation.

The same framework has been linked to previous studies on how communication delay affects stiffness perception [50, 51, 52, 53, 54, 55, 56], trying to understand how humans build the internal representation of the stiffness of an object [57] for perception and for action. This human-centered approach has been also tested with non-linear scenarios [58, 59], and considering grip forces in the picking up of both rigid and deformable objects [60, 61]. Placing the human at the center allows for designing systems that are not perfect but provide the user with a natural operation experience, In this sense, the human-centered approach is bio-inspired, as the biological control systems suffer from many non-ideal characteristics such as internal delays, low communication bandwith, noise, low gains, and yet, out of these embodied intelligence emerges.

This approach emphasizes intrinsic mechanisms at the base of human sensorimotor control. These mechanisms result in several recently reported observations, such as the difference between force and position control during interaction with elastic objects in the presence of force feedback delay [57], the proximodistal gradient of joint-driven perception of stiffness [51] and bimanual differences in perception [56], the neural mechanisms subserving the perception of stiffness [55], and grip force modulation [60, 54, 61, 62]. Another important aspect is the dissociation between

IOP Conf. Series: Materials Science and Engineering 1261 (2022) 012014 doi:10.1088/1757-899X/1261/1/012014

internal representations that are used for perception and for action, as demonstrated in the context of illusions [63, 64, 65] or psychophysical laws [66]. Perception and action are tightly coupled in physical interaction with the world: perception of the mechanical properties of the environment is important in planning future actions, and at the same time, natural haptic exploration of the environment is active – we move and probe the environment to create haptic perception. Nevertheless, recent results provide evidence for dissociation between the different information processing streams also in physical human-robot interaction. One such useful result is that while delay in force feedback biases perception of stiffness in interaction with elastic objects, the predictive control of grasping forces during that same interaction becomes accurate within a few probing movements in its timing and in the estimated stiffness [54, 56]. This result is an example of the complexity and richness of the formed internal representations, and demonstrates that it is possible to selectively tune perceptual and motor transparency even for a system that would be considered not transparent in the classical system-centered analysis.

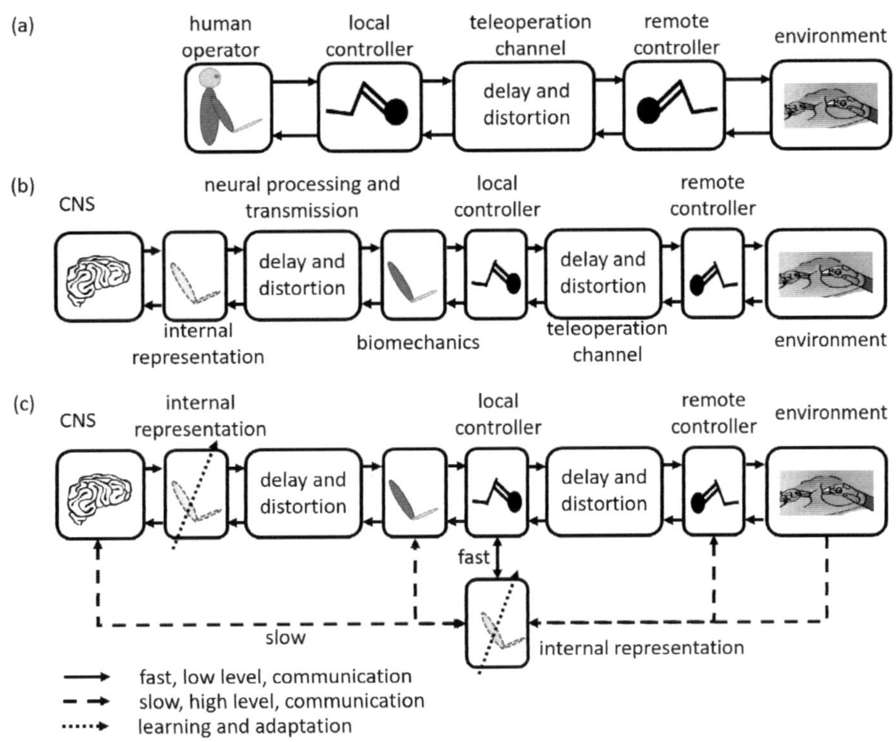

Figure 2. Schematic representation of a simplified teleoperation architecture (a), a corresponding simplified human-centered view of the same architecture (b), and a physical human-robot orchestration view of the same architecture (c). In (a), a single transmission channel is the source of potential instability and distortions. In (b), a second communication channel, representing neural processing and transmission, ads delay and distortion and may create indirect instabilities, but the internal representation and additional mechanisms may also stabilize the instabilities of the first transmission channel and correct for the distortions that are caused by it. In (c) there is an added internal representation on the robot side, a slow communication channel (dashed lines) in parallel with the fast teleoperation control (solid lines) that allows the biological (surgeon) and artificial (robot) systems to exchange high level information, and update the internal representations (dotted arrows).

IOP Conf. Series: Materials Science and Engineering 1261 (2022) 012014 doi:10.1088/1757-899X/1261/1/012014

This path will eventually benefit RAMIS, but several challenges exist [67, 11]. The most important one is that models of complex surgical movements and internal representations that can be used in RAMIS control are still lacking. RAMIS requires the mastering of complex sensorimotor behavior, viz., sensory information integration, hand-eye coordination, and fine manipulation. Even the simplest surgical procedure may involve perception and decision making (e.g. for identification of anatomy and choosing the action to be executed), and manual control (e.g. for precise manipulation of tissue during dissection and suturing); more complex procedures would require several iterations of decision making, perception, and action. Nevertheless, we have made the first steps and employed our human-centered transparency in RAMIS. We started from the simple case of perception and action in needle insertion, that can be easily approximated by a one-deminsional interaction with a highly nonlinear elastic object [58]. We proved that communication delay produces motor underestimation of the stiffness, but does not affect the perceptual estimation. In other words, the human operator is not able to correctly insert the needle, but she is fully aware of delayed induced change in perception. In another study we used the adherence of perception but not action to fundamental psychophisical law - the Weber Law [66] as an indicator of the human-centered transparency of grasping at the reach-to-grasp stage. According to the Weber Law, the human perceptual sensitivity to differences between stimuli is proportional to the magnitude of the stimuli [68]. However, this logarithmic dependence does not hold for motor variability [66]. With this approach, we proved that as long as the scaling between the local and remote gripper was not too small, in RAMIS without delay, the system was human-centered transparent [48]. This was in contrast to a similar experiment with an industrial grasping system that was not human-centered transparent, likely due to delay [69].

In our recent studies of surgical needle-driving [17] we investigated the role of the feedback channel in training of inexperienced operators. The aim was to compare the effects in learning while experiencing no force feedback, position-based force feedback and directly recorded haptic feedback. While our results failed to show benefit of force feedback to this task, we believe that these results are at least partially due to the limited transparency of the force feedback we tried, and better human-centered, force feedback algorithms have the potential to improve users' learning performance. This result also highlights another challenge in our current human-centered transparency approach. It relies heavily on lengthy psychophysical experiments, which is acceptable for assessment of human-centered transparency of a given system, but does not provide a solution for an iterative design for human-centered transparency. To make this effort feasible we need to develop experimental approaches that can identify the gradient of a human-centered transparency in the design parameters space.

2.2. Human-centered stability

Transparency is meaningless without assuring stability, which can be formally defined from the conrol theory perspective, but for our practical consideration, stability means that none of the signals in the teleoperation system grows too much and without a specific intention. Achieving transparency without compromising stability is particularly challenging [46], even when considering the simplified system, when we consider the communication delay and distortion [70, 71, 72]. Examples for such practical tradeoffs between transparency and stability range from healthcare applications to construction robotics [73, 74, 75, 76]. Instabilities can occur directly in the automatic controller in bilateral teleoperation with force feedback [46]. However, the human operator can either overcompensate for perceived errors due to delayed responses, giving rise to operator-induced oscillations [77], or contribute to the stabilization of delay-induced instabilities [49]. While the instabilities originating from the controllers can be tackled using a simplified system-centered approach, the human detrimental or beneficial contribution cannot be addressed unless we switch to a more human-centered one.

Such human-centered approach for stability has to account for neural transmission and processing, multisensory integration, internal representations, and the embodied intelligence of human biomechanics (Fig. 2(b)). It requires assuming (and testing) different models for the architecture of the human sensorimotor controller that is physically coupled to the local manipulator. Attempts to model the human sensorimotor controller go back to the early observations of natural skilled movements, such as Woodworth's observations of hand-drilling [78] and Bernstein's studies of hammering [79], which led to the discovery of organizing principles in motor control. Following many years of research fundamental theories were developed and validated with simplified tasks. resulting in a framework that combines predictive (feedforward) [80, 81, 82], reactive (feedback) [83], and biomechanical (limb stiffness) control to coordinate movement [84]. Candidate models – mathematical equations – that describe these controllers use linear feedback control [85], nonlinear control [86], optimal feedback control [87], and intermittent control [88, 61]. Optimal control models explain kinematic primitives, such as the straight path and bell-shaped velocity trajectory of reach movements [80, 82], and spatiotemporal invariants, such as the power laws of movement at equiaffine velocities [89]. The latter are kinematic invariants, hence not contributing to formal analysis of stability, but instead provide ideal trajectories for our human-centered analysis. We already know that these invariants are also observed in simple surgical maneuvers such as needle driving and suturing, and that their parameters can be indicative of the skill of the surgeon and the stage of the surgical task [31].

The integration of these models into formal analysis of stability is challenging yet rewarding. For example, in a recent study we showed that by just including in the analysis the impedance of the user, and abandoning the widely used passivity analysis [90] allows for expanding substantially the stability regions of haptic rendering with delay, and unintuitively, discovering regions where a system that is unstable for a certain delay may be stabilized by increasing the delay [49]. However, including more realistic approximations of the user in such formal analysis is still a work in progress.

2.3. Internal representations

A key feature of such approximations is incorporating feedforward controllers that use internal representation of the environment with which the user interacts. Studies of sensorimotor learning have highlighted the ability of the brain to distinguish deterministic forces and visuomotor manipulations from random disturbances. While the latter can only be compensated by feedback responses, deterministic perturbations offer the possibility to use feedforward control based on representing of the causal relations between forces and motions, or casual predictable relations between motor commands and outcomes. Such internal representation is critical given the delays of biological feedback and the limited force output and impedance of the human hands. These representations are sought to be the computational mechanism that allows for restoring centrally controlled movement invariants following adaptation to a visuomotor rotation [91], delay [92, 93, 94, 95], state-dependent forces [96, 97], and a change in the compliance of the environment [98]. Aftereffects, generalization of adaptation to different directions or different speeds, and transfer of adaptation to a different posture or a different task help to uncover the equations that describe the internal representation.

Importantly, the adaptation and update of the internal representation happens all the time, movement-by-movement [99], and even when users are explicitly guided to ignore the result [100, 101]. For many years it was believed that these updates rely on implicit, unaware processes, but recent evidence suggest that this adaptation involves multiple learning processes, including strategic explicit (aware) processes [102, 103]. Recent findings propose ways to manipulate the information and assistance that artificial systems provide to users in a way that can facilitate or attenuate these contributions [104, 101, 103]. These adaptive processes have to be taken into account in human-centered design of systems, but they could also be intertwined with adaptation on the robot control side, as we consider in the next section.

IOP Conf. Series: Materials Science and Engineering 1261 (2022) 012014 doi:10.1088/1757-899X/1261/1/012014

3. Physical human-robot orchestration and coevolution of mutual internal representations

3.1. Slow communication for mutual human-robot adaptation

The human-centred approach still provides a one-sided view of the picture, as it completely ignores the internal representations that may be used by the robotics system. This is especially important in light of the adaptive nature of human sensorimotor control. As such, a robotic system that uses models of the surgeon and her sensorimotor system must be constantly reassessing the quality of its representation of the user, and use any inconsistencies between the predictions and the observed outcome to update the models. These inconsistencies may also be fed back to the user by means of visual or robot-assisted augmentations to facilitate even more effective adaptation. We wish to characterise such mutual adaptation and collaboration as physical human-robot orchestration.

To maintain the coevolution of internal representations – that of the surgeon and that of the robotic system, an additional exchange of information is needed, and it is highlighted in a simplified schematic representation in Fig. 2(c). Importantly, this information exchange driving mutual adaptation and higher level communication is facilitated by a slow bidirectional communication channel that exchanges information between the surgeon and the robot, and we deliberately leave this illustration at a high level – it is yet to be defined exactly what information is to be transferred.

From the side of the robotic system, examples of what can be learned about the surgeon can be as simple as biomechanics: an estimation of the stiffness of the arm of the surgeon, her muscle activation level, or fatigue, or even a full dynamic system model [96]. For a specific movement or procedure, another low-level aspect that can be learned is intent detection. The skill of the surgeon can be assessed by the robotic system based on kinematic signatures [105, 31, 106, 107]. Even high level attributes such as their mental state or confidence can be estimated given rich enough information that would be exchanged between the sides. All these estimations can be used to update the predictions of the robotic system and to set its control parameters. For example, only skilled users may be allowed to use certain riskier control settings such as high gain between the movement of the local manipulator and the patient-side manipulator. Finally, because the surgeon will also be learning the system and the task at hand, an assessment of explicit or implicit learning processes of the surgeon that can be communicated to the robotic system.

From the side of the surgeon, the sensorimotor and cognitive systems will be learning and updating the internal representations continuously, and they are likely to use all possible information, including visual, kinesthetic, tactile, and auditory cues that will be available even without special architecture for mutual adaptation. Assistive forces from virtual fixtures will also likely serve as a teaching signal in case of opposing forces between the input of the surgeon and the fixture; this means that the design of the virtual fixtures will have to take into account the effect of the assistance on the learning of the surgeon. Moreover, these learning processes can be complimented with additional information, such as providing the surgeon with feedback about their estimated skill level, the quality of performance of a particular procedure, and advisory cues about how the procedure can be performed. In addition, information about the quality of the robotic system's predictions of the actions of the user can be communicated from the robotic system to the surgeon.

3.2. Relation to shared control

At first glance, the idea of physical human-robot orchestration may sound similar to concepts such as human-robot cooperation and shared control. Concerning those, the scientific interest has been exponentially growing in the last 20 years [108], especially with regards to teleoperated systems [109], and it has been shown to be a suitable substitute to human-human cooperation at times [110]. Human-robot cooperation has been investigated for the achievement of both known and unknown goals [111, 112], showcasing a more human-centered approach. The introduction to some degree of autonomy has been proven beneficial to achieve a better performance in a variety of tasks, from reaching motions to track following, and faster task execution times: prior studies show that more autonomy often brings more accuracy and precision [113]. Given such an increase in performance, shared control has also been introduced in surgical robotics to achieve higher precision and faster timings [114, 115, 116, 117]. Finally, shared control has also been shown to help to minimize the effect of sensor noise [118], and has been partially investigated as an answer to compensate delay-induced motions [119].

Then, what is still missing to achieve physical human-robot orchestration? The cohesive thinking between robot and machine. It has been proven that shared control's effectiveness is strongly limited by the extent of cooperation that the human operator is willing to allow [120]. In other words, the operator may take a different decision or ignore the robot, making the shared control effectively useless and in some cases creating a frustrating experience. Some studies have focused on how the robot could completely adapt to the user decision making, assuming that the goal is always unkown to the robot, and predicting the operator's intention by using either statistical methods [121] or neural networks [122]. Nevertheless, this process results to be entirely based on the level of accuracy of the prediction and it bypasses entirely the robot's decision making, thus limiting the true potential of human-robot orchestration. Another extreme would be a robot that continuously adapts to the changing situation, but is autonomous, and completely ignores the physical input of the user leaving the communication to the level of high level commands such as *"remove gallbladder"*, or *identify lump"*. This would however leave the surgeon out of the physical loop and is undesirable. We believe that in physical human-robot orchestration the two agents, human and robot, shall converge to the same decision and action, both at a cognitive and physical level, and learn from each other so to exploit at best their close interaction.

3.3. Relation to human-human collaboration over a teleoperated channel

To conclude, we will demonstrate the different levels of physical surgeon-robot orchestration with a few analogies from biological collaborative actions, i.e. surgeon-surgeon collaboration. For this conceptual demonstration, we replace the patient-side learning robotic system with another human surgeon. Let us assume that a remote surgeon uses a traditional bilateral teleoperation system and collaborates from a far with a local surgeon that operates on the patient, and that the tools that the local surgeon uses maybe entirely under her control or receive input also from the remote surgeon.

If both surgeons are experts, they may exchange some verbal communication following which the procedure would be performed autonomously by the local surgeon. They may also agree to collaboratively perform a certain portion of the procedure that may benefit from such collaboration. The local surgeon may also be a novice, and hence would benefit from higher weight in the control on the actions of the remote surgeon. If the entire control is given to the remote surgeon, and the local surgeon would then just attempt to passively follow the actions; this would be the most similar to a human-centered approach. It is actually very likely that the learning of the local surgeon may be very limited as he will not be given the opportunity to act and assess the quality of their actions. If the local surgeon is of intermediate experience level, the remote surgeon may instruct them on the procedure and actions to perform, and take over when assistance or correction is needed. Different weights may be assigned to the input of the local and remote surgeons, and they may change during the procedure. The key is that since both the local and remote surgeons are human, they are both adaptive, make decisions, and may choose whether to collaborate or attempt to lead the collaborative actions. After sufficient experience together both surgeons will have excellent prediction of each other actions, and have a fast and slow communication between them for smooth collaboration.

Finally, with these examples in mind, we can take an inspiration from the Turing Test [123], or its motor Turing-like version [124, 125], propose a test for the fidelity of human-robot physical orchestration. In the original Turing Test, [123], an interrogator has to determine via a telepromter whether their opponent is a human or an artificial agent. There has been much debate over the years whether this is a test of intelligence, but for our discussion this is not critical - it is sufficient that the human and the agent are indistinguishable over the telepromter to state that a certain benchmark of quality was achieved. This is the logic that was implemented in the motor version of the test [124, 125] – models of human handshake were graded based on their human-likeness by having human interrogators interact with the models or with humans via a bilateral teleoperation channel. In the conceptual test of the fidelity of human-robot physical orchastration we propose that an ideal orchestration will be achieved when the remote surgeon cannot distinguish between collaborating with a robotic system and a live local surgeon. The details of such test would have to be specified for it to be feasible and a valuable benchmark. For example, the tasks to be tested may affect the outcomes of the test. Similarly, the skill and qualification of the local surgeon would highly influence the outcome of the benchmarking. Finally, much like in the original idea [123], the actions and choices of the local surgeon may influence the test. Hence, we conclude with stating that the practical details of this and additional benchmarking approaches will have to be developed alongside with the details of the different implementations of our proposed physical human-robot orchestration.

Acknowledgements

This work was funded by the UKRI grant EP/T00519X/1 RoboPatient project, the Israeli Science Foundation (grant 327/20), the Israeli Ministry of Science and Technology, the Helmsley Charitable Trust through the Agricultural, Biological and Cognitive Robotics Initiative, and the Marcus Endowment Fund both at Ben-Gurion University of the Negev.

References

[1] Jens Rassweiler, Marcel Hruza, Dogu Teber, and Li-Ming Su. Laparoscopic and robotic assisted radical prostatectomy–critical analysis of the results. *European urology*, 49(4):612–624, 2006.

[2] K Moorthy, Y Munz, A Dosis, J Hernandez, S Martin, F Bello, T Rockall, and A Darzi. Dexterity enhancement with robotic surgery. *Surgical Endoscopy and Other Interventional Techniques*, 18(5):790–795, 2004.

[3] John C Byrn, Stefanie Schluender, Celia M Divino, John Conrad, Brooke Gurland, Edward Shlasko, and Amir Szold. Three-dimensional imaging improves surgical performance for both novice and experienced operators using the da vinci robot system. *The American Journal of Surgery*, 193(4):519–522, 2007.

[4] Tiago Leal Ghezzi and Oly Campos Corleta. 30 years of robotic surgery. *World journal of surgery*, 40(10):2550–2557, 2016.

[5] Christopher P Childers and Melinda Maggard-Gibbons. Estimation of the acquisition and operating costs for robotic surgery. *Jama*, 320(8):835–836, 2018.

[6] Intuitive surgical annual report 2018. 2018.

[7] Salvatore Gueli Alletti, C Rossitto, S Cianci, E Perrone, S Pizzacalla, G Monterossi, G Vizzielli, S Gidaro, and G Scambia. The senhance™ surgical robotic system ("senhance") for total hysterectomy in obese patients: a pilot study. *Journal of robotic surgery*, 12(2):229–234, 2018.

[8] Stephan Lang, Stefan Mattheis, Pia Hasskamp, Georges Lawson, Christian Güldner, Magis Mandapathil, Patrick Schuler, Thomas Hoffmann, Marc Scheithauer, and Marc Remacle. A european multicenter study evaluating the flex robotic system in transoral robotic surgery. *The Laryngoscope*, 127(2):391–395, 2017.

[9] Dhananjay Kelkar, Mahindra A Borse, Girish P Godbole, Utkrant Kurlekar, and Mark Slack. Interim safety analysis of the first-in-human clinical trial of the versius surgical system, a new robot-assisted device for use in minimal access surgery. *Surgical endoscopy*, 35(9):5193–5202, 2021.

[10] Lena Maier-Hein, Swaroop S Vedula, Stefanie Speidel, Nassir Navab, Ron Kikinis, Adrian Park, Matthias Eisenmann, Hubertus Feussner, Germain Forestier, Stamatia Giannarou, et al. Surgical data science for next-generation interventions. *Nature Biomedical Engineering*, 1(9):691–696, 2017.

[11] Anthony Jarc and Ilana Nisky. Application and exploration of sensorimotor coordination strategies in surgical robotics. In *Metrics of Sensory Motor Coordination and Integration in Robots and Animals*, pages 41–71. Springer, 2020.

[12] Allison M Okamura. Haptic feedback in robot-assisted minimally invasive surgery. *Current opinion in urology*, 19(1):102, 2009.

[13] Gregory Tholey, Jaydev P Desai, and Andres E Castellanos. Force feedback plays a significant role in minimally invasive surgery: results and analysis. *Annals of surgery*, 241(1):102, 2005.

[14] Akhil J Madhani, Günter Niemeyer, and J Kenneth Salisbury. The black falcon: a teleoperated surgical instrument for minimally invasive surgery. In *Proceedings. 1998 IEEE/RSJ International Conference on Intelligent Robots and Systems. Innovations in Theory, Practice and Applications (Cat. No. 98CH36190)*, volume 2, pages 936–944. IEEE, 1998.

[15] Ulrich Hagn, Rainer Konietschke, Andreas Tobergte, Mathias Nickl, Stefan Jörg, Bernhard Kübler, Georg Passig, Martin Gröger, Florian Fröhlich, Ulrich Seibold, et al. Dlr mirosurge: a versatile system for research in endoscopic telesurgery. *International journal of computer assisted radiology and surgery*, 5(2):183–193, 2010.

[16] Bernhard Weber and Sonja Schneider. The effects of force feedback on surgical task performance: a meta-analytical integration. In *International Conference on Human Haptic Sensing and Touch Enabled Computer Applications*, pages 150–157. Springer, 2014.

[17] Lidor Bahar, Yarden Sharon, and Ilana Nisky. Surgeon-centered analysis of robot-assisted needle driving under different force feedback conditions. *Frontiers in Neurorobotics*, 13(January):1–22, 2020.

[18] Zhan Fan Quek, Samuel B Schorr, Ilana Nisky, William R Provancher, and Allison M Okamura. Sensory substitution and augmentation using 3-degree-of-freedom skin deformation feedback. *IEEE transactions on haptics*, 8(2):209–221, 2015.

[19] Zhan Fan Quek, Samuel B Schorr, Ilana Nisky, Allison M Okamura, and William R Provancher. Augmentation of stiffness perception with a 1-degree-of-freedom skin stretch device. *IEEE Transactions on Human-Machine Systems*, 44(6):731–742, 2014.

[20] Zhan Fan Quek, Samuel B Schorr, Ilana Nisky, William R Provancher, and Allison M Okamura. Sensory substitution using 3-degree-of-freedom tangential and normal skin deformation feedback. In *2014 IEEE Haptics Symposium (HAPTICS)*, pages 27–33. IEEE, 2014.

[21] Mor Farajian, Raz Leib, Hanna Kossowsky, Tomer Zaidenberg, Ferdinando A Mussa-Ivaldi, and Ilana Nisky. Stretching the skin immediately enhances perceived stiffness and gradually enhances the predictive control of grip force. *Elife*, 9:e52653, 2020.

[22] Karlin Bark, William McMahan, Austin Remington, Jamie Gewirtz, Alexei Wedmid, David I Lee, and Katherine J Kuchenbecker. In vivo validation of a system for haptic feedback of tool vibrations in robotic surgery. *Surgical endoscopy*, 27(2):656–664, 2013.

[23] Leonardo Meli, Claudio Pacchierotti, and Domenico Prattichizzo. Sensory subtraction in robot-assisted surgery: fingertip skin deformation feedback to ensure safety and improve transparency in bimanual haptic interaction. *IEEE Transactions on Biomedical Engineering*, 61(4):1318–1327, 2014.

[24] Zhan Fan Quek, William R Provancher, and Allison M Okamura. Evaluation of skin deformation tactile feedback for teleoperated surgical tasks. *IEEE transactions on haptics*, 12(2):102–113, 2018.

[25] Mor Farajian, Raz Leib, Hanna Kossowsky, and Ilana Nisky. Visual feedback weakens the augmentation of perceived stiffness by artificial skin stretch. *IEEE Transactions on Haptics*, 14(3):686–691, 2021.

[26] Hanna Kossowsky, Mor Farajian, and Ilana Nisky. The effect of kinesthetic and artificial tactile noise and variability on stiffness perception. *IEEE Transactions on Haptics*, 2022.

[27] Dénes Ákos Nagy, Tamás Dániel Nagy, Renáta Elek, Imre J Rudas, and Tamás Haidegger. Ontology-based surgical subtask automation, automating blunt dissection. *Journal of Medical Robotics Research*, 3(03n04):1841005, 2018.

[28] Qingchuan Ma, Etsuko Kobayashi, Junchen Wang, Kazuaki Hara, Hideyuki Suenaga, Ichiro Sakuma, and Ken Masamune. Development and preliminary evaluation of an autonomous surgical system for oral and maxillofacial surgery. *The International Journal of Medical Robotics and Computer Assisted Surgery*, 15(4):e1997, 2019.

[29] Nicola Preda, Federica Ferraguti, Giacomo De Rossi, Cristian Secchi, Riccardo Muradore, Paolo Fiorini, and Marcello Bonfè. A cognitive robot control architecture for autonomous execution of surgical tasks. *Journal of Medical Robotics Research*, 1(04):1650008, 2016.

[30] HE Russell, Lene K Harbott, Ilana Nisky, Selina Pan, Allison M Okamura, and J Christian Gerdes. Motor learning affects car-to-driver handover in automated vehicles. *Science Robotics*, 1(1):eaah5682, 2016.

[31] Yarden Sharon and Ilana Nisky. Expertise, teleoperation, and task constraints affect the speed-curvature-torsion power law in ramis. *Journal of Medical Robotics Research*, 2018.

[32] Ilana Nisky, Yuhang Che, Zhan Fan Quek, Matthew Weber, Michael H Hsieh, and Allison M Okamura.

Teleoperated versus open needle driving: Kinematic analysis of experienced surgeons and novice users. In *2015 IEEE International Conference on Robotics and Automation (ICRA)*, pages 5371–5377. IEEE, 2015.

[33] Jacopo Buzzi, Elena De Momi, and Ilana Nisky. An uncontrolled manifold analysis of arm joint variability in virtual planar position and orientation telemanipulation. *IEEE Transactions on Biomedical Engineering*, 66(2):391–402, 2019.

[34] R Mirnezami and A Ahmed. Surgery 3.0, artificial intelligence and the next-generation surgeon. *The British journal of surgery*, 105(5):463–465, 2018.

[35] Louis B Rosenberg. Virtual fixtures: Perceptual tools for telerobotic manipulation. In *Proceedings of IEEE virtual reality annual international symposium*, pages 76–82. IEEE, 1993.

[36] Jake J Abbott, Panadda Marayong, and Allison M Okamura. Haptic virtual fixtures for robot-assisted manipulation. In *Robotics research*, pages 49–64. Springer, 2007.

[37] Ming Li, Masaru Ishii, and Russell H Taylor. Spatial motion constraints using virtual fixtures generated by anatomy. *IEEE Transactions on Robotics*, 23(1):4–19, 2007.

[38] Marie-Aude Vitrani, Cecile Poquet, and Guillaume Morel. Applying virtual fixtures to the distal end of a minimally invasive surgery instrument. *IEEE Transactions on Robotics*, 33(1):114–123, 2016.

[39] MA Fehlberg, I Nisky, AJ Doxon, and WR Provancher. Improved active handrest performance through the use of virtual fixtures. *IEEE Transactions on Human Machine Systems*, pages 1–15, 2014.

[40] Stuart A Bowyer, Brian L Davies, and Ferdinando Rodriguez y Baena. Active constraints/virtual fixtures: A survey. *IEEE Transactions on Robotics*, 30(1):138–157, 2013.

[41] Nima Enayati, Giancarlo Ferrigno, and Elena De Momi. Performance metrics for guidance active constraints in surgical robotics. *The International Journal of Medical Robotics and Computer Assisted Surgery*, 14(1):e1873, 2018.

[42] Pauline Maurice, Meghan E Huber, Neville Hogan, and Dagmar Sternad. Velocity-curvature patterns limit human–robot physical interaction. *IEEE robotics and automation letters*, 3(1):249–256, 2017.

[43] Nathanaël Jarrassé, Jamie Paik, Viviane Pasqui, and Guillaume Morel. How can human motion prediction increase transparency? In *2008 IEEE International Conference on Robotics and Automation*, pages 2134–2139. IEEE, 2008.

[44] Jakob Fröhner, Philipp Beckerle, Satoshi Endo, and Sandra Hirche. An embodiment paradigm in evaluation of human-in-the-loop control. *IFAC-PapersOnLine*, 51(34):104–109, 2019.

[45] José Ramón Medina, Tamara Lorenz, and Sandra Hirche. Considering human behavior uncertainty and disagreements in human–robot cooperative manipulation. *Trends in Control and Decision-Making for Human–Robot Collaboration Systems*, pages 207–240, 2017.

[46] Dale A Lawrence. Stability and transparency in bilateral teleoperation. In *[1992] Proceedings of the 31st IEEE Conference on Decision and Control*, pages 2649–2655. IEEE, 1992.

[47] Ilana Nisky, Ferdinando A Mussa-Ivaldi, and Amir Karniel. Analytical study of perceptual and motor transparency in bilateral teleoperation. *IEEE Transactions on Human-Machine Systems*, 43(6):570–582, 2013.

[48] Amit Milstein, Tzvi Ganel, Sigal Berman, and Ilana Nisky. Human-centered transparency of grasping via a robot-assisted minimally invasive surgery system. *IEEE Transactions on Human-Machine Systems*, 48(4):349–358, 2018.

[49] Reut Nomberg and Ilana Nisky. Human-in-the-loop stability analysis of haptic rendering with time delay by tracking the roots of the characteristic quasi-polynomial: The effect of arm impedance. *IEEE Robotics and Automation Letters*, 6(4):7564–7571, 2021.

[50] Ilana Nisky, Ferdinando A. Mussa-Ivaldi, and Amir Karniel. A regression and boundary-crossing-based model for the perception of delayed stiffness. *IEEE Transactions on Haptics*, 1(2):73–83, 2008.

[51] Ilana Nisky, Pierre Baraduc, and Amir Karniel. Proximodistal gradient in the perception of delayed stiffness. *Journal of Neurophysiology*, 103(6):3017–3026, 2010.

[52] Assaf Pressman, Ilana Nisky, Amir Karniel, and Ferdinando A Mussa-Ivaldi. Probing virtual boundaries and the perception of delayed stiffness. *Advanced Robotics*, 22(1):119–140, 2008.

[53] Assaf Pressman, Leah J Welty, Amir Karniel, and Ferdinando A Mussa-Ivaldi. Perception of delayed stiffness. *The International Journal of Robotics Research*, 26(11-12):1191–1203, 2007.

[54] Raz Leib, Amir Karniel, and Ilana Nisky. The effect of force feedback delay on stiffness perception and grip force modulation during tool-mediated interaction with elastic force fields. *Journal of neurophysiology*, 113(9):3076–3089, 2015.

[55] Raz Leib, Firas Mawase, Amir Karniel, Opher Donchin, John Rothwell, Ilana Nisky, and Marco Davare. Stimulation of ppc affects the mapping between motion and force signals for stiffness perception but not motion control. *Journal of Neuroscience*, 36(41):10545–10559, 2016.

[56] Raz Leib, Inbar Rubin, and Ilana Nisky. Force feedback delay affects perception of stiffness but not

action, and the effect depends on the hand used but not on the handedness. *Journal of neurophysiology*, 120(2):781–794, 2018.

[57] Ilana Nisky, Raz Leib, Amit Milstein, and Amir Karniel. Perception of Stiffness with Force Feedback Delay. pages 167–185, 2014.

[58] Ilana Nisky, Assaf Pressman, Carla M Pugh, Ferdinando A Mussa-Ivaldi, and Amir Karniel. Perception and action in teleoperated needle insertion. *IEEE transactions on haptics*, 4(3):155–166, 2011.

[59] Raz Leib, Ilana Nisky, and Amir Karniel. Perception of stiffness during interaction with delay-like nonlinear force field. In *International Conference on Human Haptic Sensing and Touch Enabled Computer Applications*, pages 87–92. Springer, Berlin, Heidelberg, 2010.

[60] Amit Milstein, Lital Alyagon, and Ilana Nisky. Grip Force Control during Virtual Interaction with Deformable and Rigid Objects Via a Haptic Gripper. *IEEE Transactions on Haptics*, 14(3):564–576, 2021.

[61] Olivier White, Amir Karniel, Charalambos Papaxanthis, Marie Barbiero, and Ilana Nisky. Switching in feedforward control of grip force during tool-mediated interaction with elastic force fields. *Frontiers in neurorobotics*, 12:31, 2018.

[62] Mor Farajian, Raz Leib, and Ilana Nisky. The effect of vision on the augmentation of perceived stiffness by adding artificial skin stretch to kinesthetic force. In *IEEE International Conference on Robotics and Automation workshop on Integration Multidisciplinary Approaches to Advance Physical Human-Robot Interaction*, 2020.

[63] Melvyn A Goodale and A David Milner. Separate visual pathways for perception and action. *Trends in neurosciences*, 15(1):20–25, 1992.

[64] J Randall Flanagan and Michael A Beltzner. Independence of perceptual and sensorimotor predictions in the size–weight illusion. *Nature neuroscience*, 3(7):737–741, 2000.

[65] Tzvi Ganel and Melvyn A Goodale. Visual control of action but not perception requires analytical processing of object shape. *Nature*, 426(6967):664–667, 2003.

[66] Tzvi Ganel, Eran Chajut, and Daniel Algom. Visual coding for action violates fundamental psychophysical principles. *Current Biology*, 18(14):R599–R601, 2008.

[67] Anthony Michael Jarc and Ilana Nisky. Robot-assisted surgery: an emerging platform for human neuroscience research. *Name: Frontiers in Human Neuroscience*, 9:315, 2015.

[68] Lynette A Jones and Hong Z Tan. Application of psychophysical techniques to haptic research. *IEEE transactions on haptics*, 6(3):268–284, 2012.

[69] Omri Afgin, Nir Sagi, Ilana Nisky, Tzvi Ganel, and Sigal Berman. Visuomotor resolution in telerobotic grasping with transmission delays. *Frontiers in Robotics and AI*, 4:54, 2017.

[70] Sarmad Mehrdad, Fei Liu, Minh Tu Pham, Arnaud Lelevé, and S. Farokh Atashzar. Review of advanced medical telerobots. *Applied Sciences (Switzerland)*, 11(1):1–47, 2021.

[71] Sotiris Avgousti, Eftychios G. Christoforou, Andreas S. Panayides, Sotos Voskarides, Cyril Novales, Laurence Nouaille, Constantinos S. Pattichis, and Pierre Vieyres. Medical telerobotic systems: Current status and future trends. *BioMedical Engineering Online*, 15(1):1–44, 2016.

[72] Astrid Buvik, Einar Bugge, Gunnar Knutsen, Arvid Småbrekke, and Tom Wilsgaard. Quality of care for remote orthopaedic consultations using telemedicine: A randomised controlled trial. *BMC Health Services Research*, 16(1):1–11, 2016.

[73] Xuewei Tao, Jian Li, and Qi Huang. Stability analysis of time-varying delay bilateral teleoperation system with state prediction. In *2020 6th International Conference on Control, Automation and Robotics (ICCAR)*, pages 272–276. IEEE, 2020.

[74] William R. Ferrell. Remote manipulation with transmission delay. *IEEE Transactions on Human Factors in Electronics*, HFE-6(1):24–32, 2013.

[75] Robert J. Anderson and Mark W. Spong. Bilateral Control of Teleoperators with Time Delay. *IEEE transaction of automated control*, 34(5):494–501, 1989.

[76] Lingtao Huang, Hironao Yamada, Tao Ni, and Yanan Li. A master-slave control method with gravity compensation for a hydraulic teleoperation construction robot. *Advances in Mechanical Engineering*, 9(7):1–11, 2017.

[77] Christian Smith and Patric Jensfelt. A predictor for operator input for time-delayed teleoperation. *Mechatronics*, 20(7):778–786, 2010.

[78] Robert Sessions Woodworth. Accuracy of voluntary movement. *The Psychological Review: Monograph Supplements*, 3(3):i, 1899.

[79] Nikolai Bernstein. The co-ordination and regulation of movements. *The co-ordination and regulation of movements*, 1966.

[80] Tamar Flash and Neville Hogan. The coordination of arm movements: an experimentally confirmed mathematical model. *Journal of neuroscience*, 5(7):1688–1703, 1985.

[81] Raz Leib and Amir Karniel. Minimum acceleration with constraints of center of mass: a unified model for arm movements and object manipulation. *Journal of Neurophysiology*, 108(6):1646–1655, 2012.
[82] Shay Ben-Itzhak and Amir Karniel. Minimum acceleration criterion with constraints implies bang-bang control as an underlying principle for optimal trajectories of arm reaching movements. *Neural Computation*, 20(3):779–812, 2008.
[83] Robert A Scheidt and Claude Ghez. Separate adaptive mechanisms for controlling trajectory and final position in reaching. *Journal of neurophysiology*, 98(6):3600–3613, 2007.
[84] Daniel M Wolpert, Jörn Diedrichsen, and J Randall Flanagan. Principles of sensorimotor learning. *Nature Reviews Neuroscience*, 12(12):739–751, 2011.
[85] Joseph McIntyre and Emilio Bizzi. Servo hypotheses for the biological control of movement. *Journal of motor behavior*, 25(3):193–202, 1993.
[86] Andrew G Barto, Andrew H Fagg, Nathan Sitkoff, and James C Houk. A cerebellar model of timing and prediction in the control of reaching. *Neural computation*, 11(3):565–594, 1999.
[87] Emanuel Todorov and Michael I Jordan. Optimal feedback control as a theory of motor coordination. *Nature neuroscience*, 5(11):1226–1235, 2002.
[88] Raz Leib, Marta Russo, Andrea d'Avella, and Ilana Nisky. A bang-bang control model predicts the triphasic muscles activity during hand reaching. *Journal of Neurophysiology*, 124(1):295–304, 2020.
[89] Tamar Flash, Yaron Meirovitch, and Avi Barliya. Models of human movement: Trajectory planning and inverse kinematics studies. *Robotics and Autonomous Systems*, 61(4):330–339, 2013.
[90] Jee-Hwan Ryu, Dong-Soo Kwon, and Blake Hannaford. Stability guaranteed control: Time domain passivity approach. *IEEE Transactions on Control Systems Technology*, 12(6):860–868, 2004.
[91] John W Krakauer, Zachary M Pine, Maria-Felice Ghilardi, and Claude Ghez. Learning of visuomotor transformations for vectorial planning of reaching trajectories. *Journal of Neuroscience*, 20(23):8916–8924, 2000.
[92] Chen Avraham, Guy Avraham, Ferdinando A Mussa-Ivaldi, and Ilana Nisky. Neglect-like effects on drawing symmetry induced by adaptation to a laterally asymmetric visuomotor delay. *Frontiers in human neuroscience*, 12:335, 2018.
[93] Chen Avraham, Mor Dominitz, Hana Khait, Guy Avraham, Ferdinando A Mussa-Ivaldi, and Ilana Nisky. Adaptation to laterally asymmetrical visuomotor delay has an effect on action but not on perception. *Frontiers in human neuroscience*, 13:312, 2019.
[94] Ali Farshchian, Alessandra Sciutti, Assaf Pressman, Ilana Nisky, and Ferdinando A Mussa-Ivaldi. Energy exchanges at contact events guide sensorimotor integration. *Elife*, 7:e32587, 2018.
[95] Guy Avraham, Raz Leib, Assaf Pressman, Lucia S Simo, Amir Karniel, Lior Shmuelof, Ferdinando A Mussa-Ivaldi, and Ilana Nisky. State-based delay representation and its transfer from a game of pong to reaching and tracking. *eneuro*, 4(6), 2017.
[96] Reza Shadmehr and Ferdinando A Mussa-Ivaldi. Adaptive representation of dynamics during learning of a motor task. *Journal of neuroscience*, 14(5):3208–3224, 1994.
[97] Guy Avraham, Firas Mawase, Amir Karniel, Lior Shmuelof, Opher Donchin, Ferdinando A Mussa-Ivaldi, and Ilana Nisky. Representing delayed force feedback as a combination of current and delayed states. *Journal of neurophysiology*, 118(4):2110–2131, 2017.
[98] Maura Casadio, Assaf Pressman, and Ferdinando A Mussa-Ivaldi. Learning to push and learning to move: the adaptive control of contact forces. *Frontiers in computational neuroscience*, 9:118, 2015.
[99] Robert A Scheidt, Jonathan B Dingwell, and Ferdinando A Mussa-Ivaldi. Learning to move amid uncertainty. *Journal of neurophysiology*, 86(2):971–985, 2001.
[100] J Ryan Morehead, Jordan A Taylor, Darius E Parvin, and Richard B Ivry. Characteristics of implicit sensorimotor adaptation revealed by task-irrelevant clamped feedback. *Journal of cognitive neuroscience*, 29(6):1061–1074, 2017.
[101] Guy Avraham, J Ryan Morehead, Hyosub E Kim, and Richard B Ivry. Reexposure to a sensorimotor perturbation produces opposite effects on explicit and implicit learning processes. *PLoS biology*, 19(3):e3001147, 2021.
[102] Pietro Mazzoni and John W Krakauer. An implicit plan overrides an explicit strategy during visuomotor adaptation. *Journal of Neuroscience*, 26(14):3642–3645, 2006.
[103] Samuel D McDougle and Jordan A Taylor. Dissociable cognitive strategies for sensorimotor learning. *Nature communications*, 10(1):1–13, 2019.
[104] Scott T Albert, Jihoon Jang, Hannah R Sheahan, Lonneke Teunissen, Koenraad Vandevoorde, David J Herzfeld, and Reza Shadmehr. An implicit memory of errors limits human sensorimotor adaptation. *Nature human behaviour*, pages 1–15, 2021.
[105] Yarden Sharon, Anthony M Jarc, Thomas S Lendvay, and Ilana Nisky. Rate of orientation change as a new metric for robot-assisted and open surgical skill evaluation. *IEEE Transactions on Medical Robotics and*

Bionics, 3(2):414–425, 2021.

[106] Sean Estrada, Cassidy Duran, Daryl Schulz, Jean Bismuth, Michael D Byrne, and Marcia K O'Malley. Smoothness of surgical tool tip motion correlates to skill in endovascular tasks. *IEEE Transactions on Human-Machine Systems*, 46(5):647–659, 2016.

[107] Anna French, Thomas S Lendvay, Robert M Sweet, and Timothy M Kowalewski. Predicting surgical skill from the first n seconds of a task: Value over task time using the isogony principle. *International journal of computer assisted radiology and surgery*, 12(7):1161–1170, 2017.

[108] Arash Ajoudani, Andrea Maria Zanchettin, Serena Ivaldi, Alin Albu-Schäffer, Kazuhiro Kosuge, and Oussama Khatib. Progress and prospects of the human–robot collaboration. *Autonomous Robots*, 42(5):957–975, 2018.

[109] Jing Luo, Wei He, and Chenguang Yang. Combined perception, control, and learning for teleoperation: key technologies, applications, and challenges. *Cognitive Computation and Systems*, 2(2):33–43, 2020.

[110] Ekaterina Ivanova, Gerolamo Carboni, Jonathan Eden, Jorg Kruger, and Etienne Burdet. For Motion Assistance Humans Prefer to Rely on a Robot Rather Than on an Unpredictable Human. *IEEE Open Journal of Engineering in Medicine and Biology*, 1(April):133–139, 2020.

[111] Shervin Javdani, Henny Admoni, Stefania Pellegrinelli, Siddhartha S. Srinivasa, and J. Andrew Bagnell. Shared autonomy via hindsight optimization for teleoperation and teaming. *International Journal of Robotics Research*, 37(7):717–742, 2018.

[112] Kris Hauser. Recognition, prediction, and planning for assisted teleoperation of freeform tasks. *Autonomous Robots*, 35(4):241–254, 2013.

[113] Katharina Muelling, Arun Venkatraman, Jean Sebastien Valois, John E. Downey, Jeffrey Weiss, Shervin Javdani, Martial Hebert, Andrew B. Schwartz, Jennifer L. Collinger, and J. Andrew Bagnell. Autonomy infused teleoperation with application to brain computer interface controlled manipulation. *Autonomous Robots*, 41(6):1401–1422, 2017.

[114] Fanny Ficuciello, Guglielmo Tamburrini, Alberto Arezzo, Luigi Villani, and Bruno Siciliano. Autonomy in surgical robots and its meaningful human control. *Paladyn*, 10(1):30–43, 2019.

[115] Aleks Attanasio, Bruno Scaglioni, Elena De Momi, Paolo Fiorini, and Pietro Valdastri. Autonomy in Surgical Robotics. *Annual Review of Control, Robotics, and Autonomous Systems*, 4(1):1–29, 2021.

[116] Alaa Eldin Abdelaal, Prateek Mathur, and Septimiu E. Salcudean. Robotics In Vivo: A Perspective on Human–Robot Interaction in Surgical Robotics. *Annual Review of Control, Robotics, and Autonomous Systems*, 3(1):221–242, 2020.

[117] G. P. Moustris, S. C. Hiridis, K. M. Deliparaschos, and K. M. Konstantinidis. Evolution of autonomous and semi-autonomous robotic surgical systems: A review of the literature. *International Journal of Medical Robotics and Computer Assisted Surgery*, 7(4):375–392, 2011.

[118] Christopher Schultz, Sanket Gaurav, Mathew Monfort, Lingfei Zhang, and Brian D. Ziebart. Goal-predictive robotic teleoperation from noisy sensors. *Proceedings - IEEE International Conference on Robotics and Automation*, pages 5377–5383, 2017.

[119] Shintaro Inoue, Kazutaka Toyoda, Yo Kobayashi, and Masakatsu G. Fujie. Autonomous avoidance based on motion delay of master-slave surgical robot. *Proceedings of the 31st Annual International Conference of the IEEE Engineering in Medicine and Biology Society: Engineering the Future of Biomedicine, EMBC 2009*, (1):5080–5083, 2009.

[120] Leone Costi, Luca Scimeca, Perla Maiolino, Thilina Lalitharatne, Thrishantha Nanayakkara, Ryman Hashem, and Fumiya Iida. Comparative analysis of model-based predictive shared control for delayed operation in object reaching and recognition tasks with tactile sensing. *Frontiers in Robotics and AI*, 8:313, 2021.

[121] Ajay Kumar Tanwani and Sylvain Calinon. A generative model for intention recognition and manipulation assistance in teleoperation. *IEEE International Conference on Intelligent Robots and Systems*, 2017-Septe:43–50, 2017.

[122] Siddharth Reddy, Anca D. Dragan, and Sergey Levine. Shared autonomy via deep reinforcement learning. *arXiv*, 2018.

[123] Alan M Turing. Computing machinery and intelligence. *Mind*, 59(236):433–460, 1950.

[124] Guy Avraham, Ilana Nisky, Hugo L Fernandes, Daniel E Acuna, Konrad P Kording, Gerald E Loeb, and Amir Karniel. Toward perceiving robots as humans: Three handshake models face the turing-like handshake test. *Haptics, IEEE Transactions on*, 5(3):196–207, 2012.

[125] Ilana Nisky, Guy Avraham, and Amir Karniel. Three alternatives to measure the human likeness of a handshake model in a turing-like test. *PResence: Teleoperators and Virtual Environments*, 21(2):156–182, 2012.

IOP Conf. Series: Materials Science and Engineering 1261 (2022) 012015 doi:10.1088/1757-899X/1261/1/012015

Robot swarms as embodied extensions of humans

Jonas D Rockbach[1,2], **Maren Bennewitz**[2]

[1]Human-Machine-Systems, Fraunhofer Institute for Communication, Information Processing and Ergonomics FKIE, Wachtberg, Germany

[2]Humanoid Robots Lab, Computer Science VI, University of Bonn, Germany

E-mail: jonas.rockbach@fkie.fraunhofer.de

Abstract. In order to harness the potential of robot swarms, humans must be able to control decentralized and self-organizing multi-agent systems. The field of human-swarm interaction investigates how humans with their centralized information and control demands can be paired with the decentralized architectures of robot swarms. Recently, we have proposed an architecture which aims at balancing the trade-offs between centralized and decentralized control and communication in human-swarm interaction while joining the relative capabilities of human cognition and swarm intelligence. In this article, we expand on this architecture by discussing a list of design requirements that must be met in order to make robot swarms embodied extensions of humans that act and feel like natural body parts. We also conceptually show how some of the design requirements could be implemented by simple rules when we take the world's structure into account.

'[...] I believe that the most potent near-future technologies will be those that offer integration and transformation without implants or surgery: human-machine merges that simply bypass [...] the old biological borders of skin and skull.'

Andy Clark [1, p.24]

1. Introduction

Swarm robotics [2, 3, 4, 5] is embodied swarm intelligence. In swarms, the agents exclusivity rely on local inter-robot communication for goal achievement without reference to a global leader. From an applied view, robot swarms are expected to be of great value in future application scenarios such as search and rescue [6]. From the view of fundamental research, robot swarms help us to better understand natural swarm intelligence, which in turn helps us to develop better robot swarms. This mutual benefit is well-known in bio-inspired artificial intelligence [7], and has more generally been called Pasteur's quadrant [8].

The practitioner who wants to engineer robot swarms is faced with a range of challenges [2, 4]. From a technical perspective, one must decide what behaviours the swarm should display in order to solve the tasks we build the swarm for. Then, one must design local robot controllers that achieve the desired macrolevel behaviours with robustness and scalability as general requirements. If the robot swarm should be deployed in the real world, one must also take into account how human operators can control the swarm due to technical, ethical and legal issues. Human-swarm interaction (HSI) investigates how the capabilities and demands of human operators should be joined with the decentralized control architectures of swarms [9, 10].

Recently, we have proposed a mindset on how this design challenge of HSI can be approached [9]. We think that it is essential to take into account three perspectives when building swarms; the technical perspective (swarm engineering), the human perspective (human-centred design) as well as how human and swarm can be combined into a joint human-swarm loop. In real-life, it is neither the swarm nor the human in isolation who copes with the task objectives, but human and swarm as one joint loop [11] (or one biotechnological matrix [1], or one intelligent agent [12]). Building on this view, we have further elaborated that HSI can also be interpreted as human-swarm intelligence, because the design goal is to build an intelligent (i.e., adaptive, robust and scalable) joint human-swarm loop that can deal with the world. Figure 1 summarizes this view with the triad of human-swarm intelligence which highlights its main design dimensions [9]: operator training [13], swarm engineering [2, 4] and human-swarm interface design [9, 10].

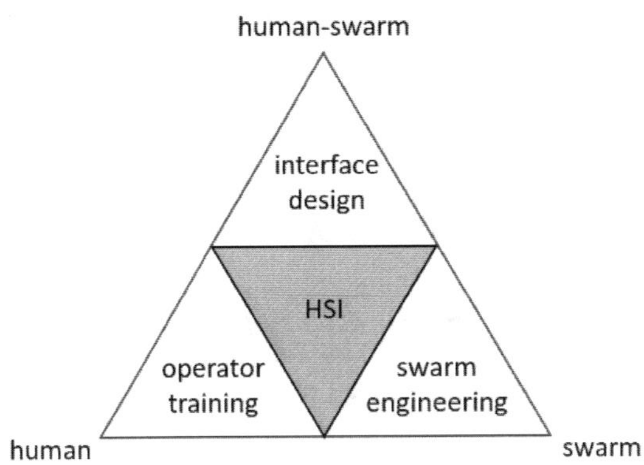

Figure 1. The triad of human-swarm intelligence shows its three main design dimensions: operator training, swarm engineering and interface design. Adapted from [9].

In our previous work, we also discussed a design theory called the swarm-amplified human [9] that could potentially guide the selection of design solutions for HSI. This proposed design theory narrows down the design solution space for each of the design dimensions (human, swarm, and human-swarm) in the triad of HSI.

Essentially, the swarm-amplified human operationalizes a robot swarm as a semi-autonomous extension of the human nervous system. For the human operator, these extensions take the form of artificial body parts. From this perspective, the swarm is not the target of the interaction, but rather a part of the situated self-regulatory human agent (i.e., part of a cybernetic organism [1, 14]). Thus, the swarm is part of the interface through which the human interacts with the world.

This perspective aims at amplifying the self-regulatory capabilities of the human operator, especially in regard to cognitive capabilities [11], by means of a robot swarm. An intelligent agent must perceive the environment, decide what to do, and act on the environment [15]. In the swarm-amplified human, a swarm self-organizes to form networks between relevant features of the environment and participating human operators, thereby amplifying the humans' natural sensation and action ranges. This extension is assumed to amplify natural cognition, given the assumption that higher-order computation depends on the body's morphology and sensory-motor capabilities [16, 17]. Figure 2 shows a simple example of how a robot swarm forms a network that connects a human h to relevant object features f_i in a two-dimensional world D. For further details, we refer the reader to [9].

In this article, we extend our previous work [9] by discussing design requirements for robot swarms as self-organizing extensions of human operators as well as showing conceptually how utilizing the world's structure can simplify the demands on the robot controllers in HSI. Table 1 provides a summary of the notation used in this work.

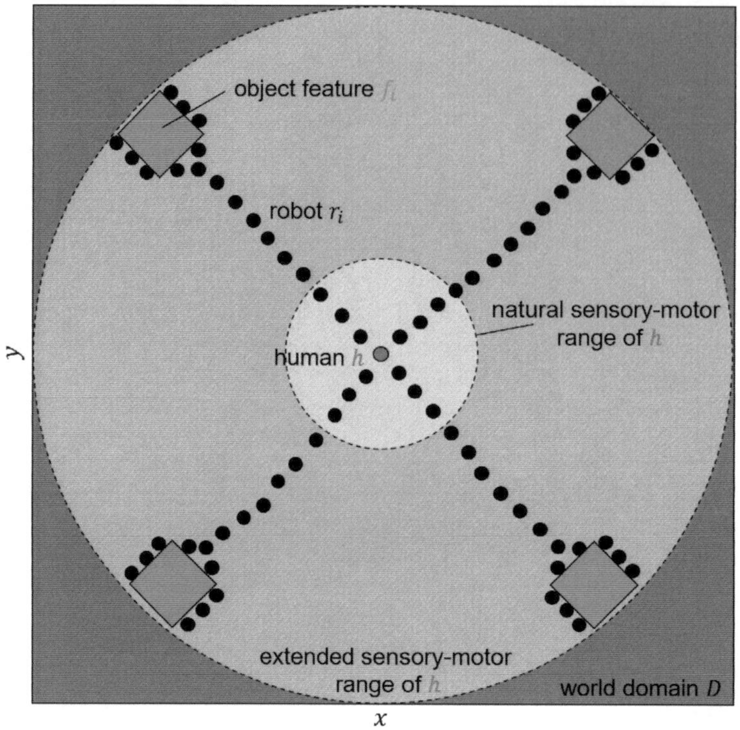

Figure 2. The human's natural sensory-motor range is extended by the swarm robots r_i that form a network between the human h and object features f_i. Thereby, a human is provided with a larger sensory-motor range that is assumed to amplify natural cognition.

2. Design requirements for swarm embodiment

As a use case, imagine that Major Tom, a member of a future search and rescue unit, is deployed to a burning warehouse[1] where an unknown number of humans are missing. A crucial part of Tom's work is his capability to interact with objects, like opening doors and moving obstacles. Figure 3 shows a sketch in which Tom manipulates an object feature F with the swarm S that is an extension of himself. Figure 4 shows the corresponding causal model. The two figures show that instead of controlling the full swarm S, Tom modulates a subset P of the swarm which in turn forms a closed loop with F, while F is assumed to be a relevant part of environment D in context of the task. The subset P has self-organized around the object feature F, driven by local rules that react to the affordances of F given the estimated state of Tom $\hat{H}$. These subsets P take the form of artificial body parts for the observing Tom. Communication between the human H and P is established by the self-organizing ad-hoc network G_C, called the connectome graph. Information about the feature state estimations $\hat{F}$ is communicated back to Tom via G_C. Figure 5 shows a sketch of Tom's head-up display.

While following the bottom-up route to developing intelligent agents [19, 20], we are motivated to identify a minimal set of abstracted design requirements for making P embodied extensions

[1] The reader could also imagine a futuristic warehouse in space.

Table 1. Notation used in this work. Adapted from [9].

Variable	Description
Domain D	The world the agents act in.
Object features F	The set of relevant object features in relation to a task, with $f_i \in F$ and $F \subseteq D$. Each f_i is defined by a state vector. While an object may have multiple features that can be of interest, in this work f_i directly corresponds to a given object for the sake of clarity.
Humans H	The set of human agents that participate in the agent network, with $h_i \in H$. Each h_i is defined by a state vector.
Swarm S	The set of robot agents that form the swarm, with $r_i \in S$. Each r_i is defined by a state vector.
Subswarms P	The set of robot agents that form local sensory-motor loops at object features f_i with $p_i \in P$ and $P \subseteq S$. A certain part p_i forms a behavioural robot cluster that is treated as an artificial body part of human h_i.
Connectome graph G_C	The graph that is formed by robot agents r_i to connect humans H to subswarms P that in turn are connected to relevant object features in F.
Estimated state of humans $\hat{H}$	Estimated state of humans H that is transmitted via G_C from H to P.
Estimated state of object features $\hat{F}$	Estimated state of object features F that is transmitted via G_C from P to H.

of a human that describe the underlying principles at work, without getting overwhelmed by the detailed complexity of the design challenge. A summary is given in Table 2 and Figure 6. We assume that the proposed list provides necessary but not sufficient requirements for making robot swarms embodied extensions. In addition, they represent our expectations about good joint human-swarm loop performance (cf. [9]) and should therefore be treated as hypotheses.

2.1. Embodiment 1: Reducing higher-order controller complexity

From the perspective of the joint human-swarm loop, we are motivated to design the swarm-amplified human as an intelligent agent that is optimized to deal with the tasks at hand. In real-life applications such as Tom's scenario, it must be remembered that the overall task is not to control the robot swarm, but rather to get the actual tasks done such as rescuing people. To do so, Tom must deal with multiple subtask in the environment, such as opening doors, moving obstacles out of the way, searching for the best way through the hostile environment, coordinating with team members, and so forth. What this perspective highlights is that the robot swarm, as a useful tool for supporting Tom, must be allocated where it needs to be. In other words, the robot swarm S should split into specialized parts P at locations of the environment that offer **affordances** (A1) to interact with the tasks that need to be solved. This perspective is deeply related to embodiment; fundamentally, an agent is capable of solving tasks in its environment by exchanging information with it *through its body*. To exchange information, the body must be located near the physical mechanisms that afford interaction. To see the sky, I must look at the sky. To grasp an object, I must move my hand to that object.

IOP Conf. Series: Materials Science and Engineering 1261 (2022) 012015 doi:10.1088/1757-899X/1261/1/012015

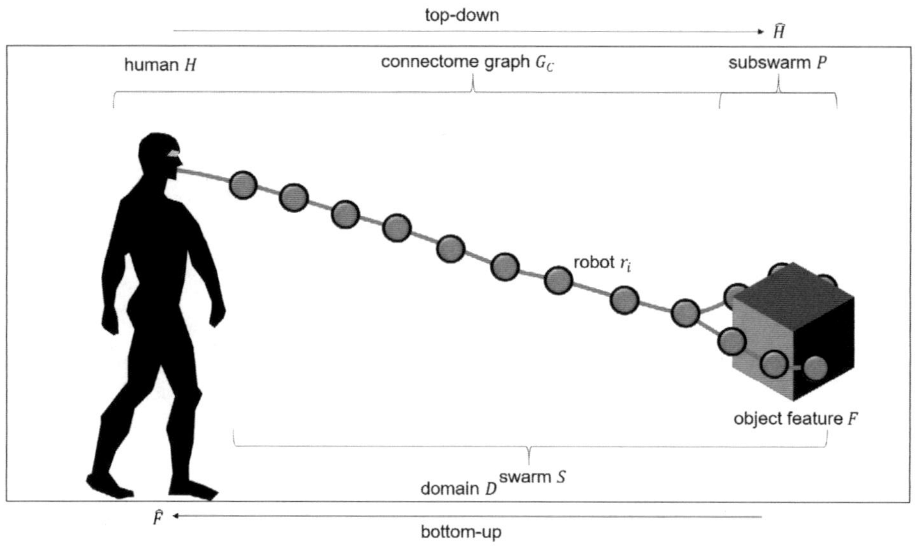

Figure 3. Sketch of the future search and rescue specialist Major Tom H who is supported by a self-organizing robot swarm S that acts as an embodied extension of himself. The swarm has self-organized into a subswarm P, that Tom interprets as an artificial body part, around the object feature F while being connected to it via the connectome graph G_C.

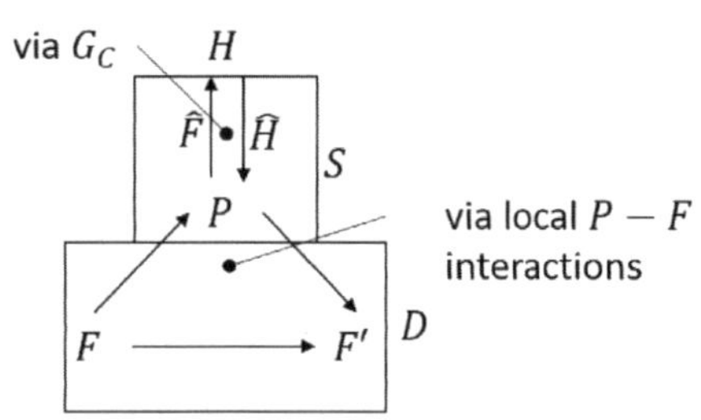

Figure 4. Causal model of the the swarm-amplified human. Adapted from [9] and [18]. See Table 1 for the notation.

Thus, a robot swarm is embodied swarm intelligence in the sense that a physical agent is situated in the real world and therefore must deal with real-world complexities [18, 21, 22]. This embodiment comes both with a weakness and a strength. The physical nature constraints the agent's possibilities [15, 23]; the agent's morphology, sensory and motor apparatus, and controllers as well as its energy supply define an upper limit for the agent's capabilities. Even the stunning size of a human's state space has a theoretical upper limit if the human's morphology is not changed or not supported by tools. On the other hand, well-known work such as passive walkers [24] demonstrate how the physical body can perform computation on its own, thereby simplifying the requirements for a higher-order controller.

Outsourcing tasks to lower-order systems, such as the body's mechanical nature, reflex pathways and low-order sensory-motor systems, can be seen as a special case of a hierarchical control architecture[2] that is tuned to deal with complex scenarios. In the real world, critical situations often demand a fast reaction [21]. If I trip over a stone, my body must be stabilized fast - there simply is no time to harness the full potential of higher-order cognition. Similarly, a mouse that can not quickly classify a cat and therefore does not implement the behaviour 'run

[2] Where elements on a microlevel self-organize into hierarchically interacting subunits on a macrolevel [9, 25]

Figure 5. Sketch of Tom's head-up display while he interacts with two objects via the swarm extension.

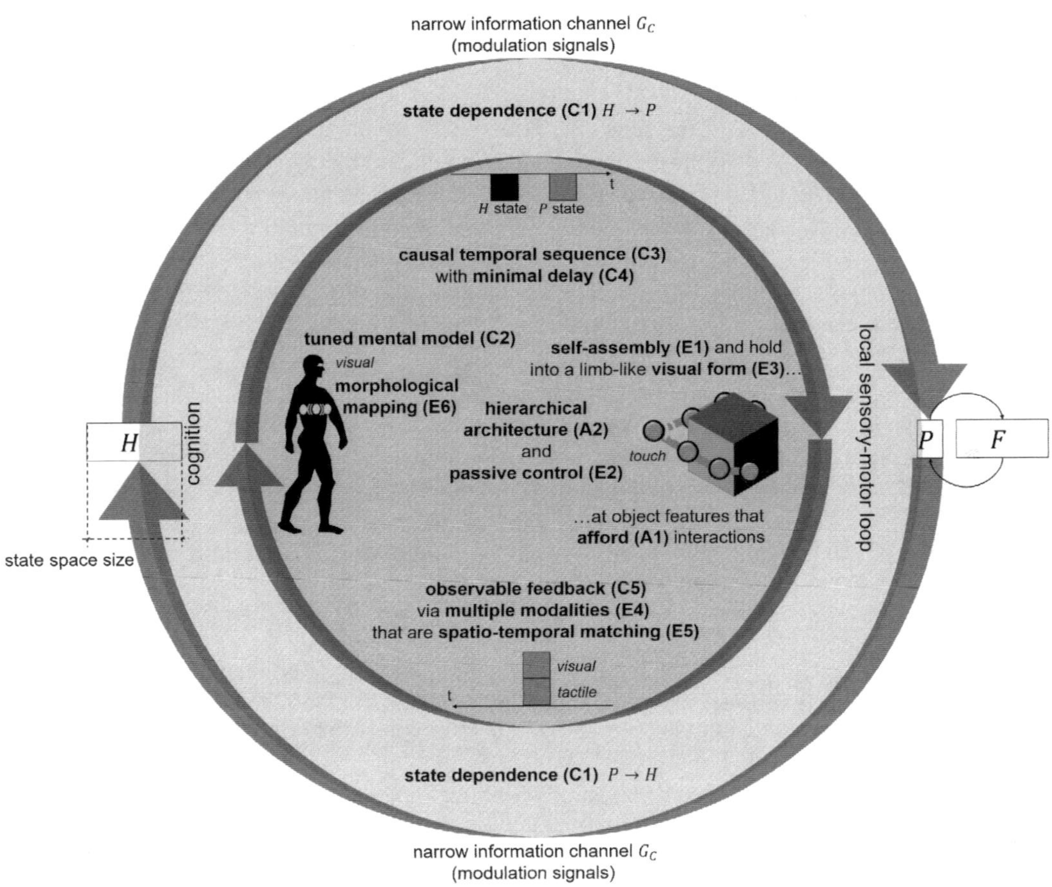

Figure 6. Illustration of the necessary design requirements for making swarms embodied extensions of humans. See Table 2 for the corresponding list.

Table 2. Necessary design requirements for making swarms embodied extensions of humans. A: Requirements for **a**rchitecture; C: Requirements for swarm **c**ontrollability; E: Requirements for swarm **e**mbodiment.

Name	Description
A1: Affordance	The swarm S must split into specialized subswarms P at locations of the environment that afford possibilities for task interactions with F.
A2: Hierarchical architecture	The subswarms P must be capable to handle some situations while being modulated by the state of the human H.
C1: State dependence	The states of the subswarms P must at least partly depend on the human state H.
C2: Tuned mental models	The human's mental models must be tuned to swarm control.
C3: Causal temporal sequence	A change of the subswarms' state P' must occur after a change of the human's state.
C4: Minimal delay	A change of the subswarms' state P' must occur shortly after a change of the human's state.
C5: Observable feedback	A change of the subswarms' state P' must be presented to the human.
E1: Self-assembly	The subswarms P must form and hold into an organization that is separate from the rest of the world.
E2: Passive control	The states of subswarms P must be influenced by the estimated human state $\hat{H}$ without the human needing to consciously supervising the subswarms while conscious control should still be possible.
E3: Limb-like visual form	The subswarms P should look similar to natural body parts, such as a hand, although non-humanoid morphologies may be possible.
E4: Multiple modalities	Feedback to the human H should be provided via multiple modalities.
E5: Spatio-temporal matching	Multiple sensory cues must be spatio-temporally matching.
E6: Morphological mapping	The morphology of the subswarms P should be mapped to the human's morphology.

away' will not survive very long [26]. An autonomous underwater vehicle that can not handle water currents on its own will fail its purpose given that operator control is not always possible underwater due to communication limitations [27].

This **hierarchical architecture** (A2) is the underlying design principle of the swarm-amplified human; the higher-order human H with its adaptive and cognitive capabilities modulates semi-autonomous lower-order embodied swarms P (Figure 4). Another way to put it is to say that the human's state space is much larger than compared to the state space of the robot swarm (outer loop in Figure 6). The subswarms may be driven by behaviour-based principles [28] on the microlevel; they can react fast[3], but will fail in more complex situations.

[3] Of course, self-organizing swarms do not always react 'fast', especially if the swarm switches between attractor states such as in distributed decision making [29]. The point here is when an attractor is found, such as a particular robot assemble [30] at the location of the task object, the swarm can directly interact with that object.

The human's cognitive system on the other hand can deal with a large situation space, which however often requires time. As demonstrated by passive walkers, a cognitive system must only modulate the lower-order system, without controlling the details. This reduces the information that needs to be exchanged between H and P, or rather H and F, via G_C.

2.2. Embodiment 2: Swarms as controllable extensions of the human body

From the perspective of distributed and extended cognition [1, 31, 32], every object that is used by an active agent becomes part of the agent's information loop through which a particular goal is accomplished. But there seems to be a phenomenological difference between something which does not belong to my body, such as a hammer, and part of my body, such as my leg. The notion of integrating, or embodying, artificial body parts P we use here is something in between [33]: we want the swarm to interface with the nervous system and be accepted by it as part of the body, because being part of my body comes with benefits such as a feeling of trust and controllability. On the other hand, we do not want human operators to be disabled if the swarm is lost, as would be the case with a real body part. In Tom's scenario, one therefore may think of swarms forming themselves into artificial body parts if the task objectives require this, while Tom feels that his body has changed and the artificial body parts are now part of himself. When the artificial body parts are not needed any more, they dissolve without disadvantages for Tom who now solely feels his biological body.

Although we do not know in detail how we can robustly trick the nervous system into believing that an artefact like a robot swarm belongs to the biological agent's body, while some neuro-cognitive experiments and clinical conditions provide motivation that this should be possible [34, 35, 36, 37, 38, 39, 40, 41, 42], we can start by formulating some obvious necessary conditions that must be met. In the neuro-cognitive sciences, evidence, although not conclusive, has been accumulated for a double dissociation between a sense of agency (i.e., a feeling that I am the author of a state, here called 'swarm controllability'), and a sense of ownership (i.e., a feeling that something belongs to my body, here called 'swarm embodiment') [35]. In general, the supervision of an artificial agent by a human operator, such as the swarm in the swarm-amplified human, requires a feeling of controllability due to practical constraints in most real-world applications. It would be hard for Tom to trust and accept the automated swarm if he does not feel that he has some form of control over it, at least if the swarm is expected to fall short of full autonomy. On top of swarm controllability, swarm embodiment should make the swarm more natively useable for Tom, which reduces swarm supervision demands and thereby leaves potential for attributing valuable cognitive resources to more demanding work tasks.

2.2.1. Swarm controllability To be a part of a dynamical system, there must be a **state dependence** (C1) between the components, i.e., the state of one component restricts the state of the other component. This, of course, requires that information can be exchanged between the components. In general, H and P can either be linked via the human's sensory-motor capabilities, e.g., if the human operator controls a swarm via a joystick, or by directly linking the swarm to the state of the nervous system, i.e, via brain-computer interfacing [43]. Given that natural body parts are not controlled via joysticks but depend on the direct electro-physiological state of the nervous system, the latter may be preferred in the long run. However, for now we can attempt to formulate design principles that allow us, as Andy Clark puts it, to *'bypass [...] the old biological borders of skin and skull.'* [1, p.24]. For example, we may classify the human's cognitive state based on physiological measurements [44], such as whether it is in a stable or critical state, and utilize this classification as input for P given the assumption that the physiological state is also a map of the neural state.

Four additional necessary requirements should be stated in order to establish controllability over the swarm. First, the human must be aware how his own actions or states (neurally, psychologically, and physiologically) relate to the state space of the swarm. If Tom does not know how he can influence the swarm, he may misattribute the swarm's state to other external factors. This highlights the importance of experience with the swarm, or operator training, that tunes the human's **mental model** (C2) to the swarm [13]. It also underlines that accessible human states should be used as control inputs and that a simple mapping between the state of the human and the swarm should be established. Second, the human's control input must be in a **causal temporal sequence** (C3) before, and preferably with **minimal delay** (C4), to the triggered swarm's action. From Tom's perspective, it would be unlikely that he is the author of a swarm state if it occurs one minute before he actually implemented a control action. Given the self-organizing nature of robot swarms, establishing fast reactions of the robot swarm is not always straight forward and must be addressed. Third, the triggered state change of the swarm must be made available in the form of **observable feedback** (C5). Presumably the simplest feedback would be the visual observation of a swarm's state change, e.g., a change in formation.

2.2.2. Swarm embodiment In order to trick the nervous system into accepting P as part of its body, we must first of all make sure that P provides the possibility that it can be experienced as one spatio-temporal entity. Body parts are components of living agents which have the fundamental capability to hold into an organization that is separate from the environment [45], thus not to disintegrate over varying situations and over time. Therefore, artificial body parts P must form and hold, i.e, **self-assemble** (E1), into a particular form that the human operator can perceive as something which exists on its own, separated from the environmental background.

We can also ask how body control feels in daily life. Most of the time, we do not explicitly think about body movement. When walking down the street, it is rare that you explicitly think about your leg movement, yet alone the detailed muscle control of your legs [9]. Therefore, the state dependence of P on the human state H should take the form of **passive control** (E2), i.e., classifying the human state and using this estimated state $\hat{H}$ as control inputs for P. However, we can also choose to consciously control our body, for example when we want to learn a new motor pattern for mastering a musical instrument. Therefore, the desired implementation should also allow the human to switch to a more conscious form of control (e.g., [46]).

The rubber hand illusion is presumably the best known experiment that demonstrates how the brain can be tricked into perciving an artificial object as part of the body [34]. A rubber hand is placed on a table, while the participant is instructed to place hers or his hand under the table at the same position as the rubber hand. The participant now observes some part of the hand being stroked, while the participant's real hand is in synchrony stroked at the same location under the table, which is not seen by the participant. This synchronized multimodal feedback seems to be sufficient to trick the brain into accepting that the rubber hand is the participant's actual hand. Predictive coding has been put forward as a general mechanisms of the brain, which fundamentally states that the brain generates hypotheses about expected sensory inputs and selects the most probable explanation given a model, in our discussion of what objects belong to the agent's body [40, 47, 48]. By identifying which sensory signals (cues) and related mechanisms the brain uses for updating its estimation of body ownership, we may identify possible leverage points that could be considered for embodying swarms into humans. Thus, the goal of swarm embodiment is to augment the swarm with convincing evidence that it is part of the human body.

It seems that the **visual form** (E3) of an object is a dominant cue the brain utilizes and has been suggested as the first filter the brain deploys for deciding body ownership [35, 49]. This means that the subswarms P should self-organize into forms that are similar to body parts, such as a hand. However, it seems that non-human forms, such as tails [41] and wings [37], can also be accepted by the brain as parts of the body [42]. Whether non-humanoid body parts can indeed be accepted by the brain as parts of the body relates to asking if the brain compares sensory inputs with models of the body that can be overwritten and is a matter of debate [35, 50].

For real body parts, proprioceptive and posture signals as well as touch sensations seem also to be of interest for the brain [35, 49]. In order to generate convincing information for the brain that the swarm is part of the body, one should therefore utilize **multiple modalities** (E4) that must **spatio-temporally match** (E5). In the rubber hand illusion, the perception of ownership becomes less likely with an increase in distance between the rubber hand and the real hand, which produces a mismatch between proprioceptive and visual cues [35].

In general, swarm embodiment could be approached from two sides which correspond to different experimental paradigms that define whether the extension is designed after the human morphology or not. For a general overview of human augmentation see [36]. First, as exemplified by the rubber hand illusion, we could strive to make the swarm as similar as possible to a real body part and then attempt to synchronize the different cues (visual, posture, and touch) with each other. This approach comes with a list of challenges, the most relevant one being the restriction to the human morphology. The main benefit of swarms are their scalable self-organizational nature, i.e., their capability to scalable form new functional structures during runtime. This strength is obviously inhibited if the robot swarm must conform to a predefined form. Efforts that extend the human morphology with prosthetic-like body parts, such as a third arm [39] or a third thumb [38], relate to this approach[4]. Second, as exemplified by augmenting humans with tails in a VR environment [41], we could attempt to train the human to remap his body representation by **mapping** (E6) non-humanoid morphologies onto the body [42]. For example, the whole body posture could be utilized [41] or we may use mapped tactile feedback on the skin [37]. Tom may be equipped with a vibrating feedback device on his chest that provides position-specific information about tactile and proprioceptive feedback. This latter approach seems more practical for swarm embodiment, although controversial (see above).

3. Exploiting the world's structure for swarm embodiment

While having the above use case of Major Tom (Figures 3 and 5) in mind, we implemented a simple simulation in MATLAB 2020b in order to illustrate how even simple rules can be sufficient for self-assembly (E1) and limb-like visual form (E3) based on derived affordances (A1) when we take the world's structure into account. The task for the robot swarm S is to split into body parts P with limb-like forms at the object features F. We placed two types of agents as point masses onto a 100x100 grid domain D:

- Human: One human h was placed at a fixed position on the lower part of the grid.
- Robots: A number of swarm robots r_i were randomly placed on the lower part of the grid. At each iteration of the simulation loop, a robot moves on the cell it is facing or randomly selects one of the eight surrounding cells if the next cell is not free. Each robot has the capability to detect when an object feature F is on a neighbouring cell.

Two box-shaped object features F were placed in the upper two corners of the grid domain. Figure 7 shows an exemplary simulation run with 200 swarm robots. Without the application of positive or negative feedback loops [3], the artificial body parts P form with two limbs, or fingers, 'grasping' the objects F based on the naive rule 'if I sense an object, I stop'. Of course, this would require that the robots are capable to identify affordances [51] such as where to attach. Importantly, it is the exploitation of the objects' geometry and position in reference to the environment [52], here the placement of box-shaped objects into the corners, that allows us to greatly minimize the controller complexity of our embodied swarm (Section 2.1).

Such an exploitation of world structures may be combined with approaches to self-assembly [53, 54, 55, 56] on suitable robot platforms like swarmanoid robots [30] that can

[4] It may also be argued that they belong to the second approach.

IOP Conf. Series: Materials Science and Engineering 1261 (2022) 012015 doi:10.1088/1757-899X/1261/1/012015

self-assemble into capable morphologies. For details about the connectome graph G_C between humans H and body parts P, the reader is referred to [9]. In sum, even naive rules in combination with a structured environment may be utilized for designing swarm embodiment - an insight, which is at the very heart of behaviour-based robotics (*'The world is its own best model'*, [21, p.15]), and embodied intelligence.

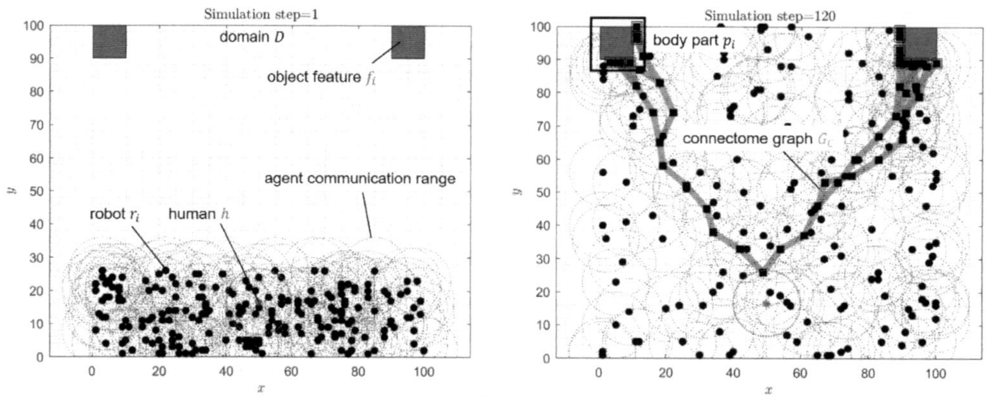

Figure 7. Left: At the start of the simulation, the robot swarm spreads out into the domain in search of object features. Right: With time, artificial body parts with two limbs, or fingers, have formed by the robots that attached to the object features.

4. Conclusion

In this contribution, we have discussed design requirements that we expect to be necessary though not sufficient for making robot swarms embodied extensions of humans. We have also shortly elaborated on how even simple solutions may be applied to produce some desired behaviours. The discussed design requirements should be seen as hypotheses in the quadrant of both applied science and fundamental science. For the former, the requirements form hypotheses about what would be a good design blueprint for making robot swarms embodied extensions of humans. This in turn is expected to result into good performance of the joint human-swarm loop. For the latter, the requirements form hypotheses about how the brain controls the world and forms its perception of body ownership. Robot swarms may therefore not only be used for understanding biological swarm intelligence, but also for understanding neural intelligence. Indeed, from a high-level standpoint, swarms and neural systems may not be so different after all [9, 57]. Human-swarm interaction intersects both with the neuro-cognitive sciences and swarm robotics and could therefore be utilized to investigate self-organizing intelligence across matters and scales.

A range of technical and ethical challenges must be solved in order to implement the swarm-amplified human in a real-life scenario. For example, a question which is currently addressed is how robot swarms can form into capable morphologies. Of particular interest in relation to the swarm-amplified human is the mergeable nervous systems [56], which demonstrates how a robot swarm can form into different morphologies with a hierarchical architecture. From an ethical perspective, we must ask ourselves whether we want to extend our bodies [1], even if this modification is reversible (Section 2.2), given the assumption of embodied intelligence that our higher-order capabilities and first-order experiences are based on our bodies' morphologies. We therefore close this article by inviting the reader to consider a variation of Nagel's question [16]:

What is it like to be a swarm-amplified human?

Acknowledgments
We would like to thank the anonymous reviewers for providing valuable feedback on the first version of this article.

5. References

[1] Clark A 2003 *Natural-Born Cyborgs: Minds, Technologies, and the Future of Human Intelligence* (Oxford university press)
[2] Barca J C and Sekercioglu Y A 2013 *Robotica* **31** 345–359 ISSN 0263-5747
[3] Bonabeau E, Dorigo M, Marco D d R D F, Theraulaz G and Théraulaz G 1999 *Swarm intelligence: from natural to artificial systems* (Oxford university press) ISBN 0195131592
[4] Brambilla M, Ferrante E, Birattari M and Dorigo M 2013 *Swarm Intelligence* **7** 1–41 ISSN 19353812
[5] Dorigo M, Theraulaz G and Trianni V 2020 *Science Robotics* **5** ISSN 2470-9476
[6] Penders J, Alboul L, Witkowski U, Naghsh A, Saez-Pons J, Herbrechtsmeier S and El-Habbal M 2011 *Advanced Robotics* **25** 93–117 ISSN 01691864
[7] Floreano D and Mattiussi C 2008 *Bio-inspired artificial intelligence: theories, methods, and technologies* (MIT press) ISBN 0262303914
[8] Stokes D E 2011 *Pasteur's quadrant: Basic science and technological innovation* (Brookings Institution Press) ISBN 0815719078
[9] Hasbach J D and Bennewitz M 2021 *Adaptive Behavior* ISSN 1059-7123 URL `http://journals.sagepub.com/doi/10.1177/10597123211017550`
[10] Kolling A, Walker P, Chakraborty N, Sycara K and Lewis M 2016 *IEEE Transactions on Human-Machine Systems* **46** 9–26 ISSN 21682291
[11] Hollnagel E and Woods D D 2005 *Joint Cognitive Systems. Foundations of Cognitive Systems Engineering* (CRC Press) ISBN 0-8493-2821-7
[12] Hasbach J D and Witte T E 2021 Human-Machine Intelligence: Frigates are Intelligent Organisms *2021 IEEE International Conference on Systems, Man, and Cybernetics (SMC)*
[13] Hasbach J D, Witte T E F and Bennewitz M 2020 On the Importance of Adaptive Operator Training in Human-Swarm Interaction *International Conference on Human-Computer Interaction* (Springer) pp 311–329
[14] Clynes M E and Kline N S 1960 *Aeronautics* **Spring** 26–34
[15] Russell S and Norvig P 2016 *Artificial Intelligence: A Modern Approach* 3rd ed (Pearson)
[16] Nagel T 1974 *Readings in philosophy of psychology* **1** 159–168
[17] Pfeifer R and Bongard J 2006 *How the body shapes the way we think: a new view of intelligence* (MIT press) ISBN 0262288524
[18] Pfeifer R, Lungarella M and Iida F 2007 *science* **318** 1088–1093 ISSN 0036-8075
[19] Braitenberg V 1986 *Vehicles: Experiments in synthetic psychology* (MIT press) ISBN 0262521121
[20] Franklin S 1995 *Artificial minds* (MIT Press)
[21] Brooks R A 1991 Intelligence without reason Tech. rep. MIT
[22] Cangelosi A, Bongard J, Fischer M H and Nolfi S 2015 Embodied intelligence *Springer Handbook of Computational Intelligence* (Springer) pp 697–714
[23] Simon H A 1955 *The quarterly journal of economics* **69** 99–118 ISSN 1531-4650
[24] Collins S, Ruina A, Tedrake R and Wisse M 2005 *Science* **307** 1082–1085 ISSN 0036-8075
[25] Heylighen F 2001 *The encyclopedia of life support systems* **5** 253–280
[26] Ashby W R 1961 *An introduction to cybernetics* (Chapman & Hall Ltd)
[27] Sheridan T B 1992 *Telerobotics, automation, and human supervisory control* (MIT press) ISBN 0262193167
[28] Arkin R C 1998 *Behaviour-Based Robotics* (MIT Press)
[29] Hamann H 2018 *Swarm robotics: A formal approach* (Springer) ISBN 9783319745282
[30] Dorigo M, Floreano D, Gambardella L M, Mondada F, Nolfi S, Baaboura T, Birattari M, Bonani M, Brambilla M, Brutschy A, Burnier D, Campo A, Christensen A L, Decugniere A, Di Caro G, Ducatelle F, Ferrante E, Förster A, Gonzales J M, Guzzi J, Longchamp V, Magnenat S, Mathews N, Montes De Oca M, O'Grady R, Pinciroli C, Pini G, Rétornaz P, Roberts J, Sperati V, Stirling T, Stranieri A, Stützle T, Trianni V, Tuci E, Turgut A E and Vaussard F 2013 *IEEE Robotics and Automation Magazine* **20** 60–71 ISSN 10709932
[31] Clark A and Chalmers D 1998 *analysis* **58** 7–19 ISSN 0003-2638
[32] Hutchins E 1995 *Cognition in the Wild* (MIT press) ISBN 0262581469
[33] Flemisch F and Baltzer M C A 2022 Are Rider-Horse or Centaurs intelligent Human Systems Integration? First Sketch of reversible and non-reversible human technology/machine/AI Symbiosis *Intelligent Human Systems Integration (IHSI 2022) Integrating People and Intelligent Systems*
[34] Botvinick M M and Cohen J 1998 *Nature* **391** 756 ISSN 0028-0836

[35] Braun N, Debener S, Spychala N, Bongartz E, Sörös P, Müller H H and Philipsen A 2018 *Frontiers in Psychology* **9** 1–17 ISSN 16641078
[36] Di Pino G, Maravita A, Zollo L, Guglielmelli E and Di Lazzaro V 2014 *Frontiers in systems neuroscience* **8** 109 ISSN 1662-5137
[37] Egeberg M C S, Lind S L R, Serubugo S, Skantarova D and Kraus M 2016 Extending the human body in virtual reality: Effect of sensory feedback on agency and ownership of virtual wings *Proceedings of the 2016 Virtual Reality International Conference* pp 1–4
[38] Kieliba P, Clode D, Maimon-Mor R O and Makin T R 2021 *Science Robotics* **6** eabd7935 ISSN 2470-9476
[39] Penaloza C I and Nishio S 2018 *Science Robotics* **3** ISSN 2470-9476
[40] Seth A K 2013 *Trends in cognitive sciences* **17** 565–573 ISSN 1364-6613
[41] Steptoe W, Steed A and Slater M 2013 *IEEE transactions on visualization and computer graphics* **19** 583–590 ISSN 1077-2626
[42] Won A S, Bailenson J, Lee J and Lanier J 2015 *Journal of Computer-Mediated Communication* **20** 241–259 ISSN 1083-6101
[43] Karavas G K, Larsson G, Artemiadis P, Larsson D T and Artemiadis P 2017 *IEEE/RSJ International Conference on Intelligent Robots and Systems* 5065–5070
[44] Schwarz J and Fuchs S 2017 Multidimensional real-time assessment of user state and performance to trigger dynamic system adaptation *Lecture Notes in Computer Science (including subseries Lecture Notes in Artificial Intelligence and Lecture Notes in Bioinformatics)* (Springer) ISSN 16113349
[45] Maturana H R and Varela F J 1980 *Autopoiesis and cognition: The realization of the living* (Reidel Pub. Co)
[46] Tsykunov E, Agishev R, Ibrahimov R, Labazanova L, Tleugazy A and Tsetserukou D 2019 *IEEE transactions on haptics* **12** 363–374 ISSN 1939-1412
[47] Friston K 2010 *Nature reviews neuroscience* **11** 127–138 ISSN 1471-0048
[48] Hohwy J 2013 *The predictive mind* (Oxford University Press) ISBN 0199682739
[49] Tsakiris M 2010 *Neuropsychologia* **48** 703–712 ISSN 0028-3932
[50] Haans A and IJsselsteijn W A 2012 *Interacting with Computers* **24** 211–218 ISSN 0953-5438
[51] Gibson J J 1979 *The ecological approach to visual perception* (Houghton-Mifflin) ISBN 1317579380
[52] Eppner C, Deimel R, Alvarez-Ruiz J, Maertens M and Brock O 2015 *The International Journal of Robotics Research* **34** 1021–1038 ISSN 0278-3649
[53] Liu W and Winfield A F T 2010 Autonomous morphogenesis in self-assembling robots using IR-based sensing and local communications *International Conference on Swarm Intelligence* (Springer) pp 107–118
[54] O'Grady R, Christensen A L and Dorigo M 2009 *IEEE Transactions on Robotics* **25** 738–743 ISSN 1552-3098
[55] Rubenstein M, Cornejo A and Nagpal R 2014 *Science* **345** 795–799 ISSN 0036-8075
[56] Mathews N, Christensen A L, O'Grady R, Mondada F and Dorigo M 2017 *Nature Communications* **8** 1–7 ISSN 20411723 URL `http://dx.doi.org/10.1038/s41467-017-00109-2`
[57] Trianni V, Tuci E, Passino K M and Marshall J A 2011 *Swarm Intelligence* **5** 3–18 ISSN 19353812

IOP Conf. Series: Materials Science and Engineering 1261 (2022) 012016 doi:10.1088/1757-899X/1261/1/012016

Embodiment in Dialogue: Daily Dialogue Android Based on Multimodal Information

Takahisa Uchida
Graduate School of Engineering Science, Osaka University, 1-3, Machikaneyamamachi, Toyonaka City, Osaka, Japan
E-mail: uchida.takahisa@irl.sys.es.osaka-u.ac.jp

Takashi Minato
Guardian Robot Project, RIKEN, 2-2-2, Hikaridai, Keihanna Science City, Kyoto, Japan
E-mail: takashi.minato@riken.jp

Hiroshi Ishiguro
Graduate School of Engineering Science, Osaka University, 1-3, Machikaneyamamachi, Toyonaka City, Osaka, Japan
E-mail: ishiguro@irl.sys.es.osaka-u.ac.jp

Abstract. The research and development of robots that can have daily dialogues autonomously with humans have become increasingly important. Most of them, however, have been limited to studies on speech recognition and interfaces. As the result, they still cannot build relationships with humans through long-term daily dialogue. In this background, we have developed a robot that can understand human relationships with their intentions and desires, and can interact with them in human-like ways through various modalities. In this chapter, we first describe an android platform for multimodal interaction based on intentions and desires. Here, both the user's and the robot's subjective preferences and opinions are essential. Then, we introduce a function to model the user's preference, which is necessary for building human relationships. We also discuss the relationship between dialogue and the opinions that can be attributed to the android considering its embodiment. Finally, based on the above studies, we discuss the embodiment in dialogue and future work.

1. Introduction

The research and development of robots that can have daily dialogue autonomously with humans have become increasingly important. Most of them, however, have been limited to studies on speech recognition and interfaces. As the result, they still cannot build relationships with humans through long-term daily dialogue. Speech recognition technology has been advanced recently, and it can now recognize speech as well as a human if its user speaks clearly. Taking the advantage of this capability, smart speakers (e.g., Google Home [1]) have been rapidly used in our lives. Their use is, however, limited to simple question-and-answer dialogue like order, that is, they still cannot realize human-like interaction. A reason for this is that the ambiguity of spoken language makes it difficult to convey commands accurately. For humans, dialogue is the easiest way to communicate. In the near future, they will interact with autonomous systems (e.g., home appliances) in human-like ways. To achieve this, they should have functions to provide appropriate services to humans even if the verbal commands are ambiguous. Additionally, they should communicate with humans not only through utterance but also through various modalities such as facial expressions and gestures. Hence, the knowledge of dialogue robots that are developed in human-robot interaction is necessary.

In this background, we have developed robots that can understand human relationships with their intentions and desires, and can interact with humans in human-like ways through various modalities. In this chapter, we introduce an autonomous daily dialogue android with a human-like appearance discussing the functions essential to interact with humans in more human-like ways. Such robots will become more important in the future. The functions of autonomous systems that support human activities in daily life will be further enhanced. For example, microwave ovens or air conditioners will not merely heat food or cool rooms, but will also appropriately control various factors such as temperature and humidity with the aim of managing human health. Here, their intentions and desires to manage human health are necessary. They will provide services in accordance with these intentions and desires. We believe that they can provide various and sophisticated services by the intention and desire to provide services to the users.

At the same time, to develop a dialogue robot, we should also consider the question "What is daily dialogue?" Humans have a strong desire to share their understanding of themselves, others, and society [2]. Vygotsky argues that one's cognitive activities are associated with others and are given meaning in society through this dialogue [3]. Here, communication plays an important role in this sharing of cognition such as understanding [4]. In particular, linguistic communication is an essential element for transmitting one's understanding [5, 6]. Okada et al. [7] have also argued that the purpose of daily dialogue is to understand others. We, therefore, suppose that daily dialogue is a means to understand others and our relationships with them. Then, why do we need to understand relationships between ourselves and others? Echterhoff et al. [8] have stated that there is a motivation to accurately understand the relations between oneself and the outside world (cognitive motivation). This allows us to adapt to the environment. Humans are social animals. Our ancestors lived in groups with other individuals in order to protect themselves from predators and to ensure that they had better opportunities to obtain food and mates [4]. We can assume, therefore, that people have benefited from belonging to groups and maintaining social bonds (e.g., [9]). Seeking favorable relations with others (i.e., a need to belong) is considered an advantage for adaptation [10, 11]. In addition, isolation and rejection from others cause fundamental anxiety [10]. Understanding human relationships are, therefore, an important and fundamental requirement for humans.

Our purpose is to develop an autonomous dialogue robot that has its own intentions and desires to provide services to humans understanding human relationships. We believe that they can realize a symbiotic relationship with humans, where both have their own intentions and desires and infer each other's ones through dialogue. In this chapter, we first describe an android platform for multimodal interaction based on intentions and desires. Here, both the user's and the robot's subjective preferences and opinions are essential to achieve the purpose. We introduce a function to model the user's preference, which is necessary for building human relationships. We also discuss the relationship between dialogue and the opinions that can be attributed to the android considering its embodiment. Finally, based on the above studies, we discuss the embodiment in dialogue and future work.

2. Multimodal Dialogue Android Based on Intentions and Desires

In this section, we introduce a multimodal dialogue android based on intentions and desires. The research and development of autonomous robots began with Shake [12], an autonomous mobile robot developed at AI Center of SRI International between 1966 and 1972. It was developed to achieve general-purpose artificial intelligence. It can recognize rooms and objects with its onboard camera and execute commands given to it. However, its movements were slow, and it was far from robots of today. The development of Shakey revealed problems to be solved and created various research fields related to robotics and artificial intelligence, such as computer vision, planning, and motion control. The core function of Shakey was a planning function called STRIPS. It generated a chain of observations and actions to execute given commands. In other words, it was a primitive cognitive model of movement. The study of this cognitive model was developed by SOAR [13] and ACT-R [14]. SOAR is a cognitive architecture developed by Laird, Newell, and Rosenbloom based on Newell's Unified Theory of Cognition [15]. It uses procedural knowledge, declarative knowledge, episodic knowledge, and representational knowledge to handle various problems. ACT-R (Adaptive Control of Thought-Rational), proposed by Anderson in 1993, is a cognitive architecture based on the findings of cognitive neuroscience. Its origins go back to a model of human memory [16] proposed by Anderson and Bower in 1973. The most important assumption of ACT-R is that human knowledge consists of declarative memory and procedural memory. Based on this assumption, various modules have been incorporated referring to findings from cognitive neuroscience. The development of these modules is still ongoing [17].

These studies are based on the findings of cognitive science and cognitive neuroscience, and aim to reproduce the functions of the human brain. Similar to SOAR and ACT-R, BDI model is used in artificial intelligence research. It is based on the theory of intentions proposed by the philosopher Bratman [18] in 1987, and is designed to realize agents that act rationally. It consists of beliefs (B), desires (D), and intentions (I), and it is a model in which an agent selects a goal to be achieved from its beliefs and desires and the means (plan) to achieve it, forms and maintains an intention to execute the plan selected as the means, and acts according to the intention. This model has also been studied in robotics [19]. Since it attempts to model human rational thought, it is difficult to implement reflexive behavior without using beliefs and intentions. Brooks' subsumption architecture [20] realizes a robot that is like a creature by implementing reflexive behavior. It consists of multiple layers designed for different purposes, with the upper layers having more abstract purposes. For example, the lowest layer has the goal of avoiding objects, and the higher layer has the goal of roaming. The problem with this architecture is that the achievable actions are mainly generated by a chain of reflexive actions. To achieve rational behavior, we need a way to integrate it with BDI model and so on.

2.1. Intention Desire Architecture

Figure 1 shows the architecture for reproducing human autonomous behavior, which combines BDI model and subsumption architecture [21]. Here, the network of reflexive behavior, which is mainly handled by the subsumption architecture, is at the bottom, and the hierarchy of action/speech, intention, and desire/concept are above it. In each hierarchy, the relations between modules are described in the form of graphs. Though they can be described by rules as in other systems, we adopt the graph representation as the most general one. The reaction(R) and behavior(B) nodes are the parts that interact with the real world, and related behaviors and reactions are selected based on the input from it. Conversely, some of them can affect it. In particular, some behaviors have an ordering property, that is, some behaviors can be chained together. The behavior can be affected by intentions (I), and intentions can by concepts/desires (C). There is a possibility of chaining between the intentions, and also between the concepts (desires). The internal representation of each level memorizes the physical world, intentions, and desires/concepts, respectively. At the lowest action/speech level, the memory of sensory actions is stored through the actions and speech, and the physical world is modeled based on it. At the intention level, the memory of the intentions of self and others is stored as procedural memory. At the desire/concept level, the memory of the concepts of self and others is stored as declarative memory.

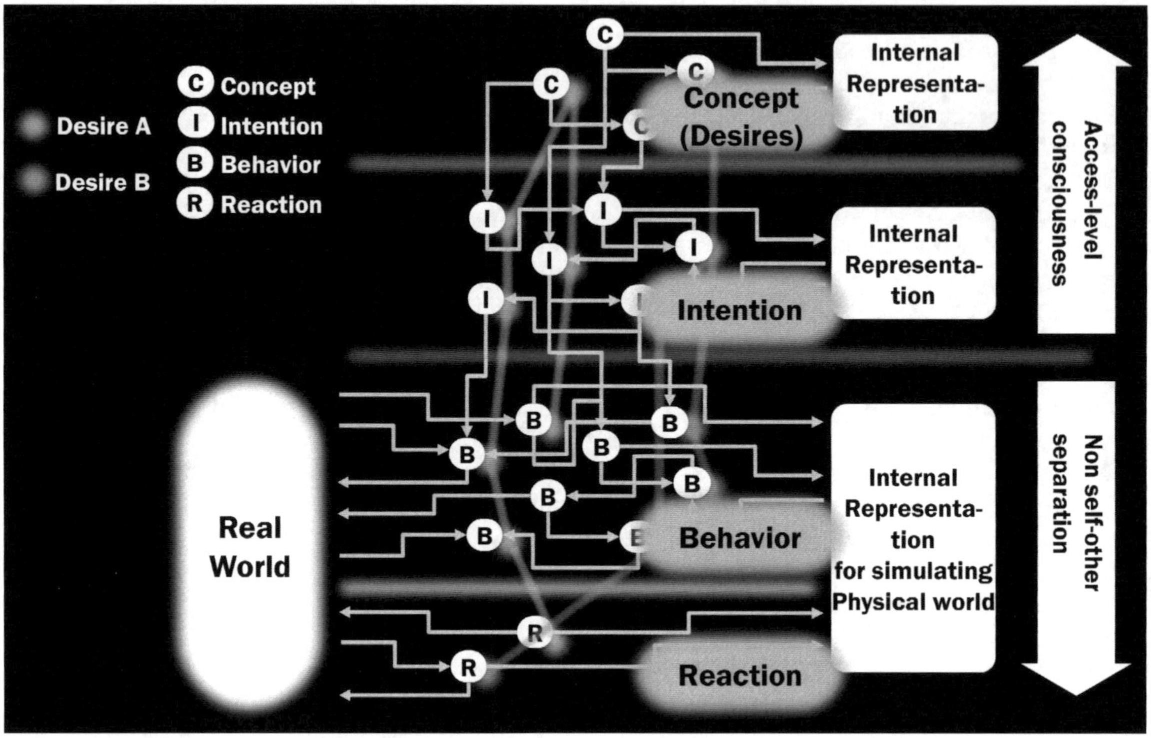

Figure 1. Intention desire architecture

What makes this architecture most different from previous architectures is that desires are expressed both at the conceptual level and at the reflexive behavioral level. Neither human infants nor animals have an explicit understanding of desires. They are embedded in reflexive behavior as instincts. For instance, they eat food when they are hungry, or they cry when their mothers disappear, and these behaviors are expressed without understanding why they do that. As they grow up, they become aware of the desire of appetite or desire of missing their mother on a conceptual level, and they can rationally select behaviors to satisfy their desires. Furthermore, they will be able to understand not only their own desires but also the ones of others at the conceptual level. This dual structure of desires, that is, desires as instincts embedded in reflexive behavior and conceptual-level desires that contribute to rational selection of behavior, can characterize human beings [22]. We suppose that this dual structure of desires is necessary to realize robots with complex functions. On the other hand, this function is quite hard to implement. Alternatively, if the robot is an adult robot, it does not necessarily need to be aware of its own instincts embedded in its reflexive behavior and model them on its own at the conceptual level. In other words, if the robot is an adult, the designer can start the research by giving it the desire at the conceptual level.

2.2. Functions of Autonomous Dialogue System Based on Intention And Desire

Based on the intention desire architecture, we have developed an autonomous dialogue android ERICA [23] (Figure 2), which can interact with a new acquaintance for 10-20 minutes. It sits in the waiting room of a research institute and talks with them in a daily dialogue situation. It was developed as an autonomous dialogue platform, which has a human-like appearance and 44 degrees of freedom in its upper body, and is capable of expressing human-like gaze, lip movements, head movements, facial expressions, and gestures during dialogue. Its voice is generated by VOICE TEXT ERICA [24], a speech synthesis software. As a recognition system, we have developed a sensor network system consisting of multiple depth sensors, microphone arrays, and cameras placed around the android to perform human detection and tracking, speaker recognition, speech separation and recognition for each speaker, and facial expression recognition (Figure 2). In our previous research, we have found that not only human-like expressions of each modality but also synchronous expressions between modalities as humans' are important for multimodal androids. In the developed system, we incorporated a system that automatically generates lip movements [25], head movements [26], facial expressions [27, 28], and gestures [29] to synchronize with the speech of the android.

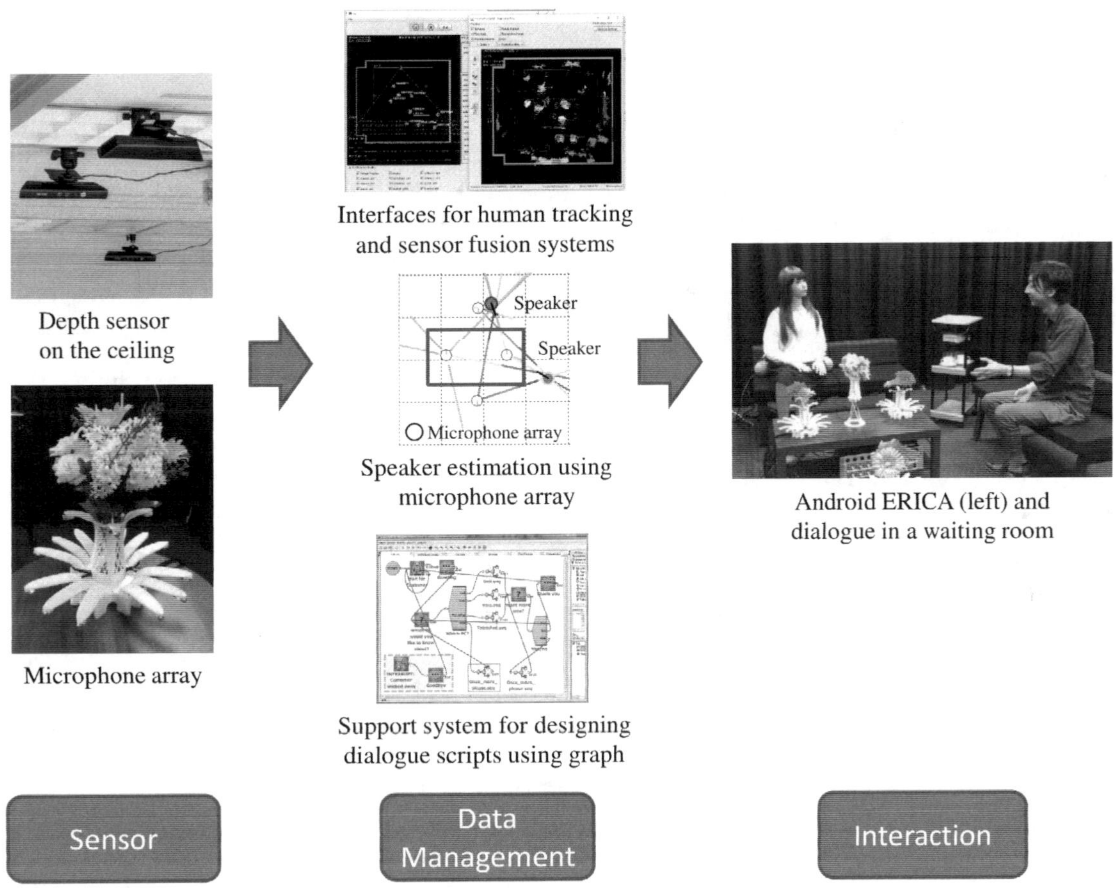

Figure 2. Android ERICA and recognition system

The purpose of daily dialogue is empathy, sharing of mutual experiences, understanding of preferences, and so on. In dialogues in a first meeting, communication partners build relationships through them. To achieve this with autonomous robots, it needs not only abstract and generalized memories (semantic memories) but also subjective memories (episodic memories). In the developed system, a designer prepares scripts consisting of a few to a dozen turns of dialogue (equivalent to Schank's Script [30]) in advance. The android selects topics according to its own internal state (e.g., desires and emotions) and the state of the user. In the script, its utterances are controlled based on keyword-matching recognition. For instance, if it selects a script about "hometown," it asks the user where they are from. After that, it gives impressions of the place if it knows, or asks the user the details of where they are from if it does not know, and so on. The information obtained and the experience of the android changes depending on what the users say. The log of the selected topic becomes an episodic memory of it.

When talking about their experiences, without the appropriate listening behavior of the android, the user cannot feel they shared their experience. This kind of behavior is important in daily dialogue. We have developed a system that can generate appropriate backchannels [31], rephrasing the focal word in the user's utterance, and in-depth questions [32] with machine learning technologies based on databases of human-human dialogue. Error recovery is another important function of a dialogue robot. Dialogue breakdowns, such as overlapping (collision) between the speeches or inappropriate responses caused by misrecognition of the system due to the limitations in natural language processing, discourage the users from interacting with the android. When a speech collision is detected, it takes or gives the turn to speak, depending on the strength of its argument on the current topic. There is also a function for handling dialogue breakdowns related to inappropriate responses [33]. Understanding the preferences and perspective of values of the user is a necessary function for building empathy and relationships in daily dialogue. We also explain the details of this function in the following section.

We implement a dialogue control module of the android based on the intention desire architecture with the functions described above. The dialogue strategy based on this model can be summarized as follows.

- To satisfy individual desires, the android actively talks to the preferable user and selects topics with more self-disclosure.
- To satisfy social desires, it becomes considerate of the user and lets them take the initiative and the criteria for selecting the topic (select a topic that they would like). In addition, to resolve an awkward atmosphere with a new acquaintance, it tries to estimate the relationship by talking to them. It also decides whether to continue or terminate the dialogue, taking into account the likability for the user and relationship.
- It tries to equalize the priority of personal desires and social desires (the degree of priority of self and others) according to reciprocity in the first meeting [34].

The android has a preference for topics as its personality and likability for the user as its mental state. The latter changes according to prepared rules, depending on what the user says and its own preferences. It also estimates the mental state of the user. The user's interest in the android is estimated from the content of their speech, facial expressions, gaze, and so on. Each script has a predefined level of self-disclosure depending on its content (e.g., low self-disclosure for talking about hometown, and high for complaining). In human-human communication, we self-disclose when we like the interlocutor [35]. In the same way, the android selects a script with the level of self-disclosure that corresponds to its likability for the user. If the android is trying to prioritize the user, however, it selects a topic that they seem to like. Whether it actively talks to the user depends on the likability of the user, the user's estimated interest in the android, and the priority of self and the other. The priority of self and the other depends on the frequency with which the android has taken the initiative. For example, the more often it speaks to the user, the higher the priority is given to the user. It also estimates the degree of self-disclosure of the user. Here, we simply assume that if they speak more, they disclose more about themselves. It recognizes the relationship between itself and the user based on degrees of its own self-disclosure and the user's one.

IOP Conf. Series: Materials Science and Engineering 1261 (2022) 012016 doi:10.1088/1757-899X/1261/1/012016

3. Modeling the User's Preference

Modeling (understanding) the user is important for understanding human relationships. In daily dialogue, people often talk about preferences and experiences [36]. Hence, we focus on preferences as a target of user modeling. In this section, we explain a dialogue robot that estimates others' preference models [37]. To improve the user's satisfaction and motivation, we demonstrate a robot's ability to understand a user's opinions. This study focuses on user preferences (likes and dislikes) in a certain category and explores a method that demonstrates the robot's ability to understand the user's preferences.

For estimating preferences, Kobayashi et al. [38] developed a dialogue system that estimates user's preferences in dialogues by recognizing the polarity of user utterances using syntax, surface case, and deep case analysis. Other methods estimate user interests by combining linguistic and non-linguistic information [39, 40]. Although estimating user preferences for the current item or topic is possible, no relation between the preferences for other items, that is, the concept of user's preferences, is estimated. Collaborative filtering [41], which is commonly used to estimate user preferences in recommendation systems, estimates the preferences of a target user based on the information of many other users. However, it does not work effectively when insufficient user information is available (sparsity problem [42]). If it is applied in a dialogue system, many questions must be asked to get user information. This requirement decreases motivation to interact with it. An information provision system [43] can quickly estimate the user's concept in the dialogues. It is enabled by using the concept spaces of multiple people acquired in advance. On the other hand, it focuses on task-oriented dialogues (i.e., information provision), and it simply asks questions one after another about the user's preferences. Too many questions would decrease their motivation to talk with the system. Therefore, the user's information must be collected effectively while encouraging user motivation.

3.1. Preference Modeling System

In this study, modeling the user's preference, we propose a dialogue strategy that shows the robot's comprehension ability and willingness to understand the user's concept of preferences. Here the concept of preferences consists of the user's subjective similarity between items (similarity model) and preferences for them (preference model). When people try to understand another person's opinion through dialogues while constructing a model of their thought (concept) based on the obtained information, they ask questions that are likely to update the model. Therefore, in a scenario that tries to understand user preferences, we first construct a method that updates them and the similarity models from the user information obtained through dialogues. Next with the models, we propose a method that efficiently collects the data of the user's subjective similarity among preferences through dialogues. The preference and similarity model is represented by a table.

3.1.1. Table Complementation Rule We show the table complementation rules as shown in Figure 3. In this figure, 1 means like/similar, 0 does not like/not similar, and the number with <> represents the estimated values that are complemented. Assume that the system has a user's preference model and its own model (the system's preference table is predefined). Since people refer to their own model when estimating their partner's model (i.e., preference/similarity), this system also has its own model. There are four possible complements: from preference table to preference table, from preference table to similarity table, from similarity table to preference table, and from similarity table to similarity table. If the estimated values are affirmed by the data confirmation utterances in the dialogue, the values are treated as observed values. If the complemented data are inconsistent with the obtained data within each rule, no complementation occurs. Each complementation rule is explained below.

IOP Conf. Series: Materials Science and Engineering 1261 (2022) 012016 doi:10.1088/1757-899X/1261/1/012016

- Complementation of preference table from preference table: This rule complements a preference table based on the similarity in preferences. If the system and the user partially share identical preferences in a certain attribute, the preference table is complemented so that both share identical preferences. If the coincidence degree of the observed preference data in a certain attribute is greater than or equal to a threshold, and if the number of known items in the preference data is greater than or equal to a threshold, the same value of the preference data is complemented as the estimated values. This rule represents speculation that if the preferences for some items in a certain attribute match between the system and the user, both will have identical preferences for the other items in the attribute.
- Complementation of similarity table from preference table: This rule complements a similarity table based on the similarity in preferences. If the preferences of two items in a certain attribute match, the rule estimates that those items have similar attributes. Here, it is assumed that those with identical preferences have similarities.
- Complementation of preference table from similarity table: This rule complements a preference table based on the data of a similarity table. If two items are similar for an attribute, the rule estimates that they have the same preferences in the attribute. If the observed similarity between two items of an attribute is similar and if only one item's preference data is missing, the rule complements the preference of this item with the same value of the known preference data. If not similar and if only one item's preference data are missing, it complements the preference of this item with the opposite value of the known preference data.
- Complementation of similarity table from similarity table: This rule complements a similarity table based on the data of a similarity table. If the coincidence degree of the observed similarity data in a certain attribute is greater than or equal to a threshold, and if the number of known items in the similarity data is greater than or equal to a threshold, the same value of similarity data is complemented as the estimated values. This rule represents speculation that the similarity for an attribute will match on some items if the similarity for the attribute matches on the other items.

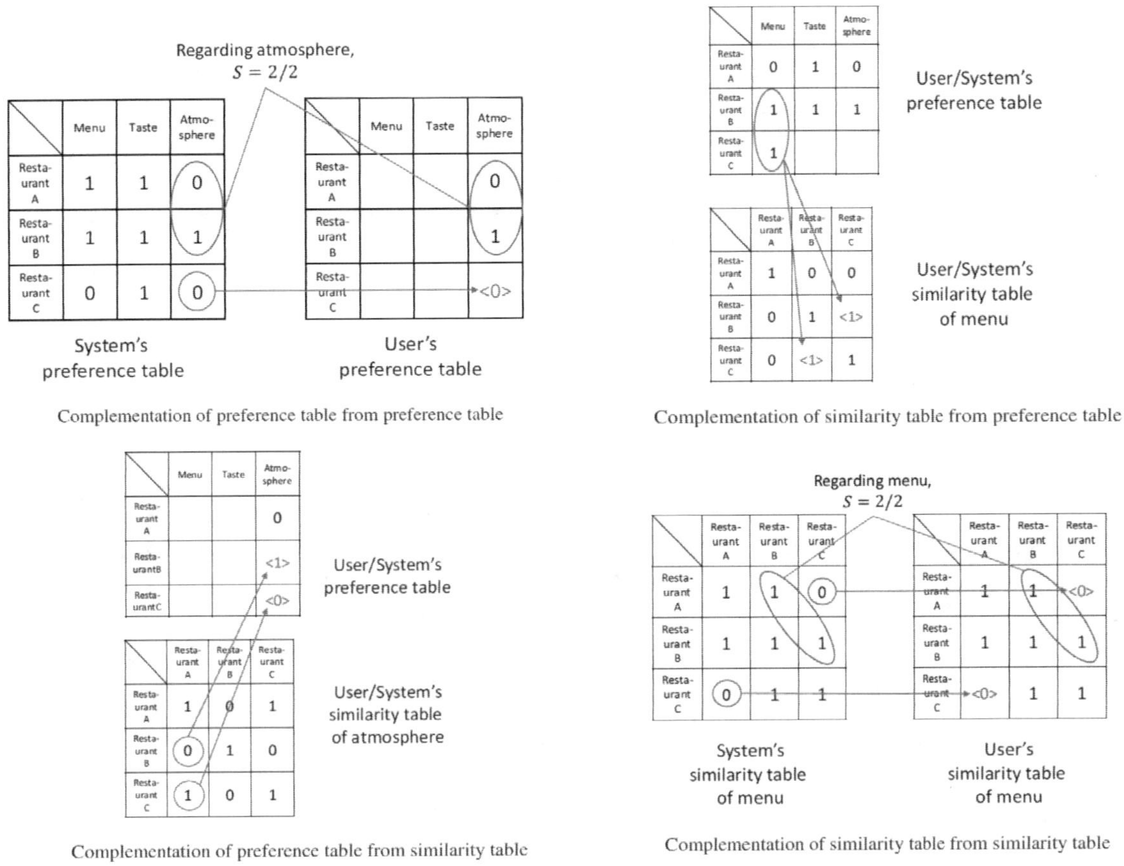

Figure 3. Table complementation rules

3.1.2. Utterance Generation Rule We describe the method to generate dialogue with the user. First of all, the system fixes one topic (e.g., tourist destinations) that the robot and its user would talk about. Within the fixed topic, the system generates utterances below. The system first makes an initial utterance, e.g., "Which Y do you like from the viewpoint of X?," where X is an attribute and Y is a topic. If the user's initial utterance contains an element in the item database, the preference data is stored in the user's preference table. If the initial utterance is not in the item database, the system repeats it for the randomly selected item: e.g., "Do you like Y's X?," where X is an attribute and Y is an item. After that, the system checks whether the complementation rules can be applied to the user's tables. When complementation occurs, select the last complemented data and it confirms whether the complementation of the data is correct (Data confirmation utterance). For example, when the user's preference for A (e.g., the appearance of a golden retriever) is complemented, it says, "You like A, don't you?" If the user agrees, the complemented data are adopted. If they disagree, the complemented data are rejected, and the obtained data are stored in the user's preference table; if the user disagrees with "Do you like A?," then it stores the data that indicate that they do not like A. Data acquisition utterances are performed when complementation rules cannot be applied by selecting one cell that has no preference/similarity data. The data obtained by the data acquisition utterances are stored in the table; the user likes A or thinks A resembles B. After that, it checks the above complementation rules and performs the same processes.

IOP Conf. Series: Materials Science and Engineering 1261 (2022) 012016 doi:10.1088/1757-899X/1261/1/012016

3.1.3. Dialogue Strategy for Effective Data Acquisition Utterance Here, we propose a dialogue strategy for acquiring data efficiently. It is desirable for the system to efficiently complete the preference and similarity tables through dialogues while asking questions that are likely to contribute to model updating more. Therefore, we consider a method that efficiently completes the tables from dialogues with users.

The items whose cells have not been filled with data are candidates for data acquisition utterances. When the system obtains the user's preference for an item or the similarity between two items after the data acquisition utterance, 1 (like or similar) or 0 (unlike or not similar) is stored in the cell. Let N^0 be the number of cells that are complemented by the complementation rules after 0 is assumed to be stored in the target cell, and let N^1 be the number of cells that are complemented by the complementation rules after 1 is assumed to be stored. Expected complementation amount (ECA) E_{ij} for the cell in the ith row of the jth column is calculated by the following formula:

$$E_{ij} = (N^0 + N^1)/2. \tag{1}$$

ECA is the expected amount of complementation after the system asks about the user's preference for the ith item in the jth viewpoint or about the user's thought of similarity between the ith and jth items in the kth viewpoint. ECA is calculated for every cell that is not filled in the preference or similarity tables, and the cell with the maximum ECA is adopted as the reference cell for the data acquisition utterance. By using this method, the system can effectively complete the table and acquire the user's concept of preferences.

3.2. Experiment

We conducted an experiment to verify the effect of the proposed method. Sixteen Japanese people (6 men, 10 women, average age 20.3 (SD 1.72)) participated in two conditions: generating data acquisition utterances based on ECA (ECA condition) and generating them randomly by selecting a cell (control condition). In both conditions, the same complementation rules were applied. We evaluated the amount of data estimated by the system. For the amount of estimated data, we adopted the complementation rate as a scale. It is the ratio of the number of complemented cells in the preference/similarity table to the number of cells in the preference/similarity table (averaged viewpoints mentioned in the dialogue). The complementation rate of the proposed condition (44.3%) is higher than that of the control condition (21.7%). Next, we describe the results of the impression evaluations. We compared the proposed and control conditions for comprehension ability, willingness to understand, dialogue satisfaction, dialogue motivation, and suitability of comments by using a seven-point-scale questionnaire. A Wilcoxon signed-rank-sum test revealed a significant tendency ($p = 0.0708$) in the willingness to understand and a significant difference ($p = 0.0295$) in dialogue satisfaction.

These experimental results suggest that the proposed method efficiently obtains user preference models and enhances user dialogue satisfaction. In addition, the degree of satisfaction is positively correlated with both the android's ability to understand and willingness to understand. On the other hand, the degree of motivation (i.e., long-term willingness to communicate with the android) is positively correlated only with the android's willingness to understand. The experimental results not only showed the effectiveness of the proposed method but also suggest a design theory of dialogue robots to stimulate dialogue motivation.

4. Android's Subjective Opinion and Users' Dialogue Motivation

As explained in section 2 and 3, we have implemented the functions necessary for the android to understand human relationships based on intentions and desires, using various modalities in dialogue. In this section, we discuss the relationship between dialogue and the opinions that can be attributed to the android considering its embodiment.

4.1. Hypothesis

A previous study reported that the exchange of subjective opinions increases and that of objective facts decreases in daily dialogues [44]. Therefore, the android should exchange subjective opinions with the user in daily dialogues. Another previous study has stated that people hardly attribute subjective experiences related value (good or bad) to a robot [45]. Here, even if the robot states that food tastes good, its user is unlikely to attribute the experience of eating the food to it, and is unlikely to find that the robot actually has the opinion. In some cases, therefore, the robot's subjective opinions are not realistic, that is, the user does not feel the robot has its own opinions. We hypothesize that this may decrease the user's motivation to talk with the robot. In this study, we express "the user thinks the robot has a subjective opinion" as "the user attributes subjective opinion to the robot." We also define subjective opinion as "an evaluation made by an individual on a target." For example, if the robot evaluates pasta as tasty, the robot's subjective opinion is "the pasta is tasty." To test the hypothesis, we compare the opinions that can be attributed to the robot with those that cannot, and examine the effect of the attribution of opinions on the robot on the motivation to talk [46].

A previous study has examined the relationship between the rate at which an android (robots that look similar to humans) agrees with the user and their motivation to talk to them [47]. In human-human dialogue, if a person makes too much agreement or too much disagreement with their interaction partner, the partner's motivation to talk would decrease. The study has reported that when the android expresses subjective opinions that are difficult to attribute to it, the user's motivation to talk did not change regardless of the agreement rate. This can suggest that opinions that are difficult to attribute to the robot do not stimulate motivation in any way. However, there can be various interpretations of why the motivation to talk did not change with the agreement rate. Furthermore, in their experiment, there were two separate participants, one who judged whether the opinions were attributed to the android and the other who evaluated the motivation to talk with it. Since whether opinions are attributed to the robot depends on individuals, the study has not fully tested whether the attribution of opinion to robots affected its user's motivation to talk with it. In this study, the same participants (i.e., within-participants design) evaluate whether they can attribute various opinions (e.g., the deliciousness of the food, the quality of computer performance) to the robot and their motivation to talk with it.

4.2. Experiment

In this experiment, we verified the hypothesis that not only the user's interest in the dialogue topic but the user's attribution of subjective opinions to the android affects the motivation to talk with it. Forty-eight Japanese people (21 males and 27 females, average age 28.58, SD 7.92) participated in this experiment. To control the participants' knowledge of the android and to teach them it is a dialogue robot, we showed them an introductory video of it. In this video, it just says, "I am an android." (translated from original Japanese text). We asked them to evaluate whether the android can judge each opinion and whether they want to talk with the android about it. We prepared 100 opinions such as "taste of food," "beauty of paintings," and "difficulty of mathematics." Since the motivation to talk is also affected by their interest in the topic, we also asked them to evaluate how interested they were in the topic. To evaluate these evaluation items (i.e., dialogue motivation, opinion attribution, and interest), we used the following questions.

- Q1(Dialogue motivation): "On each topic, how much do you want to interact with the android you just saw in the video?" (0. I don't think so, 1. I think so a little, 2. I think so, 3. I extremely think so)
- Q2(Opinion attribution): "For each topic, do you think the android you just saw in the video can judge it? E.g.) Regarding "Interestingness of TV programs," please answer "Yes"

IOP Conf. Series: Materials Science and Engineering 1261 (2022) 012016 doi:10.1088/1757-899X/1261/1/012016

if you think the android can judge whether the TV program is interesting or not, and "No" if you don't think it can." (0. I don't think so, 1. I think so)

- Q3(Interest): "How much interest do you have in each topic?" (0. not interested, 1. slightly interested, 2. interested, 3. very interested)

For each topic, participants choose one of four options for motivation to talk and interest, two options for opinion attribution.

We investigated the effects of the opinion attribution to the android and the interest for the dialogue motivation. Regarding the opinion attribution (Q2), we divided the data into two groups: those with attribution group ("1. I think so") and without attribution group ("0. I don't think so"). Similarly, regarding the interest (Q3), we divided the data into two groups: with interest group ("1. slightly interested, 2. interested, 3. very interested") and without interest group ("0. not interested"). Table 1 shows the result of the experiment. A two-way ANOVA revealed an interaction between opinion attribution and interest ($F(1, 4796) = 5.53, p < 0.05$). A simple main effect test revealed that the simple main effect of interest at the without attribution group ($F(1, 4796) = 158.04, p < 0.05$), that of interest at with attribution group ($F(1, 4796) = 252.71, p < 0.05$), that of opinion attribution at the without interest group ($F(1, 4796) = 87.22, p < 0.05$), and that of opinion attribution at with interest group ($F(1, 4796) = 160.39, p < 0.05$) was significant. The results show that the dialogue motivation was significantly higher for those with attribution than for those without attribution regardless of whether the participants were interested or not. The dialogue motivation was significantly higher for those who were more interested than for those who were not interested, regardless of whether they attribute the opinion to the android. In other words, the participants feel more motivation to talk with the android not only in topics that they are interested in but in topics that they think it can have opinions about. While interest in a topic is a self-oriented factor formed by one's past experiences, opinion attribution is another oriented factor formed by the interaction partner (here, the android). From the findings, both self-oriented and other-oriented factors need to be taken into account when designing dialogue androids that stimulate the user's motivation to talk.

Table 1. Score dialogue motivation.

mean (SD)	Without attribution	With attribution
Without interest	0.399 (0.704)	0.836 (0.809)
With interest	0.987 (1.019)	1.580 (1.001)

4.3. Discussion of Opinion Attribution

As factors of opinion attribution, we discuss two aspects: topic and adjective. The experimental results show that topics related to nature, science, and communication/ computer were highly attributed to the android. This can be because the android is a state-of-the-art technology, and many participants felt that it could understand that kind of technological aspect. This may change depending on how robotic and AI technologies will evolve in the future and on the situations that robots interact with people in society. Though the attribution in the topic field of culture and relationship is low, the percentage of attribution for items such as "stylishness of clothes" in culture and "beauty of appearance" in relationships was high. This may be induced by the appearance of the androids in the experiment. To clarify this, it is necessary to evaluate the impressions of various androids and robots and to investigate the influence of the impressions on opinion attribution.

Regarding the aspect of adjectives, a comparison of the percentages of attribution between emotional and attributive adjectives revealed that there were significant differences between the two types of adjectives in some topic fields. In all of those fields, opinions with attributive adjectives were more attributed to the android than those with emotional adjectives. One possible reason is that participants believed that it is more technically feasible for the android to make evaluations with attributive adjectives such as "accuracy" and "difficulty" than with emotional adjectives such as "fun" and "fear". However, some topic fields were excluded from the analysis; they only included either emotional or attributive adjectives. Moreover, not all topic fields show significant differences. Therefore, it is not sufficient to generalize that attributive adjectives are more easily attributed to the android than emotional adjectives. Future work includes the analysis with large sizes of adjectives.

Emotional and attributive adjectives can be classified into more subcategories. Based on a previous study in Japanese [48], emotional adjectives are classified into two categories: emotion (e.g., "happy", $N = 21$) and perception (e.g., "painful", $N = 1$), and attributive adjectives are also classified into four categories: broad attributes (e.g., "same", $N = 3$), object attributes (e.g., "red", $N = 53$), person attributes (e.g., "kind", $N = 3$), and things attributes (e.g., "striking", $N = 19$). We examined the percentages of opinion attribution for each category of adjectives. The percentage of opinion attribution for "painful" belonging to perception adjectives was lowest among the emotional adjectives. One possible reason is that androids do not share the same physical functions as humans, making it particularly difficult to imagine them experiencing pain. Although other perceptions may be equally difficult to imagine, they were not included in this experiment. Therefore, it is not sufficient to generalize that opinions that involve adjectives about perception are difficult to attribute to androids among emotional adjectives. To clarify this point, it is necessary to include adjectives related to various perceptions in future investigations.

In the subcategories of attributive adjectives, the percentage of attributions about person attributes (e.g., "kind") tended to be lower than that of other attributive adjectives. Among the evaluations for attributive adjectives, it may be difficult to imagine that androids evaluate a person's kindness. However, the four kinds of attributive adjectives are not uniformly distributed among the topic fields in the 100 topics prepared in this experiment. For example, the topics that contain attributive adjectives related to people are "kindness of family", "goodness of personality", and "popularity of celebrity", and the target of opinion ("family", "personality", and "celebrity") are all related to people. In other words, it is possible that the influence of the topic fields of opinion is not related to the adjectives, but only to the topic fields of the opinion. Therefore, to confirm whether the above-mentioned tendency exists regardless of the topic field, it is necessary to prepare topics so that the adjective types are uniformly distributed among the topic fields.

5. Embodiment in Dialogue and Future Study

In this chapter, we first described the android platform for multimodal interaction based on intentions and desires. Next, we introduced a function to model the user's subjective preference, which can be a fundamental factor of building human relationships. We also discussed the relationship between the android's subjective opinions and the user's dialogue motivation. Based on them, we discuss embodiment in dialogue.

In this study, we used the android that has a human-like appearance (i.e., it shares a high degree of embodiment with humans) to realize dialogues in human-like ways. For such dialogues, the embodiment that can process and express multimodal information is useful. A previous study has reported that people find robots more fun and trustworthy than CG agents [49]. Another states that robots give the impression of being social, trustworthy, and competent compared to CG agents [50]. For the system with intentions and desires, human-like ones are important to develop a robot that is easy for humans to interact with. While it is possible to implement different ones from humans, the uses may be difficult to understand them behind the robot's behavior. At the same time, we need to discuss whether exactly the same intentions and desires as humans are appropriate for robots. Research on the intentions and desires that robots should have is future work.

When designing a task-specific robot, it becomes easier to set intentions and desires. In other words, the more specific the robot's task, the easier to determine the robot's requirements. In the research of dialogue systems, there have been many studies that focus on specific tasks (i.e., task-oriented dialogue system [51, 52]). This research aims to realize an android that can communicate with humans in human-like ways by the intention-desire architecture that has more abstract goals.

In our preference modeling system (section 3), we have focused only on the user's preferences. To realize a symbiotic relationship between humans and robots, however, robots themselves need to express their own preferences and try to find parts that they can sympathize with their user. The android's embodiment can be useful to emotionally empathize with the user's preferences because bodies are necessary for acquiring preferences. Experiences that lead to the preferences may involve more embodiment. To have a perspective of values, we also need a physical body. Therefore, the androids that shares a high degree of embodiment with humans can be useful for dialogue that share preferences, experiences, and perspectives of values. In other words, the embodied intelligence of androids in dialogue may bias the user's recognition and mind, which makes it easier for them to accept what the android is saying.

In section 4, we investigated the effect of opinion attribution on the dialogue motivation. Since the opinions referred in dialogue stem from people's daily lives, human-like embodiment can be closely related to them. Therefore, the android with human-like appearance and modalities in dialogue may be more capable of attributing opinions than other robots. Experiments using robots other than androids are necessary as a future study. It is also necessary to verify whether the relationship between opinion attribution and dialogue motivation can be generalized to other robots.

The development of the android ERICA revealed many research issues, similar to the research on Shakey described at the beginning of this chapter. Shakey was created by combining various technologies available at the time, such as computers, cameras, and motors. Building it makes clear what it could and could not do, and what kind of research is necessary in the future. This created the various research fields of robotics and artificial intelligence. The development of the android ERICA has also led to the discovery of many research issues as below.

- Dialogue with multiple users: The dialogue system described in this paper is a one-to-one dialogue between the android and a user. We should discuss what functions are necessary for dialogue with multiple users, and how to handle intentions, desires, preferences, and perspectives of values.
- Designing personalities: If an autonomous robot has intentions and desires, users will expect the robot to have its personality. This kind of robot is more likely to establish a close relationship with users. How should we design such a personality?
- Intentions and desires that robots should have: Humans have two major desires, individual preservation and species preservation, but robots may not need them. How should we design the intentions and desires that robots have? What are the reasons for these intentions and desires?
- Desires, emotions, and consciousness: Humans use positive emotional expressions when their desires are satisfied, and negative ones when they are not. In other words, desires and emotions are closely related to each other. Furthermore, people who express desires

and emotions have a strong sense of consciousness. We need to clarify the relation between intention, desire, emotion, and consciousness.

These are just a few examples of new research themes. More research is necessary to realize autonomous dialogue robots that understand human relationships with intentions and desires.

Acknowledgements

This work was supported by JST Moonshot R&D Grant Number JPMJMS2011 (discussion of embodiment in dialogue), JST ERATO Ishiguro Symbiotic Human Robot Interaction Project Grant Number JPMJER1401 (summarizing studies of multimodal dialogue robot), and JSPS KAKENHI Grant Number 19H05693 (summarizing studies of user modeling).

References

[1] Hey Google, `https://assistant.google.com/intl/en_uk/platforms/speakers/` (25/2/2022)
[2] Hardin C D and Higgins E T 1996
[3] Vygotsky L S 1962 *Thought and language (translated by Shibata Y)*
[4] Koji M 2012 *Social Cognition and Affect (Theories and Applications of Cognitive Psychology 6) (in Japanese)* (Kitaoji Shoboh)
[5] Tomasello M 2003 *Language in mind: Advances in the study of language and thought* 47–57
[6] Sanders R E 1988 *Language in Society* **17** 604–609
[7] Okada M, Suzuki N, Ishii K and Altman E 1997 *IPSJ SIG technical reports* vol 1997-SLP-017 pp 39–44 (in Japanese)
[8] Echterhoff G, Higgins E T and Levine J M 2009 *Perspectives on Psychological Science* **4** 496–521
[9] Buss D M 1990 *Motivation and Emotion* **14** 265–286
[10] Baumeister R F and Leary M R 1995 *Psychological bulletin* **117** 497
[11] Smith E R, Murphy J and Coats S 1999 *Journal of personality and social psychology* **77** 94
[12] Shakey, `http://www.ai.sri.com/shakey/` (25/2/2022)
[13] Laird J E, Newell A and Rosenbloom P S 1987 *Artificial intelligence* **33** 1–64
[14] Anderson J R 1993 *Rules of the mind* (Lawrence Erlbaum Associates)
[15] Newell A 1990 *Unified theories of cognition* (Harvard University Press)
[16] Anderson J R and Bower G H 1973
[17] ACT-R, `http://act-r.psy.cmu.edu/` (25/2/2022)
[18] Bratman M 1987
[19] Gottifredi S, Tucat M, Corbatta D, García A J and Simari G R 2008 *XIV Congreso Argentino de Ciencias de la Computación*
[20] Brooks R 1986 *IEEE journal on robotics and automation* **2** 14–23
[21] Ishiguro H, Minato T and Koyama T 2019 *Journal of the Robotics Society of Japan* **37** 312–317 (in Japanese)
[22] Koyama T and Ishiguro H 2017 *The 34th Annual Meeting of the Japanese Cognitive Science Society* pp 380–388 (in Japanese)
[23] Glas D F, Minato T, Ishi C T, Kawahara T and Ishiguro H 2016 *2016 25th IEEE International symposium on robot and human interactive communication (RO-MAN)* (IEEE) pp 22–29
[24] ReadSpeaker, `http://voicetext.jp/` (25/2/2022)
[25] Ishi C T, Liu C, Ishiguro H and Hagita N 2012 *2012 IEEE/RSJ International Conference on Intelligent Robots and Systems* (IEEE) pp 2377–2382
[26] Sakai K, Minato T, Ishi C T and Ishiguro H 2016 *2016 25th IEEE International Symposium on Robot and Human Interactive Communication (RO-MAN)* (IEEE) pp 232–239
[27] Ishi C T, Funayama T, Minato T and Ishiguro H 2016 *2016 IEEE/RSJ International Conference on Intelligent Robots and Systems (IROS)* (IEEE) pp 3327–3332
[28] Ishi C T, Minato T and Ishiguro H 2017 *IEEE Robotics and Automation Letters* **2** 1748–1754
[29] Ishi C T, Machiyashiki D, Mikata R and Ishiguro H 2018 *IEEE Robotics and Automation Letters* **3** 3757–3764
[30] Schank R C 1996 *Tell me a story (translated by Nagao K and Nagao K)* (Hakuyosya)
[31] Kawahara T, Yamaguchi T, Inoue K, Takanashi K and Ward N G 2016 *Interspeech* pp 2890–2894
[32] Ishida M, Inoue K, Nakamura S, Takanashi K and Kawahara T 2018 *SIG-SLUD* **5** 7–12 (in Japanese)
[33] Uchida T, Minato T, Koyama T and Ishiguro H 2019 *Frontiers in Robotics and AI* **6** 29
[34] Ogawa K 2000 *Bulletin of the Graduate School of Education and Human Development. Psychology and human developmental sciences* **47** 173–183 (in Japanese)
[35] Altman I and Taylor D A 1973 *Social penetration: The development of interpersonal relationships.* (Holt, Rinehart & Winston)

[36] Hirano T, Kobayashi N, Higashinaka R, Makino T and Matsuo Y 2016 *Transactions of the Japanese Society for Artificial Intelligence* **31** DSF–B 1–10
[37] Uchida T, Minato T, Nakamura Y, Yoshikawa Y and Ishiguro H 2021 *International Journal of Social Robotics* 1–18
[38] Kobyashi S and Hagiwara M 2016 *Transactions of the Japanese Society for Artificial Intelligence* **31**
[39] Inoue K, Lala D, Takanashi K and Kawahara T 2018 *APSIPA Transactions on Signal and Information Processing* **7**
[40] Yu Z, Ramanarayanan V, Lange P and Suendermann-Oeft D 2019 *Advanced social interaction with agents* (Springer) pp 199–207
[41] Shardanand U and Maes P 1995 *Proceedings of the SIGCHI conference on Human factors in computing systems* pp 210–217
[42] Good N, Schafer J B, Konstan J A, Borchers A, Sarwar B, Herlocker J, Riedl J *et al.* 1999 *Aaai/iaai* **439** 1–8
[43] Sumi K, Sumi Y, Mase K, Nakasuka S and Hori K 2000 *Systems and Computers in Japan* **31** 41–55
[44] Tokuhisa R and Terashima R 2006 *Proceedings of the 7th SIGdial Workshop on Discourse and Dialogue* pp 161–167
[45] Sytsma J and Machery E 2010 *Philosophical studies* **151** 299–327
[46] Uchida T, Minato T and Ishiguro H 2019 *Transactions of the Japanese Society for Artificial Intelligence* **34** B–I62 1–8
[47] Uchida T, Minato T and Ishiguro H 2016 *Proceedings of the Fourth International Conference on Human Agent Interaction* pp 187–192
[48] Nishio T 1972 *A descriptive study of the meaning and uses of Japanese adjectives* vol 44 (in Japanese) (Syuei Publishing)
[49] Kidd C D and Breazeal C 2004 *2004 IEEE/RSJ International Conference on Intelligent Robots and Systems (IROS)(IEEE Cat. No. 04CH37566)* vol 4 (IEEE) pp 3559–3564
[50] Powers A, Kiesler S, Fussell S and Torrey C 2007 *Proceedings of the ACM/IEEE international conference on Human-robot interaction* pp 145–152
[51] Komatani K, Adachi F, Ueno S, Kawahara T and Okuno H G 2003 *Proceedings of the Fourth SIGdial Workshop of Discourse and Dialogue* pp 87–96
[52] Seneff S and Polifroni J 2000 *ANLP-NAACL 2000 Workshop: Conversational Systems*

Benefits and Challenges of Morphological Adaptation in Real-world Robots

Tønnes F. Nygaard[1,2] and Kyrre Glette[2]

[1] Oslo Metropolitan University, Norway
[2] University of Oslo, Norway

E-mail: tonnesfr@oslomet.no

Abstract. For many years, the main target of robotics research has been how to improve control and behavior to allow robots to solve more challenging tasks. Recently, the concept of embodied intelligence has shifted this focus from pure control to a more holistic approach. The mind, body, environment, and the interactions between all these are all highly important for the performance of a robot system. This has previously been investigated in virtual robots in simulation or very simple physical robots in the lab, but the benefits for more advanced and capable real-world robots are still to be determined. We present a case study with several examples of real-world adaptation of morphology, and discuss the benefits and challenges as this direction becomes more important in the field of robotics. Taking an embodied approach promises more adaptive and robust robots, but there are many challenges that need to be addressed before wide-spread adoption in the robotics community.

1. Introduction

Robotics and computer science research has long focused on the brain of a robot, its controller, when it comes to optimization and adaptability. In recent years, however, a somewhat broader and more holistic approach has gained traction, taking inspiration from psychology. The concept of *embodied cognition* states that the brain of an organism is not its sole source of cognition, but that rather the brain, body, the environment it inhibits, and the interaction between these all serve as sources of cognition and problem solving [1]. The body of a robot has mostly been viewed as a design choice an engineer had to choose before building it, but as embodied cognition states, it can also be modified actively by the robot itself for better problem solving during operation.

There are many examples of experiments on morphology adaptation using virtual robots, allowing exploration of a large variety of designs using a physics simulator. The challenge with using virtual robots is that they might act very differently when tested in the real world. The difference in performance and behavior between simplified simulations and the physical world is referred to as the *reality gap*, (or more recently, sim-to-real gap in the reinforcement learning domain), and is considered one of the main challenges of robotics today [2, 3, 4]. There are many approaches to reducing this reality gap, but they have only had moderate success, and that is only considering simple robots, simple tasks, and simple environments. As expectations of what robots can perform, and in which environments they could operate in, rise, so will the reality gap problem. As embodied approaches become more widespread, we believe incremental efforts to reduce the reality gap through better simulators or surrogate modeling will not be enough to accurately produce robots that can exploit the natural noise and richness of interaction with

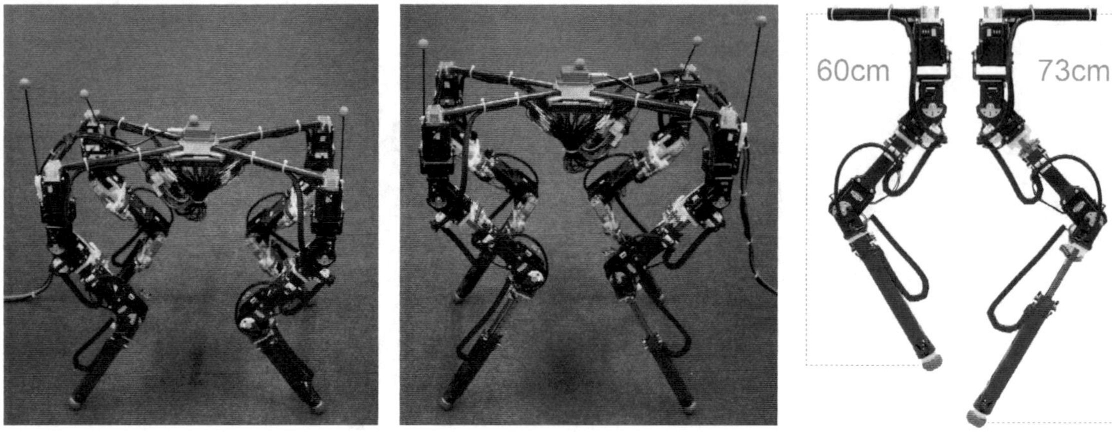

Figure 1: The DyRET robotic platform for exploring dynamic morphology adaptation. These pictures shows the minimum and maximum length of the legs of the robot. The robot stands 13cm taller when fully extended, and each link can be controlled individually.

real world environments. The only approach to completely bypass this issue is to work with the robots exclusively in hardware. Though that does come with a myriad of other challenges [5], we believe this is the right direction for future morphologically adaptive robots.

One of the sub-fields of robotics that has the most experience in embodied approaches is evolutionary robotics. This field takes inspiration from natural evolution to optimize and improve the mind (control) and body (morphology) of robots [6, 7]. A multi-objective approach optimizing both at the same time reduces the chance of premature convergence and allows more finely-tuned solutions to be found. Most work that optimizes the morphology, however, only does so in simulation [2, 3]. There are examples where the evolution of morphology is done in simulation before a select few individuals are built and tested in the real world [8], but this often comes with a large decrease in performance due to the reality gap. There are a few examples of work that combines predefined modules in new ways to build robot bodies in the real world [9, 10]. This allows adapting the bodies of robots to the environment they operate in, but this is still a bit limited due to the simple nature of the robots, relative slow process of building, and need for human intervention.

In this paper, we discuss real-world morphological adaptation in robotics in the context of artificial evolution and learning. Through our case study, we show that the concept is not only feasible with current techniques and technology, but is already at a point where it yields good results on real-world robots. Furthermore, we identify some of the benefits and challenges of a hardware-only approach, and potential ways forward.

2. Our approach

The goal of our project has been to investigate and demonstrate morphological adaptation in the real world on a capable physical robot. To do this, we designed and built DyRET, the DYnamic Robot for Embodied Testing, seen in Figure 1. It can control the length of the two lower links of its legs, the femur and tibia, which allows it to change its morphology during operation. Compared to other approaches that manufacture a population of different robots to test different morphologies, our approach simplifies the process significantly while still allowing a range of different morphologies and controllers to be tested.

Our first attempt at morphological adaptation looked into performing offline adaption to changing internal conditions in the robot. This allowed us to investigate the role of control and

Figure 2: The robot on the four different walking surfaces is was tested on in our work on adapting to the operating environment in the lab [12] to the left. We showed that evolution was able to utilize both control and morphology to adapt to different environments. We also showed the results could generalize to new and unseen environments. The right shows the robot during testing of our online adaptation algorithm in outdoor unstructured terrains [13]. Here, the robot was able to exploit limited training in the lab to quickly and efficiently adapt its body to new and unseen environments.

morphology without having to consider the external environment to a large degree. We applied a co-evolutionary approach of morphology and control to adapt to different battery voltages that would be present during operation, and found that the evolutionary algorithm was able to exploit both the control and morphology in its adaption [11]. The next step was to introduce the environment, and we did a similar evolutionary adaptation to different walking surfaces. There, we observed that the system was able to exploit both morphology and control to adapt to its external environment, and that the results generalized well to previously unseen walking surfaces as well [12]. Even though demonstrating that the results generalize bodes well for its applicability to real-world problems, it was still an offline approach. Even though it works in hardware, there is still a gap in performance and robot behavior between the lab environment and its final operating conditions outside. In our next experiment, we implemented an online approach to learning where minimal training in controlled environments inside helped bootstrap the learning process. We found that this allowed the robot to adapt its morphology while walking in challenging unstructured outdoor environments and continuously improve its model as it walked [13].

Through our use of the DyRET robotic platform, we have seen that morphological adaptation is not only possible to carry out in a practical way, but also leads to highly beneficial adaptations to internal and external changes. It would be natural to think that other systems implementing morphological change have the potential for similar benefits, and maybe even more so with materials and actuators allowing for more flexible changes.

3. Benefits and challenges of morphological adaptation in the real world

Benefits: Taking a hardware-only approach to morphological adaptation comes with several benefits, and we enumerate what we think are the most important ones below:

- **Bypassing the reality gap.** Working solely on real-word robots completely bypasses the reality gap problem, which is considered one of the biggest challenges in evolutionary robotics and robotic reinforcement learning today. This is not only about simplified simulations giving reduced performance in the real world, but also the algorithms exploiting the simulators and focusing on simulations that are far from realistic, often rendering solutions from simulation useless when transferred to reality.

- **Embodied intelligence.** Working with real-world robots allows investigation into and exploitation of the principles of embodied intelligence. The robot can fully utilize the natural noise and richness of the physical world, which can potentially contribute much to its performance and problem solving ability. Our own results from adapting to internal power conditions illustrate this [11]. Other robots that rely on complex dynamics [14], or interactions with the environment [15] would also be hard to exploit fully using simplified simulations.
- **Real-world platform.** Compared to the simple virtual robots typically used in simulation, we have demonstrated that this works on a highly capable robot that is applicable to a wide range of different problems and tasks. This increases the impact of the work, and might help increase the adoption of the techniques to a much higher degree as it starts being used in practical robot systems.

Challenges: Working exclusively on real-world robots also introduces some challenges, which need to be considered before initiating experiments. However identifying these challenges will also serve as a guideline for where to focus research in improving future efficiency of real-world morphology adaptation:

- **Search efficiency.** One of the main issues in the context of evolutionary optimization and other approaches to learning is the limited evaluation budget allowed by the hardware-only platform, as such algorithms typically perform optimally with a large number of evaluations. Even though the robot is able to operate on its own, our experiments have all required some form of human intervention. In addition, the evaluations typically need to be done sequentially, and necessarily in real-time. Compared to simulation experiments that can run faster than real-time, be highly parallelized, and run cheaply in the cloud, this severely limits the degree of adaptability that can be explored in hardware [16].
- **Physical wear and damage.** Another issue is damage to the robot, increasing risk significantly. Having to repair, maintain and initially build the robots also takes considerable effort, regardless of the strategy considered. This includes automated solutions like real-world assembly by an external robot or self-adaptation mechanisms. During our experiments our robot broke in many ways, and the design had to be improved and parts re-manufactured, introducing delays to the experimentation process [5].
- **Design and scale of experiments** The above-mentioned repair and design changes may be a challenge for maintaining consistency across experiments, but there are also other factors challenging the experimental design. During one single experiment, the characteristics of the robot may change, such as motors gradually heating up and leading to reduced performance. If trials are always carried out in the same order, such factors may introduce unwanted and difficult-to-identify effects in the overall results [5]. Limiting the number of individuals and replications in evolutionary experiments also reduces the applicability of the experiments for investigations into the biological principles behind evolution, which is one of the main motivations behind work in the ALIFE field [17, 18].

4. Conclusion and the way forward

In this paper, we have discussed the benefits of actively using the morphology as a part of the problem-solving space for real-world robots, grounded in our concrete application of a quadrupedal robot with dynamic leg lengths. Our previous work leveraged the dynamic morphology both with evolutionary and machine learning approaches, building upon ideas and principles from the field of embodied cognition.

While we have pointed out some challenges of real-world morphology adaptation, such as limited evaluation budgets and risk of damage, we believe that the prospect of rich real-world interaction should outweigh these, and that new materials and algorithms could alleviate existing problems. In particular, approaches which generate internal models from the real-world interactions could be a promising way to guide the real-world exploration and save trials. This would involve making or learning a model of the robot and its environment that could be used for further optimization without relying on the physical robot directly, potentially drastically reducing the number of hardware evaluations needed. Moreover, soft and compliant approaches to materials and actuators could also reduce the mechanical wear and tear of testing potentially taxing solutions.

We hope to inspire further exploration into the role of adaptive morphology for real-world robotic problem-solving and adaptation. Embodied intelligence has gained increased interest in the robotics community in the last few years, but few pursue hardware-based platforms with a dynamic morphology. The development of this concept, both in terms of methods and new materials and actuators, could lead to highly adaptive robots capable of meeting new and unforeseen situations effectively.

References

[1] Wilson A D and Golonka S 2013 Embodied cognition is not what you think it is vol 4 (Frontiers) p 58
[2] Silva F, Duarte M, Correia L, Oliveira S M and Christensen A L 2016 Open issues in evolutionary robotics vol 24 (MIT Press) pp 205–236
[3] Mouret J B and Chatzilygeroudis K 2017 20 years of reality gap: a few thoughts about simulators in evolutionary robotics *Proceedings of the Genetic and Evolutionary Computation Conference Companion* pp 1121–1124
[4] Chebotar Y, Handa A, Makoviychuk V, Macklin M, Issac J, Ratliff N and Fox D 2019 Closing the sim-to-real loop: Adapting simulation randomization with real world experience *2019 International Conference on Robotics and Automation (ICRA)* (IEEE) pp 8973–8979
[5] Nygaard T F, Nordmoen J, Ellefsen K O, Martin C P, Tørresen J and Glette K 2019 Experiences from Real-World Evolution with DyRET: Dynamic Robot for Embodied Testing *Symposium of the Norwegian AI Society* (Springer) pp 58–68
[6] Doncieux S, Mouret J B, Bredeche N and Padois V 2011 Evolutionary robotics: Exploring new horizons *New horizons in evolutionary robotics* (Springer) pp 3–25
[7] Bongard J C 2013 Evolutionary robotics vol 56 (ACM) p 74–83 ISSN 0001-0782
[8] Lipson H and Pollack J B 2000 Automatic design and manufacture of robotic lifeforms vol 406 (Nature Publishing Group)
[9] Jelisavcic M, de Carlo M, Hupkes E, Eustratiadis P, Orlowski J, Haasdijk E, Auerbach J E and Eiben A E 2017 Real-world evolution of robot morphologies: A proof of concept vol 23 pp 206–235
[10] Vujovic V, Rosendo A, Brodbeck L and Iida F 2017 Evolutionary developmental robotics: Improving morphology and control of physical robots (MIT Press)
[11] Nygaard T F, Martin C P, Samuelsen E, Torresen J and Glette K 2018 Real-world evolution adapts robot morphology and control to hardware limitations *Proceedings of the Genetic and Evolutionary Computation Conference* (ACM) ISBN 978-1-4503-5618-3
[12] Nygaard T F, Martin C P, Howard D, Torresen J and Glette K 2021 Environmental Adaptation of Robot Morphology and Control Through Real-world Evolution (MIT Press) ISSN 1063-6560
[13] Nygaard T F, Martin C P, Torresen J, Glette K and Howard D 2021 Real-world embodied AI through a morphologically adaptive quadruped robot (Nature Publishing Group)
[14] Katiyar S A, Gouwanda D, Iida F and Nurzaman S G 2021 Power-efficient adaptive behavior through a shape-changing elastic robot (SAGE Publications Sage UK: London, England)
[15] Howison T, Hughes J and Iida F 2021 Morphological sensitivity and falling behavior of paper v-shapes pp 1–16 ISSN 1064-5462
[16] Nygaard T F, Martin C P, Torresen J and Glette K 2019 Evolving Robots on Easy Mode: Towards a Variable Complexity Controller for Quadrupeds *Applications of Evolutionary Computation* (Springer International Publishing)
[17] Aguilar W, Santamaría-Bonfil G, Froese T and Gershenson C 2014 The past, present, and future of artificial life vol 1 (Frontiers) p 8
[18] Roberts S F, Hirokawa J, Rosenblum H G, Sakhtah H, Gutierrez A A, Porter M E and Long Jr J H 2014 Testing biological hypotheses with embodied robots: adaptations, accidents, and by-products in the evolution of vertebrates vol 1 (Frontiers) p 12

IOP Conf. Series: Materials Science and Engineering 1261 (2022) 012018 doi:10.1088/1757-899X/1261/1/012018

Taking Shape: A Perspective on the Future of Embodied Cognition and a new Generation of Evolutionary Robotics

David Howard[1*], Jack Collins[1,2], and Nicole Robinson[3]

[1]Robotics and Autonomous Systems Group, CSIRO, QCAT, 1 Technology Court Pullenvale 4069 QLD Australia

[2] School of Electrical Engineering and Robotics, Queensland University of Technology, Australia

[3] Department of Electrical and Computer Systems Engineering, Turner Institute for Brain and Mental Health, Monash University, Australia

E-mail: *david.howard@csiro.au

Abstract. The philosophy of Embodied Cognition (EC) presents an intuitive lens with which to consider a variety of problems that attempt to optimise or refine some physically-grounded combination of form and function. Conceptualised as a Venn diagram, with circles of 'body', 'brain', and 'environment', the prevailing notion is that moving to the centre of the diagram is the ultimate goal - intertwining and leveraging all three components to generate adaptive solutions. Philosophically, at least, this makes sense: nature abounds with examples of life that extol the virtues of tightly-coupled embodiment and hint at the possibilities attainable when designing entities through EC principles. The algorithmic basis for following this approach is similarly intuitive: extra degrees of freedom to the design process combined with explicit consideration of the environment allow for a wider range of interesting, useful behaviours. However, moving from philosophy to concrete algorithmic implementation presents a number of pitfalls and barriers that have prevented EC from being more ubiquitously applied as a mainstream problem-solving methodology. In this comment, we speculate on one possible avenue for the future of EC wherein fuller implementations of EC are enabled through adoption of algorithmic advances from the neighbouring field of Machine Learning. Further, we suggest to re-frame evolutionary robotics as a model learning problem, wherein the end goal is to generate an accurate design landscape through the application of high-throughput techniques and tightly coupled digital-experimental systems. Combined, these techniques offer the possibility to reinvent the state of the art and hint at a bright future for evolutionary robotics.

1. Introduction

Embodied Cognition (EC) [1] is a vital and relevant philosophy; a persuasive and powerful lens through which we may solve a plethora of problems. As a framing device, it encourages us to employ systems thinking; to consider the whole rather than the parts, and to encourage beneficial interactions and co-adaptation as the predominant problem-solving mechanism [2]. To date, EC has predominantly been applied in domains where an adaptable body, brain, and environment have direct analogues in the problem space [3], for example in Evolutionary Robotics (ER) [4].

Evolutionary Robotics and EC are strongly coupled. A rich vein of research shows the benefits of this pairing (e.g., [5]), and drives a far-future vision of robots fully employing EC to

IOP Conf. Series: Materials Science and Engineering 1261 (2022) 012018 doi:10.1088/1757-899X/1261/1/012018

autonomously persist, adapt, and thrive in challenging environments [6]. For robotics, EC-plus-ER is the natural, and perhaps only, solution to this challenge.

Despite the aforementioned strong alignment, there is a pervasive feeling that the potential of this pairing is yet to be fully realised. Although philosophically aligned, a full, concrete, and complete mapping of EC principles onto ER is as yet unrealised; held back by algorithmic and fabrication challenges.

We can define three clearly delineated generations of ER (Figure 1). First generation ER was historically done as co-evolution (in simulation, so with no physical instantiation) [7], or traditional robot learning (in reality, but with no morphological adaptation) [8]. Second generation ER brought in 3D printing as a means to generate adaptable bodies, but typically only the final robots were physically realised, partially attaining 'True EC' but being subject to the reality gap as no feedback exists between reality and simulation [9, 10].

We define the third generation ER as a full and complete mapping of EC frameworks into the ER space. Practically, this is enabled through continual physical instantiation with the end goal of maintaining a 'living' dynamic design model that fully captures that complex interrelationships between body, brain, and environment and harnesses them to generate powerful adaptive designs.

For clarity, the body is the robot's morphology; it's sensor and actuator configuration and placement, its constituent materials and their arrangement and properties (coefficients of friction, restitution, Young's modulus, Shore value, and so one). The brain is the controller that takes sensory input and provides actuating output to the body. The environment defines the operating context of the robot, as well some as sensory input channels and modulation of output behaviours. Note that each of these components may be represented physically or virtually, i.e., in simulation. In fact, the third generation of ER relies on the interchangeable nature of the two methods of instantiation.

In this comment we provide two concrete steps to enable this 'third generation' of evolutionary robotics.

- First, we highlight several rapidly-advancing sub-fields of Machine Learning that enable a more direct and complete mapping of EC principles into ER algorithms.
- Second, we suggest that adoption of high-throughput techniques originally pioneered in materials science [11] can generate powerful, expressive, and accurate models of robotic design spaces [12, 13], including material, morphology, and control. Such techniques allowing for scalable, expansive search as bulk of the design work to be done digitally.

It is our contention that the end goal of ER is to fully implement EC. This means body, brain, and environment are all adaptive and interact to contribute to the generation of natural behaviours. Through the steps delineated above, we look towards a bright future for evolutionary robotics where strong EC is consistently and fully applied as a codified design methodology, and where strong EC can start to look outwards to application in other high-impact fields.

2. Embodied Cognition and the future of Evolutionary robotics

The intertwining and co-adaptation of body and brain to a dynamic environment is a compelling, logical solution to the problem of designing robots for a plethora of currently-unattainable tasks and deployment scenarios (Figure 1).

However, significant issues arise during the algorithmic implementation of the EC philosophy[1]. Computational co-evolution for robotics has well-known issues with dimensionality and tractability [14]. Primarily this is because the design space of two varied adaptive elements is huge, and the resulting behaviour depends on strong coupling, which negates the application

[1] This is not something that natural EC systems need to worry about.

IOP Conf. Series: Materials Science and Engineering 1261 (2022) 012018 doi:10.1088/1757-899X/1261/1/012018

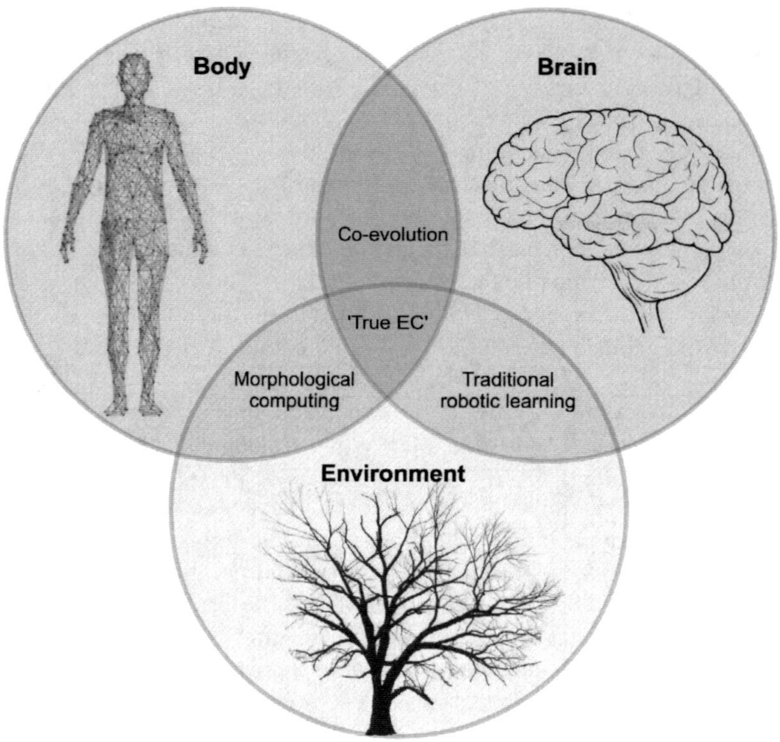

Figure 1. Delineation of the body/brain/environment triad central to Embodied Cognition, with well-established research fields at the intersections. First generation evolutionary robotics was historically done as co-evolution (in simulation), or traditional robot learning (in reality). Second generation ER used 3D printing to generate bespoke bodies, but typically only the final robots were printed, partially attaining 'True EC'. Our third generation ER, postulated here, involves continual physical instantiation via high-throughput techniques to update a 'living' dynamic design model, which is in turn used to generate environmentally fit robots.

of reductionist approaches as seen in other fields. Note that this is just for co-evolution, and doesn't cover the ability we have to computationally adapt an environment to provide beneficial learning experiences – a 'true EC' implementation should arguably consider adaptive/dynamic environments as well as bodies and brains, further reducing algorithmic tractability.

This coupling means that robot behaviour is often brittle and difficult to optimise. Change the body even a little and the brain no longer works as intended, change the brain a little and completely different control may result. Additionally, indirect representations, or those based on neural networks, can be sensitive to small changes in the genome, so even an apparently conservative mutation can entail a huge impact on behaviour which makes local optimisation tricky. Add in an environment which also requires some degree of adaptation, and the upshot is that EC is often viewed as a problem to be solved, rather than a problem-solving framework. Despite promoting synergy and coupling, what we frequently see is a *trade-off* between brain and body development [15], rather than the promised land of a co-evolutionary free lunch [16].[2]

[2] Interestingly, the concept of promoting adaptive solutions over brittle ones is the underlying concept seen in

2.1. Step 1: Map Embodied Cognition more fully onto Evolutionary Robotics through Cutting-edge Machine Learning Algorithms

All is not lost however, and the tools and techniques required to fully map EC to evolutionary robotics are starting to become available. This is mostly because EC and ER are tightly linked to Reinforcement Learning (RL), which has recently benefitted from a notable increase in popularity and research activity [18]. RL also benefits from advances in Machine learning and deep learning more generally. Problems that arise in RL are similar to those experienced by ER and EC, and so solutions are (at least partially) transferrable. Adopting contemporary techniques from ML can drive the development of 'full EC' systems; breaking down computational, accuracy, and scalability barriers to allow a more thorough and complete application of EC principles to evolving robotic systems.

Modern ML can more fully implement EC to promote three key goals of ER, namely; (i) we want to evolve in reasonable time, (ii) we want to evolve capable behaviours suitable for real-world problems, and (iii) we want our eventual robots to be real rather than simulated, so they can be used to solve real problems. Here are some practical steps for achieving these goals through adaption of modern ML:

Evolving in reasonable time implies the use of faster-than-realtime, parallelisable simulation or modelling. Differentiable simulation [19] is a recently-popular field that reduces computational complexity by turning (part of) the robot design into a gradient-based optimisation, and is applicable to soft robotics [20], which is a high-value target domain for EC. Differentiable simulation additionally offers a computationally frugal approach to system identification, allowing us to add accurate sensor and actuator models to simulators and cross the reality gap (the non-differentiable version of learning-based system ID is known to cross the reality gap [21]). EC+ER would be much faster if each generation's controllers didn't have to be learned from scratch, which has previously been described in terms of Lamarkian inheritance [22] but can also be thought of as transfer learning [23], which is a much larger, more contemporary, and more active area of research.

As well as evolving in reasonable time, we also need to **evolve sufficiently complex behaviours**. EC frames the environment as an adaptive tool to drive learning. Curricula provide a straightforward implementation of this, allowing increasingly complex behaviours to be incrementally learned through environmental modulation, promoting interesting interactions that scale with environmental complexity [24] . Curricula naturally couple with procedural content generation, which has recently been shown to increase (learned) behavioural generality [25]. To expand on this concept, multi-task learning [26] of curriculum-learned behaviours offers a route towards rich, flexible behavioural repertoires [27]. As a bio-inspired design paradigm, EC will also benefit from bio-inspired learning paradigms including self-supervision [28], attention, and artificial curiosity [29], to create necessarily complex behaviours through targetted environmental interactions. Algorithms may be tailored to strongly resemble biological form and function [30], opening up new design spaces in a principled way to encourage the discovery of useful behaviours.

Now that we have our behaviours we need to make them **work in reality**. The reality gap is an old foe for EC, which relies on high-fidelity interactions between robot and environment. Interestingly, both differentiable simulation and curriculum learning are beneficial for gap-crossing. The future of EC for ER relies on injection of real-world data as a means of crossing the gap. System-identified sensor and actuator models have been shown to be critical to crossing the gap, can be learned by simple neural networks, and can be readily deployed in most simulation environments [31] with minimal engineering effort [21]. Aside from tuning just the elements that are hardest to simulate, we can tune parameters of the underlying physics engine to experimental

the field of 'Resilience Engineering'; see e.g., [17], for a holistic overview of this discipline.

observation through Domain Adaptation. Taken together, continuing developments in these recent ML methods will allow ER to more easily realise 'True EC' systems, through a reduction in computational complexity, simplification of the co-design problem, or provision of an adaptive environment to drive learning and reduce reality gap effects.

2.2. Step 2: Get on the High-Throughput Highway

The High Throughput Highway [11] is a computational-experimental framework for accelerating materials discovery. High throughput in this case means that data is produced in such quantity that algorithmic interventions are required to analyse it effectively. Automated synthesis and characterisation facilities[3] generate huge amounts of experimental data and provide physcially-grounded veracity to predictive materials models. The resulting 'megalibraries' [32] are subsequently explored to find promising new candidate materials. High throughput experimentation has revolutionised materials discovery, generating novel battery materials, insulin, and catalysts [33], amongst others. A typical instantiation of a high throughput system includes:

- An *experimental characterisation facility* - e.g., a Cartesian gantry robot suitably equipped with a suite of sensors.
- Arrayed underneath the robot are *a variety of samples* (of materials/chemicals/catalysts...) - the gantry robot proceeds sequentially across the array, gathering sensor information that characterises each sample.
- The information is used to seed *a predictive model* and tie it closely to experimental data. The model can be used to (i) find interesting materials, and (ii) suggest experiments (new arrays of materials) that maximally improve the quality of the model.

The Materials Genome Initiative [34] is a closely related international collaborative effort to accelerate the pace of new materials discovery, with the aim of mapping entire *genomes* — complete performance spaces of a particular class of material. This is a daunting challenge; material properties depend on a huge amount of factors including a plethora of synthesis conditions [4], and the final product may vary according to time and storage conditions. Moreover, the discoverable materials space is practically infinite.

If these issues of developing effective digital genomes, and designing in huge and highly complex search spaces sound familiar, it's because they are identical to those discussed above for ER. Based on successes reported in the MGI, we contend that the future of evolutionary robotics involves creating a 'a high throughput highway' for soft robotics, a concept first introduced in [12]. This would work for rigid robots as well, however soft robots are a highly relevant, possess powerful embodiments, and are an unsolved design challenge, and so we use them as our case study.

The central idea to such an approach relies on creation of a 'living model' that is strongly coupled to the physical experiment [12, 13]: 3D printers produce many soft robots, generating sufficient data points through physical testing [35] to learn a predictive model of the design space that is strongly coupled to reality. As the model improves, printing becomes less and less necessary, and the bulk of the design work can be transferred to the (now accurate) cheap, parallelisable model. Scale can then be brought to bear to explore challenging soft robotic design spaces. Note this is philosophically different from the 'Evolution of Things' [36] paradigm, as the end goal is model-based design, rather than performing all evolution in reality.

Fabrication is still expensive in person-hours and money, which can be ameliorated by adopting Design of Experiments (DoE) techniques, such that the goal of the an ER experiment

[3] e.g., CSIRO's Rapid Automated Materials and Processing (RAMP) facility
[4] that we in ER might call 'environmental'

becomes the generation of maximally accurate design models in as few physical tests as possible, rather than simply optimising a robot. To close the loop in fabrication and characterisation, the DoE-optimised experimental plans are carried out without human intervention using autonomous robotic scientists [37]; a natural progression of physical testing apparatus already used in ER [38, 35, 39]. This creates a fully autonomous robot designing system.

3. Discussion

This article discussed a potential future for EC and ER, founded on the growth and exploitation of natural links between EC, ER, and RL to solve algorithmic issues that have — to date — prevented EC frameworks from fully mapping onto ER problems.

We proposed two concrete steps to usher in the 3rd generation of ER, one where EC principles can be readily and fully applied as a problem-solving methodology. First, we encourage researchers investigate transferrable techniques from ML that can be deployed effectively in ER to implement EC. Second, we propose to reframe the evolutionary robotics problem in terms of high-throughput experimentation, using techniques from that field to create reality-coupled accurate design models to catalyse the scalable, principled exploration of robotic design spaces.

Aside from the technological leaps enabled by the MGI, another, arguably more important outcome was the development of effective, large-scale international collaborative experimentation and data sharing, together with significant investment of the prerequisite infrastructure to collectively tackle such a large problem. Our future vision must encompass both technological and collaborative advances, as well as the training required to work in this frontier, multi-disciplinary space [40]. Consider for example the powerful notion of running high-throughput soft robotics experimentation simultaneously across multiple international collaborators, each contributing their own infrastructure, expertise, models, and robotic libraries to collectively solve a problem.

Overall, robotics researchers have been running up against familiar challenges for the past 30 years. A future confluence of heightened collaboration, advanced evolutionary algorithms, cross-pollination with RL, and adoption of high throughput techniques as the de facto design tool for evolutionary roboticists heralds a bright future for EC and ER.

And after we've solved ER? The underpinning mechanism of co-adaptation and research into the interaction mechanisms rather than the co-adapting parts is compelling enough to be applied further afield (so long as the parts retain their respective meanings), and it can be convincingly argued that a wide range of different domains could benefit from application of strong embodiment frameworks, depending on how one chooses to define a body or a brain, etc. As our ability to construct powerful demonstrator EC systems in reality increases, we are likely to see the strengths of EC spread to other fields of research. Our future is likely to see application of strong EC to Human Robot Teaming (HRT), an important field that increasingly reflects the optimal deployment scenario for most robotics systems. Adopting an EC standpoint may lead to systems that more tightly couple the human and robot agents to each other, to their tasks, and to their environments. In doing so, we may see the more widespread adoption of EC frameworks across a new range of problems, and correspondingly novel solutions developed. Together with further development into concrete computational models of embodiment phenomena [41], increasing success in applying strong EC principles in evolutionary robotics will encourage other fields to think about how they could benefit from thinking *beyond the brain* [42].

References

[1] S. Gallagher, *How the body shapes the mind.* Clarendon Press, 2006.
[2] R. Pfeifer and J. Bongard, *How the body shapes the way we think: a new view of intelligence.* MIT press, 2006.
[3] T. F. Nygaard, C. P. Martin, J. Torresen, K. Glette, and D. Howard, "Real-world embodied ai through a morphologically adaptive quadruped robot," *Nature Machine Intelligence*, vol. 3, no. 5, pp. 410–419, 2021.

[4] A. E. Eiben, "Grand challenges for evolutionary robotics," *Frontiers in Robotics and AI*, vol. 1, p. 4, 2014.
[5] J. Bongard, "Morphological change in machines accelerates the evolution of robust behavior," *Proceedings of the National Academy of Sciences*, vol. 108, no. 4, pp. 1234–1239, 2011.
[6] G. Nitschke and D. Howard, "Autofac: The perpetual robot machine," *IEEE Transactions on Artificial Intelligence*, 2021.
[7] K. Sims, "Evolving 3d morphology and behavior by competition," *Artificial life*, vol. 1, no. 4, pp. 353–372, 1994.
[8] S. Nolfi and D. Floreano, *Evolutionary robotics: The biology, intelligence, and technology of self-organizing machines.* MIT press, 2000.
[9] H. Lipson and J. B. Pollack, "Automatic design and manufacture of robotic lifeforms," *Nature*, vol. 406, no. 6799, pp. 974–978, 2000.
[10] J. Auerbach, D. Aydin, A. Maesani, P. Kornatowski, T. Cieslewski, G. Heitz, P. Fernando, I. Loshchilov, L. Daler, and D. Floreano, "Robogen: Robot generation through artificial evolution," in *ALIFE 14: The Fourteenth International Conference on the Synthesis and Simulation of Living Systems.* MIT Press, 2014, pp. 136–137.
[11] S. Curtarolo, G. L. Hart, M. B. Nardelli, N. Mingo, S. Sanvito, and O. Levy, "The high-throughput highway to computational materials design," *Nature materials*, vol. 12, no. 3, pp. 191–201, 2013.
[12] D. Howard, A. E. Eiben, D. F. Kennedy, J.-B. Mouret, P. Valencia, and D. Winkler, "Evolving embodied intelligence from materials to machines," *Nature Machine Intelligence*, vol. 1, no. 1, pp. 12–19, 2019.
[13] T. Howison, S. Hauser, J. Hughes, and F. Iida, "Reality-assisted evolution of soft robots through large-scale physical experimentation: a review," *Artificial Life*, vol. 26, no. 4, pp. 484–506, 2020.
[14] S. Doncieux, N. Bredeche, J.-B. Mouret, and A. E. G. Eiben, "Evolutionary robotics: what, why, and where to," *Frontiers in Robotics and AI*, vol. 2, p. 4, 2015.
[15] A. Rosendo, M. Von Atzigen, and F. Iida, "The trade-off between morphology and control in the co-optimized design of robots," *PloS one*, vol. 12, no. 10, p. e0186107, 2017.
[16] D. H. Wolpert and W. G. Macready, "No free lunch theorems for optimization," *IEEE transactions on evolutionary computation*, vol. 1, no. 1, pp. 67–82, 1997.
[17] E. Hollnagel, D. D. Woods, and N. Leveson, *Resilience engineering: Concepts and precepts.* Ashgate Publishing, Ltd., 2006.
[18] R. S. Sutton and A. G. Barto, *Reinforcement learning: An introduction.* MIT press, 2018.
[19] T. Du, J. Hughes, S. Wah, W. Matusik, and D. Rus, "Underwater Soft Robot Modeling and Control With Differentiable Simulation," *IEEE Robotics and Automation Letters*, vol. 6, no. 3, pp. 4994–5001, 2021.
[20] Y. Hu, J. Liu, A. Spielberg, J. B. Tenenbaum, W. T. Freeman, J. Wu, D. Rus, and W. Matusik, "ChainQueen: A Real-Time Differentiable Physical Simulator for Soft Robotics," in *2019 International Conference on Robotics and Automation (ICRA)*, 2019, pp. 6265–6271.
[21] J. Hwangbo, J. Lee, A. Dosovitskiy, D. Bellicoso, V. Tsounis, V. Koltun, and M. Hutter, "Learning agile and dynamic motor skills for legged robots," *Science Robotics*, vol. 4, no. 26, 2019.
[22] M. Jelisavcic, R. Kiesel, K. Glette, E. Haasdijk, and A. Eiben, "Analysis of lamarckian evolution in morphologically evolving robots," in *Artificial Life Conference Proceedings 14.* MIT Press, 2017, pp. 214–221.
[23] S. J. Pan and Q. Yang, "A survey on transfer learning," *IEEE Transactions on knowledge and data engineering*, vol. 22, no. 10, pp. 1345–1359, 2009.
[24] N. Heess, D. TB, S. Sriram, J. Lemmon, J. Merel, G. Wayne, Y. Tassa, T. Erez, Z. Wang, S. M. A. Eslami, M. Riedmiller, and D. Silver, "Emergence of locomotion behaviours in rich environments," 2017.
[25] S. Risi and J. Togelius, "Increasing generality in machine learning through procedural content generation," *Nature Machine Intelligence*, vol. 2, no. 8, pp. 428–436, 2020.
[26] Y. Zhang and Q. Yang, "A Survey on Multi-Task Learning," *IEEE Transactions on Knowledge and Data Engineering*, 2021.
[27] X. Zheng, A. K. Qin, M. Gong, and D. Zhou, "Self-regulated evolutionary multitask optimization," *IEEE Transactions on Evolutionary Computation*, vol. 24, no. 1, pp. 16–28, 2020.
[28] D. Hendrycks, M. Mazeika, S. Kadavath, and D. Song, "Using Self-supervised Learning can Improve Model Robustness and Uncertainty," *arXiv preprint arXiv:1906.12340*, 2019.
[29] D. Pathak, P. Agrawal, A. Efros, and T. Darrell, "Curiosity-driven exploration by self-supervised prediction," in *Proceedings of the 34th International Conference on Machine Learning*, 2017, p. 2778–2787.
[30] N. Hockings and D. Howard, "New biological morphogenetic methods for evolutionary design of robot bodies," *Frontiers in Bioengineering and Biotechnology*, vol. 8, p. 621, 2020.
[31] J. Collins, S. Chand, A. Vanderkop, and D. Howard, "A review of physics simulators for robotic applications," *IEEE Access*, vol. 9, pp. 51 416–51 431, 2021.
[32] E. J. Kluender, J. L. Hedrick, K. A. Brown, R. Rao, B. Meckes, J. S. Du, L. M. Moreau,

B. Maruyama, and C. A. Mirkin, "Catalyst discovery through megalibraries of nanomaterials," *Proceedings of the National Academy of Sciences*, vol. 116, no. 1, pp. 40–45, 2019. [Online]. Available: https://www.pnas.org/content/116/1/40

[33] J. Hattrick-Simpers, C. Wen, and J. Lauterbach, "The materials super highway: integrating high-throughput experimentation into mapping the catalysis materials genome," *Catalysis Letters*, vol. 145, no. 1, pp. 290–298, 2015.

[34] J. J. de Pablo, N. E. Jackson, M. A. Webb, L.-Q. Chen, J. E. Moore, D. Morgan, R. Jacobs, T. Pollock, D. G. Schlom, E. S. Toberer *et al.*, "New frontiers for the materials genome initiative," *npj Computational Materials*, vol. 5, no. 1, pp. 1–23, 2019.

[35] H. Heijnen, D. Howard, and N. Kottege, "A testbed that evolves hexapod controllers in hardware," in *2017 IEEE International Conference on Robotics and Automation (ICRA)*. IEEE, 2017, pp. 1065–1071.

[36] A. E. Eiben and J. Smith, "From evolutionary computation to the evolution of things," *Nature*, vol. 521, no. 7553, pp. 476–482, 2015.

[37] R. D. King, K. E. Whelan, F. M. Jones, P. G. Reiser, C. H. Bryant, S. H. Muggleton, D. B. Kell, and S. G. Oliver, "Functional genomic hypothesis generation and experimentation by a robot scientist," *Nature*, vol. 427, no. 6971, pp. 247–252, 2004.

[38] V. Vujovic, A. Rosendo, L. Brodbeck, and F. Iida, "Evolutionary Developmental Robotics: Improving Morphology and Control of Physical Robots," *Artificial Life*, vol. 23, no. 2, pp. 169–185, 05 2017. [Online]. Available: https://doi.org/10.1162/ARTL_a_00228

[39] D. Howard, "A platform that directly evolves multirotor controllers," *IEEE Transactions on Evolutionary Computation*, vol. 21, no. 6, pp. 943–955, 2017.

[40] A. Miriyev and M. Kovač, "Skills for physical artificial intelligence," *Nature Machine Intelligence*, vol. 2, no. 11, pp. 658–660, 2020.

[41] G. Pezzulo, L. W. Barsalou, A. Cangelosi, M. H. Fischer, M. Spivey, and K. McRae, "The mechanics of embodiment: A dialog on embodiment and computational modeling," *Frontiers in psychology*, vol. 2, p. 5, 2011.

[42] L. Barrett, *Beyond the brain.* Princeton University Press, 2011.

IOP Conf. Series: Materials Science and Engineering 1261 (2022) 012019 doi:10.1088/1757-899X/1261/1/012019

Meta-brain Models: biologically-inspired cognitive agents

B Alicea[1,2] and J Parent[1]

[1] Orthogonal Research and Education Laboratory, Champaign-Urbana, IL

[2] OpenWorm Foundation, Boston, MA.

E-mail: bradly.alicea@outlook.com

Abstract. Artificial Intelligence (AI) systems based solely on neural networks or symbolic computation present a representational complexity challenge. While minimal representations can produce behavioral outputs like locomotion or simple decision-making, more elaborate internal representations might offer a richer variety of behaviors. We propose that these issues can be addressed with a computational approach we call meta-brain models. Meta-brain models are embodied hybrid models that include layered components featuring varying degrees of representational complexity. We will propose combinations of layers composed using specialized types of models. Rather than using a generic black box approach to unify each component, this relationship mimics systems like the neocortical-thalamic system relationship of the mammalian brain, which utilizes both feedforward and feedback connectivity to facilitate functional communication. Importantly, the relationship between layers can be made anatomically explicit. This allows for structural specificity that can be incorporated into the model's function in interesting ways. We will propose several types of layers that might be functionally integrated into agents that perform unique types of tasks, from agents that simultaneously perform morphogenesis and perception, to agents that undergo morphogenesis and the acquisition of conceptual representations simultaneously. Our approach to meta-brain models involves creating models with different degrees of representational complexity, creating a layered meta-architecture that mimics the structural and functional heterogeneity of biological brains, and an input/output methodology flexible enough to accommodate cognitive functions, social interactions, and adaptive behaviors more generally. We will conclude by proposing next steps in the development of this flexible and open-source approach.

1. Introduction

The history of Artificial Intelligence (AI) has been marked by the rise of dominant models that have been hyped as being sufficient for simulating general intelligence. The hype surrounding such models has subsequently led to the exposure of flaws in both symbolic and connectionist models, and perhaps even the idea of general intelligence itself. Perhaps a few sources of inspiration can be found through embodied behavior, biological inspiration, and the strategic application of knowledge representation. The goal of replicating the rich behavioral repertoires and flexible internal models found in biological brains has consequences for constructing more realistic and informative models of intelligence. While it is obvious that biological nervous systems must be both heterogeneous and embodied, the correct combination of these attributes is ripe for innovation. In terms of emulation of biological intelligence in a computational system, this suggests that both subsymbolic (similar to what we will refer to as representation-free) and symbolic (similar to what we will refer to as representation-rich) are crucial components. This is where meta-brain models come into play: meta-brain models enable neuro symbolic models [1] to be expressed, refined, and extended in the form of embodied hybrid models.

When we speak of representation, it refers to the interpretation of information in a manner that is decoupled from direct perception. A representation results in an internal mental state. An agent's internal state should reflect distinctive psychological functions located in different parts of a neural representation [2]. This should resemble the functional segregation of a biological brain, which is not a completely separable phenomenon [3]. The different layers of a meta-brain model thus consist of several interacting internal states that must be integrated. Representation can also be defined as the ability of one state to be transformed into something else [4].

While this has interesting applications with respect to lower degrees of representation, richer levels of representation allow for a variety of functional attributes such as causality, intentionality, and object properties [5]. As with biological brains, the meta-brain is not aware of how things in the world work, only that some attempts at solving problems work better than others [6]. We must also be careful to not ascribe too much intentionality to different levels of representation. The role of even the richest representation is not to provide conscious awareness or purpose, but to provide a joint probability map of stimulus and response [7].

Our definition of representation is flexible, but can be defined in two ways. The first type of representation is a top-down representation, an example of which involves models organized around classificatory structures or symbolic frameworks. Specific models range from labeled input data to grounded symbols. This can be contrasted with a bottom-up representation, which involves topological structure that provides information about the sensory world or environment. Bottom-up systems range from cells that encode place/sensory fields to agent-based models with cells representing units of analysis of varying complexity.

1.1 Taking an embodied and developmental perspective

Meta-brains draw from several diverse areas of inquiry such as AI, Neuroscience, Cognitive Science, Developmental Science, Cybernetics, and Cell and Molecular Biology. These fields provide a unique perspective on the core concepts most relevant to meta-brains: embodiment, developmental science, constructivism, and connectionism. Taken together, these features serve to extend meta-brain models from a simple case of a single sensory input and two layers to complex models with multiple layers, inputs, and even behavioral outputs. Throughout our discussion, we will expand upon our basic framework of layered, representational computational structures with features that make them relevant to developmental embodied agents, and eventually, to application in the broader world.

The most influential contributor to meta-brain model architectures involves embodiment. This includes both cognitive and Meta-brain models that are most effective when embodied in an agent with a body. Yet even in the disembodied case (Figure 1), the arrangement of layers and the sensory apparatus are embodied in a spatial context. Embodiment enables the information processing power of effectors, affordances, and various interfaces with the environment. Meta-brain models can also utilize the body of an agent as a mechanism for preprocessing sensory information, which would lie outside of the layered model structure. Morphological preprocessing would serve as a means to sample or filter information en route to the representation-free layer of the model.

In terms of developmental science, we can take advantage of two features of developmental agents: the generative nature of morphogenesis and compositional nature of learning. Our multilayered meta-brains and associated agent phenotypes rely upon morphogenetic growth, particularly innate rules that put these features of the model in place through a developmental process. Our conception of innateness is not simply to dictate what components a model should begin its life with, but also how it is shaped by interactions with the environment. As a complementary process, composition takes a number of forms in a meta-brain model, from the composition of a phenotype to the composition of complex behavioral outputs. Once the morphogenetic component of development is largely finished, a phenomenon called developmental freedom allows for an adaptive output to emerge from what becomes a largely static neural and morphological phenotype [8]. As the developmental process stabilizes, the ability to construct coherent outputs and the informational synchrony of a meta-brain increases.

1.2 Learning and layering

Drawing from learning itself, meta-brain agents take an unorthodox view to a traditional constructivist approach. In line with the layering principle of representational components, meta-brain models feature an innate component that provides a baseline as well as a heritability component to computational models of the brain. As a basement layer, the innate component resembles the genome in biological agents. In the spirit of creating efficient abstractions, innateness in meta-brain models consist of a transcription layer and a mutational layer. This is similar to mechanisms underlying epigenetic landscapes [9-11]. The mutational layer is similar to a genetic encoding in that it consists of binary strings encoded with basic instructions on how to construct the brain and body of the computational agent. But the real power is in the mutation (and preservation) of this encoding.

While mutations can introduce variation on a theme, our transcription layer allows for the mutational layer to be expressed in an analogue fashion. The transcription and mutational layers are generative in different ways, and provide a good balance of instructional information and response to particular environmental contexts. The lower layer(s) of a meta-brain model are considered to be representation-free, the most common and powerful of which is the connectionist model. In their most basic implementation, connectionist models are generally representation-poor, and in their basic form homogeneous in structure [12]. In meta-brain models, connectionist models can be heterogeneous in their structure, but crucially have connections to more representation-rich model types.

1.3 Layering in Biological Brains

In biological brains, laminar organization can be found in a number of contexts, with the greatest amount of characterization in the neocortex. Felleman and van Essen [13] proposed a cortical hierarchy for understanding layering in the visual cortex of Macaque. In an object lesson for computational modeling, these layers do not have sharp functional or structural boundaries. Nevertheless, laminar structure is enforced by cytophysiological constraints which act to selectively constrain neural connectivity and activity [14]. Such models have been refined to construct such as the laminar-based neural unit [15], which characterizes the many neuronal cell types found in distinct cortical lamina. Using laminar-specific neuroimaging, we can better understand both the computational role of and connectivity between different cortical layers in cognitive functions such as language processing [16], working memory [17], and predictive processing [18]. Brain imaging studies have also found that the normal formation of a six-layered cortex is disrupted during embryonic and postnatal development in disorders such as autism [19] and microcephaly [20].

2. Basic Architecture of a Meta-brain Model

Meta-brain models are hybrid models consisting of computational layers inspired by laminar anatomical structures. Each layer is a different type of computational model, and is meant to mimic structural differences inherent in many nervous systems. From a computational representation standpoint, meta-brain models provide us with neuro-symbolic capabilities while also selectively determining how neural and symbolic components interact. For example, the representation-rich component can "encase" or be "layered" on top of the representation-free component. The ability to toggle between these layers might be useful in a dynamical sense, or might allow us to better understand how our own brain selectively integrates information during cognitive processing. We have developed an open-source repository where a sizable collection of agent-based models are suggested for incorporation into our various layers of the model (https://github.com/Orthogonal-Research-Lab/Meta-brain-Models).

In Figure 1, a layered meta-brain is shown that can be encased structurally within other layers. Encasing in biological systems serves a number of diverse functions. Axons are encased by myelin, bone is encased by muscle and fatty tissue, different parts of the brain are encased by others, and germ layers define the earliest functional divisions in development. Often, encasing serves a functional role between tissues of different types, from chemical signaling between cells of distinct tissue types to more direct chemical communication using synaptic connections. Functionally, encasing emphasizes boundaries between tissue types or subsystems. In the case of meta-brains, the encasing we are approximating the outcomes of differential tissue growth, activity-dependent target tissue plasticity [21], and more general developmental processes. In meta-brains, encasing can play a role in providing multiple access points between specific spatial locations in each layer. Topographical organization, or spatially explicit representations and projections, is a common organizational principle of the nervous system [22]. Meta-brain models provide a means to simulate the structure of such systems, and perhaps leverage their functional advantages as well.

IOP Conf. Series: Materials Science and Engineering 1261 (2022) 012019 doi:10.1088/1757-899X/1261/1/012019

2.1 Embodiment and Regulation of the Meta-brain

While meta-brain models are flexible, all implementations should have a set of common principles that guide their application. One of these principles is *embodiment*, which allows us to place our hybrid models in a broader context. We can also couple embodiment and the practice of encasing to create a subsumption architecture [23]. Subsumption involves the bottom-up processing of sensory stimuli, and has been considered to be biologically-inspired. In a meta-brain model, subsumption occurs through the processing of sensory information at progressively richer levels of representation. Unlike past implementations, the coordination of layers allows for representations of different complexities to emerge from representation-poor layers.

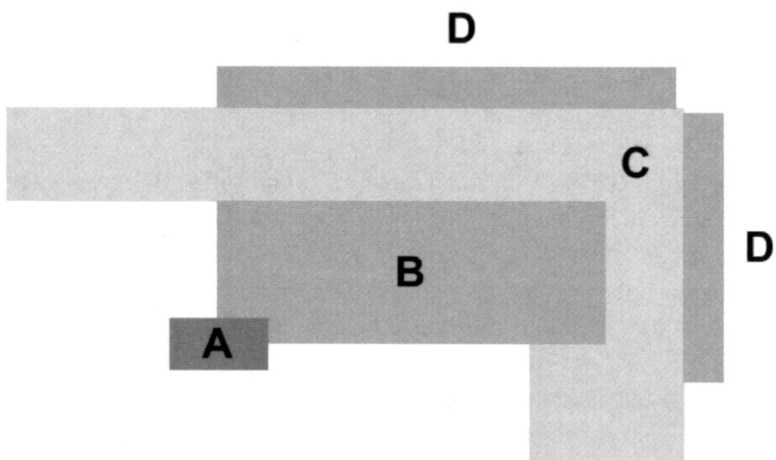

Figure 1. Structure of a meta-brain, encased in a geometry resembling a vertebrate brain. A: sensory input module. B: lowest representational layer, C: higher representational layer, D: highest representational layer.

2.2 Cybernetic regulation

Subsumption architectures can be rather limiting when we consider meta-brains from the perspective of *cybernetic regulation.* While feedback is widely used in neuroscience at the level of systems such as the sensorimotor loop, feedback and other regulatory mechanisms are also critical to the communication and coordination between two meta-brain layers. In the language of meta-brain models, cybernetic control requires that the encasing layer (regulator) has more degrees of freedom than the underlying layer (controlled system). This results in a feedforward signal projecting towards higher degrees of representation, while feedback projects towards lesser degrees of representation. Aside from the potential for feedback between representational layers, interactions between layers necessitate one system to fully map to another. This requires a matching between the complexity of each layer. If a set of coupled models are vastly different in terms of their complexity (e.g. state space) then these deficiencies will lead to either insufficient control or severe errors in communication.

We also utilize the Every Good Regulator theorem [24] to guide the composition of the various layer outputs and interactions. Meta-brains that conform to the EGRT allows for prediction and control given certain organizational principles. More generally, Ashby [25] utilizes the homeostat to propose cybernetic control for brains as the natural selection of behavior patterns. Meta-brains are similar to a homeostat in that the connectivity between layers can both fluctuate and exhibit temporary interdependence.

Applying the cybernetics approach also suggests that meta-brains constitute a teleological regulatory system [26] subject to the constraints of entropy and evolution. Although this is beyond the scope of our computational model, it is nevertheless an important principle to approximate. Connecting a meta-brain to cybernetic control also allows us to more rigorously analyze model outputs using the Free Energy Principle and related Bayesian approaches [27].

2.3 Levels of representation
Meta-brain models address the issue of connectionist-symbolic integration by combining different degrees of representation into a single system. In the human brain, the representation of color relies upon a distributed set of layers that correspond to various features of the sensory phenomenon [28, 29]. Meta-brains operate as a more general processing system which is not specialized for any specific property of the sensory world. A two-layer meta-brain might consist of a representation-free layer (connectionist model) and a representation-rich layer (object-oriented model). Meta-brains can also enable different levels of symbolism, from simple object classifications to cultural representations. In a multi-layer model, information processing can proceed orthogonally: information can be processed through the network or kernel of a single layer, while layer outputs can be utilized in a compositional manner to communicate between our layers of models. The relationship between layers can be defined as a homeomorphism, where the encasing layer is a richer model of the underlying layer.

2.4 Isometric integration
The simplest way to join two or more layers of a meta-brain is through isometric integration. Previous work on utilizing isometric integration-like organization has been implemented in the form of layered control architectures [30, 31]. As shown in Figure 2, isometric layering can be demonstrated without incorporation in an anatomical or higher geometric setting. Nevertheless, isometrically arranged layers can exchange information in a geometrically-specific manner. This arrangement also demonstrates how layers can be combined to potentially leverage geometric constraints in a way that provides meta-structure to the interactions between layers. In Figure 2A, we present a three-layer laminar model, where the sparse representation receives inputs from both the representation-rich and representation-poor layers. A different configuration is shown in Figure 2B, where two sets of two-layer laminar are topologically distinct by connecting with a bidirectional

connection. In this configuration, there is a duplicate sparse representation layer that serves to further coordinate each laminar set of layers.

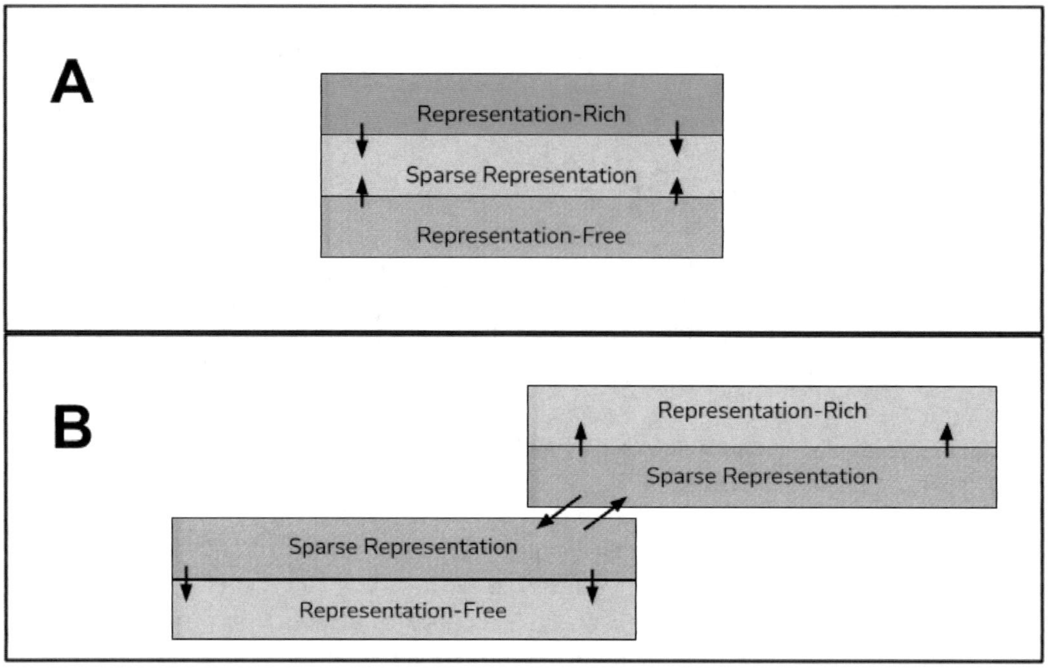

Figure 2. Basic isometric integration between the layers for two configurations of a meta-brain. A: a three-layer configuration with an arrangement that forms a representational gradient. B: an alternate meta-brain configuration with two sets of layers that form a spatially embodied and semi-discontinuous representational gradient.

3. Embeddedness and Embodied Context

Meta-brain models are incomplete without being embedded in an embodied context. By embodiment, we mean that an agent's physical body plays a significant role in shaping the cognitive capabilities of the agent. In Figures 1 and 2, meta-brain models are shown as disembodied entities. Yet embodiment can add adaptive capacity to an agent in addition to task performance utility [32, 33]. Meta-brains are thus best suited to being embedded in an anatomical context, represented by an agent with anatomical fidelity. Embodiment presumes that meta-brain will exhibit at least four properties: situatedness with respect to a particular context, time-dependence with respect to a task, the ability to offload computational requirements from the nervous system into the environment, and interactivity between a task and its environment [34].

In Figure 1, we show a component called "sensory input". In the embodied context, this component becomes important as a mediator of direct perception [35]. Direct perception is a bottom-up approach to cognition that relies upon the environment to provide computational structure to the brain. In the case of radical embodied cognition [36, 37], the case is made for a lack of computation in the brain itself, ascribing much of this capacity to direct perception. While the meta-brain perspective is not fully in accordance with this view, it does strike a balance between computationalism and non-computationalism. The lower layers of a meta-brain can lack representational capacity and be enriched by direct perception, which is in turn enabled by sensory inputs and the agent body. The higher layers of representational capacity then enable behaviors that are more intentional in nature.

This situation is ideal for things like direct learning [38], which extracts a dynamical system from the environment presented to an agent. Enactive intelligence [39, 40], or intelligence that arises through interactions between the agent's body and external environment, is also something that can be expanded upon using the meta-brain approach. In biological systems, morphology also plays a role in the development of mental representations. Body representations in the human brain act to create continuous feedback for motor control and positional awareness, as well as awareness of bodily position relative to spatial orientation [41]. Meta-brain models do not utilize bodily representations *per se*, but such representations abstracted to computational models might be useful to integrate various aspects of the global context.

3.1 Anatomical fidelity

One benefit of embodiment is the ability to incorporate the topology of anatomy into the computational model. With meta-brain models, there are two aspects of embodiment that provide anatomical fidelity to our models. The first benefit of embodiment is that the body itself engages in computation with respect to the brain-environment coupling. Even in agents, the effectors and their component parts can yield muscle synergies [42] that regulate movement behavior in addition to cognitive offloading [43] capabilities. The second benefit, perhaps more generally useful to meta-brained agents, is that these encased and layered systems replicated the natural geometry of neural connectivity. Topological complexity in terms of adjacent layers provide benefits such as shorter wiring distances and tighter integration with an embodied agent's morphology.

Anatomical fidelity comes in the form of a layered morphology where some layers exhibit isometric integration and some act as parts of the morphology. However, they are all subject to developmental constraints under which layers grow according to a scaling law [44] and can also exhibit growth, movement, translation, and rotation [45] with respect to other layers (Figure 3). Anatomical fidelity can also offer morphologically consistent affordances that enable a means for agents to change their size or shape in an adaptive manner with respect to behavior [46]. Not only is this advantageous for the development of adaptive behaviors, but it also enables heterogeneity in the agent's structure and function.

3.2. Intrinsic and extrinsic factors: heterogeneity

To move beyond the input/output methodology without losing the benefits of computational and hypothetical simplicity, we can leverage the heterogeneity of a meta-brain model to provide a means for functional specialization. For example, heterogeneous meta-brains allow us to be explicit in

IOP Conf. Series: Materials Science and Engineering 1261 (2022) 012019 doi:10.1088/1757-899X/1261/1/012019

terms of determining functional hierarchies like the kind we observe in the Primate visual system [47, 48]. We can also integrate and selectively turn off levels as needed for specific tasks. Heterogeneity also has benefits in terms of robustness and persistence of dynamical function. In general, diversity provides multiple paths to a specific function. Yet in the case of developmental meta-brain models, we can still preserve the ability to trace specific functional pathways. This type of architecture allows us to strike a balance between credit assignment [49, 50] and emergent perception. Finally, single layers should exhibit temporal and topological heterogeneity to maximize representational stability and durability as discussed previously [12].

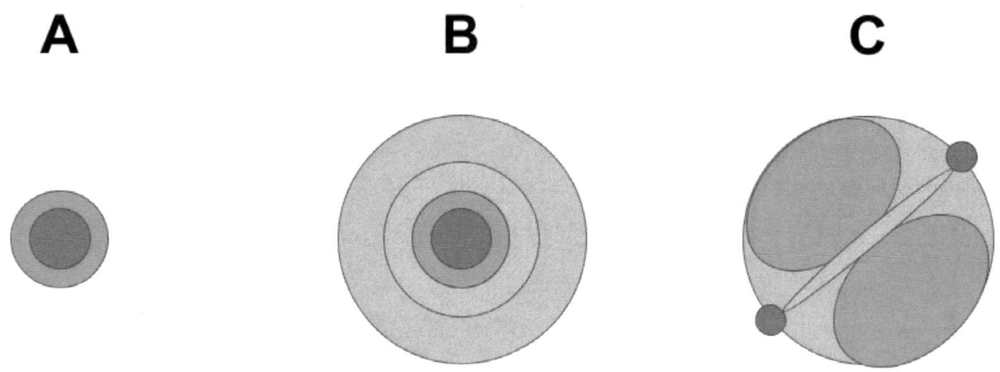

Figure 3. Example of anatomical fidelity as a developmental process. A: a two-layered embryonic agent. B: a four-layered developmental phenotype that expands in accordance with an 8/5 scaling law. C: four-layered mature phenotype with layers exhibiting growth, translation, movement, and rotation.

3.3. Intrinsic and extrinsic factors: innateness

We can add a significant innate component to a layered model. Moving forward from a homogeneous model of an artificial nervous system, we need to enable dynamic attributes such as encodings and generativity. In extending meta-brain models to the Developmental AI paradigm, we will focus on evolution and development as means to imbue our models with innate mechanisms for generating intelligent behavior. One means to incorporate innateness is to provide a genetic representation for a given meta-brained agent. A genetic representation is a computational encoding that contains a generalized specification of an agent's body, brain, and behavior. While we do not propose using a typical genetic algorithm (GA), this encoding of the morphology and meta-brain layers should draw from the principles of GAs and genetic regulatory networks (GRNs).

Using a GA/GRN architecture yields a set that can be gradually expressed throughout development and tuned to the agent's environment. Aside from selectively enabling plasticity, the GA/GRN architecture also imbues meta-brains with a basis for hereditary information as well as a source of developmental and functional invariance. In terms of our representational hierarchy, the genetic encoding is separated from the various levels of representation in the meta-brain itself, and

IOP Conf. Series: Materials Science and Engineering 1261 (2022) 012019 doi:10.1088/1757-899X/1261/1/012019

serves as a generative system that provides a baseline of information to the agent without relying on the vagaries of environment or a more restrictive blueprint.

3.4. Intrinsic and extrinsic factors: development

Meta-brains can also take inspiration from developmental (or ontogenetic) processes to provide an alternative to the so-called static adult mind [51, 52]. This provides a means to understand learning as both a compositional and morphogenetic process. One example of this is by using the example of small, emerging connectomes [53]. Development enables gradual differentiation of cells in the connectome, along with their formation of neural circuits. Meta-brains can approximate this process as simple connectomes at different times/stages of development serving as different layers and degrees of representations can model these processes coming online in an agent.

The development of meta-brains as emerging small connectomes can be described using the metaphor of Flatland. Flatland [54] is a mythical world of only two spatial dimensions. The agents who inhabit Flatland do not have any concept of a third spatial dimension. And so it is with immature meta-brains, which develop their capabilities gradually in order of representational power. Only in the ontogenetic case it is representational capacity, not dimensionality, that defines the immature agent's perspectivist shortcomings. As development proceeds, representational capacity increases, and conceptual understanding of the third dimension emerges. In terms of developing a form in which embodiment can take place, Morphogenesis might proceed by the innate component producing a precursor phenotype, which allows for the differentiation of multiple computational layers.

As the layers differentiate, individual components of each layer come online. Then these layers need to be embodied in the agent's environment, which to an Flatland-dwelling agent resembles an unfamiliar world. In a simple meta-brain with only a single representation-free layer, the richness of the agent's environment is lost, depriving the developing agent of contextual information. As additional layers form that are more representationally robust, the environment can now be represented in a fuller form. However, a problem remains: the different layers of the meta-brain are at different developmental stages, and so are still unsynchronized in their ability to integrate information. This is where the interconnections between layers become important. These interconnections can be both feedforward and feedback in orientation, and work collectively to normalize differences in developmental stages.

3.5 Intrinsic and extrinsic factors: evolution

Meta-brains can also evolve across generations of agents, and provides a means to think of meta-brain models from a dynamical, phylogenetic, and population perspective. This provides us with a means to explore the behavior of meta-brain models over long time intervals. The evolutionary trajectory of a given agent population is also influenced by its history of learning from sensory information [55]. By evolving the innate component of meta-brain models, we can create a phylogenetic representation of multiple generations. Given a mechanism for natural selection, this also allows us to produce different layers that are potentially adaptive to the problem at hand, and

IOP Conf. Series: Materials Science and Engineering 1261 (2022) 012019 doi:10.1088/1757-899X/1261/1/012019

ultimately find the right degree of representation. The incorporation of natural selection and adaptation as a way to enable long-term meta-brain morphogenesis also requires populations of meta-brains, each exhibiting a range of variation. This variation could come in the form of network topologies at the single layer level, or as different patterns of interconnections between layers.

Alternatively, we can consider the developmental expression of various behaviors and layers from a *ontophylogenetic* approach [56, 57]. Cisek [56] suggests that different parts of the brain expand in various taxonomic groups in the tree of life, which leaves a developmental signature consisting of areas that emerge early versus later in morphogenesis. A number of rudimentary behaviors shared by most species that have a brain are examined by Bennett [57]. This study reveals that many of these behaviors are implementable in agent-based systems such as Braitenberg Vehicles [58] or other agents with minimal cognition [59]. This suggests selection may have favored an additive approach to building neurobehavioral complexity. In short, evolution is an important component of configuring meta-brain models.

4. Application of Meta-brains to Specific Systems

The application of meta-brain models to specific systems also allows us to consider how developmental and evolutionary approaches might be used to build our layered heterogeneous representations in a specific environmental context. Of particular interest is the dynamic behavior of connections between each layer along with the establishment of new layers. The ability to toggle between layers, or the ability to use higher degrees of representation for some tasks and lower degrees of representation for others, might be valuable for a wide range of applications that require time-dependent processing or different levels of perspective for the same problem.

Aside from the further methodological refinement of meta-brains, there also exist a number of application domains that might be particularly amenable to this approach. These include tasks such as self-driving vehicles or agent navigation of unexplored environments. For these types of application, agents must employ a variety of behavioral and cognitive functions such as spatial cognition, learning, and phototaxis. Meta-brains might also be useful for more mundane tasks such as image processing, where segmentation and decomposition of an image can involve multiple levels of computational representation. To realize their full potential, the meta-brain approach also requires technical refinement on a number of fronts. One area that needs further technical detail is exploiting the connectivity patterns between layers. Another area to develop further is the topological nature of individual layers and the layering process more generally. A third domain for improvement is the sophistication of interactions with the environment via sensors and effectors.

4.1 Meta-brain implementation contexts

Meta-brain agents can also be implemented in populations, where they experience interactions with other meta-brain agents. There are of course a number of coordination problems that become relevant. But social interactions also pose a host of unique sociocultural issues for computational agents [60], including affordance learning and social embodiment. Affordance learning [61] enables individual agents to learn higher-level representations from how other agents interact with their surroundings. In such cases, new meta-brain layers and their representational capacity can emerge soon after model initiation (during a developmental stage).

Social embodiment is also important for meta-brain development. From a representational standpoint, biological brains feature higher-level social representations that are scaffolded from lower-level sensorimotor representations [62]. Scaffolding [31] allows for many complex functions to emerge from brain lamina via functional constraints. In this way, meta-brains can feature representations of social relationships with select connections to sensory processing. Other forms of social behavior include swarm optimization and the emergence of collective behaviors [63, 64], which enable coordinated social behaviors between agents as well as the ability to learn these representations from one another.

Lastly, the Developmental AI paradigm can be used to enable meta-brains to fully utilize the benefits of plasticity and constructivism that a set of developmental processes provides. Developmentally-inspired approaches such as Developmental Braitenberg Vehicles [65], and overarching theories such as Continual Embodied Learning [8] all provide different means to incorporate features of developmental processes into a computational agent. One application domain for developmental embodied agents is spatial learning in small artificial nervous systems. Using a meta-brain to model spatial learning might not only enable behaviors such as path integration [66], but also the acquisition of highly abstract spatial learning useful for navigating and analyzing high-dimensional, discontinuous topologies [67].

5. Future Directions

To introduce the concept, we contextualize meta-brain models in terms of the history of Artificial Intelligence and inspiration from biological systems. Then we discuss the basic architecture of meta-brain models, including the three principles that guide implementation. This is followed by a discussion of architectural design features such as anatomical fidelity, heterogeneity, and cybernetic regulation. Meta-brain models also feature an innate component, which comes in the form of a base generative layer in addition to processes that interact with agent shape and structure. For example, agents are subject to processes such as development or evolution which affect form and function at multiple time scales. We then conclude with a discussion of applying meta-brains to specific systems, particularly with relevance to the Developmental AI approach.

While it is critical to understand the nature of connectivity between layers, and regulation between layers more specifically, we must also account for the independent representational drift of models that constitute different layers of a meta-brain. Representational drift [68, 69] occurs when changes to the environment or local content challenge the current state of the internal model. This has the potential to create a persistent source of error between layers [69], which can interfere with their informational integration. Yet error can be controlled through the regulation of gain, as is the case in Xia et.al [70]. During viewing of a naturalistic stimulus that provides episodic activity, as long as gain can be dynamically regulated, the representation can remain stable. This is where the cybernetic approaches can assist us in building robust representations at multiple scales. More generally, internal representation has the potential to be seamlessly integrated with directly acquired sensory information. Through the measurement of mutual predictive information, predictions generated from an internal model and predictions extracted from environmental stimuli are not distinguishable [71].

5.1 Broader context of intrinsic and extrinsic factors
In terms of the interrelationship between morphogenesis and acquisition in development, developmental freedom is roughly analogous to neuroplasticity and connection weights in a representation-free model, but can also be captured using a form of reinforcement learning (RL). An RL perspective might be valuable at higher levels of representation, particularly in terms of overcoming the inconsistencies with plastic changes at lower representational layers. Similar to the concept of affordance learning, multi-agent RL [72, 73] would allow for meta-brain agents to acquire successful policies through the social learning of representations. Put into meta-brain terms, latent strategies [74] are acquired from fellow agents in the form of representation-rich models that influence future policy selection. Returning to the problems of representational drift, it is also critical that the relationship between units within a layer are sufficiently different from other layers. In lesser representational layers, units should connect to other units using a series of transformational operations. Meanwhile, higher representational layers are connected through descriptive linkages [75].

Although our argument has relied upon the analogy between layers in an embodied brain and layering analogies in natural systems, it is not clear that biological levels are a coherent organizing principle in biology [76]. Moreover, our division of representational levels is somewhat arbitrary, given that they rely on various types of computational models. Even the manner in which we define functional relevance and thus the relationships between levels have a much different cast when viewed from a phylogenetic perspective [56]. Another caveat we offer is that much of what we have presented has not yet been implemented or may not even be computationally rigorous. That said, we believe the component part described here can lead to models that improve upon current AI models, in addition to making such models much more biologically-inspired and robust.

References
[1] Garcez AD and Lamb LC 2020 Neurosymbolic AI: The 3rd Wave *arXiv* 2012.05876

[2] Arbib MA 1972 The metaphorical brain: an introduction to cybernetics as artificial intelligence and brain theory (New York: Wiley)

[3] Deco G Tononi G Boly M and Kringelbach ML 2015 Rethinking segregation and integration: contributions of whole-brain modeling. *Nature Reviews Neuroscience* **16** 430–439

[4] Newell A 1980 Physical Symbol Systems *Cogn Sci* **4(2)** 135-183

[5] Ramsey WM 2007 Representation Reconsidered (Cambridge, UK: Cambridge University Press)

[6] Dretske F 1988 Explaining Behavior (Cambridge, MA: MIT Press)

[7] Eliasmith C 2005 A new perspective on representational problems. *J Cogn Sci* **6** 97-123

[8] Alicea B et al 2020 Embodied continual learning across developmental time via Developmental Braitenberg Vehicles *arXiv* 2103.05753

[9] Waddington CH 1962 New Patterns in Genetics and Development (Columbia University Press: New York)

[10] Goldberg AD Allis D and Bernstein E 2007 Epigenetics: a landscape takes shape *Cell* **128(4)** 635-638

[11] Bhattacharya S Zhang Q and Andersen ME 2011 A deterministic map of Waddington's epigenetic landscape for cell fate specification *BMC Syst Biol* **5** 85

[12] Katyal K Parent J and Alicea B 2021 Connectionism, Complexity, and Living Systems: a comparison of artificial and biological neural networks *arXiv* 2103.15553

[13] Felleman DJ and Van Essen DC 1991 Distributed hierarchical processing in the primate cerebral cortex *Cereb Cortex* **1** 1-47

[14] Miller R 1996 Neural assemblies and laminar interactions in the cerebral cortex *Rev Biol Cybernet*, **75(3)** 253-261

[15] Corbitt PT Ulloa A and Horwitz B 2018 Simulating laminar neuroimaging data for a visual delayed match-to-sample task. *Neuroimage* **173** 199-222

[16] Sharoh D Van Mourik T Bains LJ Segaert K Weber K Hagoort P and Norris D 2019 Laminar specific fMRI reveals directed interactions in distributed networks during language processing *PNAS* **116(42)** 21185-21190

[17] Finn ES Huber L Jangraw DC Molfese PJ and Bandettini PA 2019 Layer-dependent activity in human prefrontal cortex during working memory *Nat Neur*sci **22(10)** 1687-1695

[18] Haarsma J Kok P and Browning M 2020 The promise of layer-specific neuroimaging for testing predictive coding theories of psychosis *Schizophr Res* S0920-9964(20)30496-5

[19] Stoner R Chow ML Boyle MP Sunkin SM Mouton PR Roy S Wynshaw-Boris A Colamarino SA Lein ES and Courchesne E 2014 Patches of Disorganization in the Neocortex of Children with Autism. *N Engl J Med* **370** 1209-1219

[20] Badea A Schmalzigaug R Kim W Bonner P Ahmed U Johnson GA Cofer G Foster M Anderson RJ, Badea C and Premont RT 2021 Microcephaly with altered cortical layering in GIT1 deficiency revealed by quantitative neuroimaging *Magn Reson Imaging*, 76, 26-38

[21] Striedter GF 2006 Principles of Brain Evolution (Sunderland, MA: Sinauer Associates)

[22] Patel GH Kaplan DM and Snyder LH 2014 Topographic organization in the brain: searching for general principles *Trends Cogn Sci* **18(7)** 351-363

[23] Brooks RA 1990 Elephants Don't Play Chess. Designing autonomous agents: theory and practice from biology to engineering and back (Cambridge, MA: MIT Press)

[24] Conant RC and Ashby WR 1970 Every good regulator of a system must be a model of that system *Int J Syst Sci* **1(2)** 89-97

[25] Ashby WR 1952 Design for a Brain (London: Butler and Tanner)

[26] Rosenblueth A Wiener N and Bigelow J 1943 Behavior, Purpose and Teleology *Philos Sci* **10** 18-24.

[27] Safron A 2021 The Radically Embodied Conscious Cybernetic Bayesian Brain: from free energy to free will and back again. *Entropy* **23(6)** 783

[28] Zeki S and Marini L 1988 Three cortical stages of color processing in the human brain *Brain* **121** 1669-1685

[29] Seymour KJ Williams MA and Rich AN 2016 The Representation of Color across the Human Visual Cortex: distinguishing chromatic signals contributing to object form versus surface color *Cereb Cortex* **26(5)** 1997–2005

[30] Prescott TJ Redgrave P and Gurney K 1999 Layered control architectures in robots and vertebrates. *Adapt Behav* **7** 99-127

[31] Wilson SP and Prescott TJ 2021 Scaffolding layered control architectures through constraint closure: insights into brain evolution and development *Philos Trans Royal Soc B* **377** 20200519

[32] Bongard J 2010 The utility of evolving simulated robot morphology Increases with task complexity for object manipulation *Artif Life* **16(3)** 201–223

[33] Bredeche N Haasdijk E and Prieto A 2018 Embodied Evolution in Collective Robotics: a review. *Frontiers in Robotics and AI* 5 18

[34] Wilson M 2002 Six views of embodied cognition *Psychon Bull Rev* **9(4)** 625-636

[35] Gibson JJ 1979 The ecological approach to visual perception (Boston, MA: Houghton Mifflin)

[36] Chemero AP 2009 Radical Embodied Cognitive Science (Cambridge, MA: MIT Press)

[37] Raja V and Anderson ML 2019 Radical Embodied Cognitive Neuroscience *Ecol Psychol* **31(3)** 166-181

[38] Jacobs DM and Michaels CF 2007 Direct learning *Ecol Psychol* **19(4)** 321-349

[39] De Loor P Manach K and Tisseau J 2009 Enaction-based Artificial Intelligence: toward coevolution with humans in the loop. *Minds Mach* **19** 319-343

[40] Froese T and DiPaolo EA 2011 The enactive approach: theoretical sketches from cell to society *Pragmat Cogn* **19(1)** 1-36

[41] Carruthers C 2008 Types of body representation and the sense of embodiment *Conscious Cogn* **17(4)** 1302-1316

[42] Alessandro C Delis I Nori F Panzeri S and Berret B 2013 Muscle synergies in neuroscience and robotics: from input-space to task-space perspectives. *Front Comput Neurosci* **7** 43

[43] Risko EF and Gilbert SJ 2016 Cognitive offloading *Trends Cogn Sci* **20(9)** 676-688

[44] West GB 1999 The origin of universal scaling laws in biology Physica A **263** 104-113

[45] Thompson DW 1917 On Growth and Form (Cambridge, UK: Cambridge University Press)

[46] Stokkermans A Chakrabarti A Wang L Moghe P Subramanian K Steenbergen P Monke G Hiiragi T Prevedel R Mahadevan L and Ikmi A 2021 Ethology of morphogenesis reveals the design principles of cnidarian size and shape development *bioRxiv* doi:10.1101/2021.08.19.456976

[47] Kruger N Janssen P Kalkan S Lappe M Leonardis A Piater J Rodríguez-Sanchez AJ and Wiskott L 2013 Deep hierarchies in the primate visual cortex: what can we learn for computer vision? *IEEE Trans Pattern Anal Mach Intell* **35(8)** 1847-1871

[48] Arcaro MJ and Livingstone MS 2017 A hierarchical, retinotopic proto-organization of the primate visual system at birth *eLife* **6** e26196

[49] Minsky M 1961 Steps toward Artificial Intelligence *Proc IRE* **January** doi:10.1109/JRPROC.1961. 287775

[50] Richards BA and Lillicrap TP 2018 Dendritic solutions to the credit assignment problem *Curr Opin Neurobiol* **54** 28–36

[51] Oyama S 2000 The Ontogeny of Information: developmental systems and evolution (Durham, NC: Duke University Press)

[52] Airenti G 2019 The place of development in the history of Psychology and Cognitive Science *Front Psych* **10** 895

[53] Alicea B 2018 The emergent connectome in *Caenorhabditis elegans* embryogenesis *BioSystems* **173** 247-255

[54] Abbott E A 1884 Flatland: a romance of many dimensions (London: Seeley and Company)

[55] Hinton GE and Nowlan SJ 1987 How learning can guide evolution *Complex Syst* **1** 495-502

[56] Cisek P 2019 Resynthesizing behavior through phylogenetic refinement *Atten, Percept, Psychophys* **81** 2265–2287

[57] Bennett MS 2021 What behavioral abilities emerged at key milestones in human brain evolution? 13 hypotheses on the 600-million year phylogenetic history of human intelligence *Front Psych* **12** 685853

[58] Braitenberg V 1984 Vehicles: experiments in synthetic psychology (Cambridge, MA: MIT Press)

[59] van Duijn M Keijzer F and Franken D 2006 Principles of Minimal Cognition: casting cognition as sensorimotor coordination *Adapt Behav* **14** 157

[60] Bolotta S and Dumas G 2022 Social Neuro AI: social interaction as the "dark matter" of AI *arXiv* 2112:15459

[61] Whiten A 2021 The burgeoning reach of animal culture *Science* **372** eabe6514

[62] Meier BP Schnall S Schwarz N and Bargh JA 2012 Embodiment in Social Psychology *Top Cogn Sci* **2012** 1–12

[63] Ha D and Tang Y 2021 Collective Intelligence for Deep Learning: a survey of recent developments *arXiv* 2111.14377

[64] Gordon G 2019 Social behavior as an emergent property of embodied curiosity: a robotics perspective *Philos Trans Royal Soc B* **374** 20180029

[65] Dvoretskii S Gong Z Gupta A Parent J and Alicea B 2022 Braitenberg Vehicles as Developmental Neurosimulation *Artif Life* **28(3)**

[66] Harootonian SK Wilson RC Hejtmanek L Ziskin EM and Ekstrom AD 2020 Path integration in large-scale space and with novel geometries: comparing vector addition and encoding-error models *PLoS Comput Biol* **16(5)** e1007489

[67] Svarverud E Gilson S and Glennerster A 2012 A demonstration of 'broken' visual space *PLoS One* **7(3)** e33782

[68] Clopath C Bonhoeffer T Hubener M and Rose T 2017 Variance and invariance of neuronal long-term representations *Philos Trans Royal Soc B* **372(1715)** 20160161

[69] Rule ME O'Leary T and Harvey CD 2019 Causes and consequences of representational drift. *Curr Opin Neurobiol* **58** 141-147

[70] Xia J Marks TD Goard MJ and Wessel R 2021 Stable representation of a naturalistic movie emerges from episodic activity with gain variability *Nat Commun* **12** 5170

[71] Candadai M and Izquierdo EJ 2020 Sources of predictive information in dynamical neural networks *Sci Rep* **10** 16901

[72] Canese L Cardarilli GC Di Nunzio L Fazzolari R Giardino D Re M and Spano S 2021 Multi-Agent Reinforcement Learning: a review of challenges and applications *Appl Sci* **11(11)** 4948

[73] Zhang K Yang Z and Basar T 2019 Multi-Agent Reinforcement Learning: a selective overview of theories and algorithms *arXiv* 1911.10635.

[74] Xie A Losey DP Tolsma R Finn C and Sadigh D 2020 Learning latent representations to influence multi-agent interaction *arXiv* 2011.06619.

[75] Barack DL and Krakauer JW 2021 Two views on the cognitive brain. *Nature Rev Neurosci* **22** 359–371

[76] Potochnik A 2021 Our World Isn't Organized into Levels In *Levels of Organization in the Biological Sciences* DS Brooks J DiFrisco and WC Wimsatt eds (Cambridge, MA: MIT Press)

IOP Conf. Series: Materials Science and Engineering 1261 (2022) 012020 doi:10.1088/1757-899X/1261/1/012020

Active inference, preference learning and adaptive behaviour

Noor Sajid[1], Panagiotis Tigas[2] & Karl Friston[1]

[1]WCHN, University College London, UK
[2]OATML, Oxford University, UK

Abstract. The ability to adapt to a changing environment underwrites sentient behaviour e.g., wearing a raincoat when walking in the rain but removing it when indoors. In such instances, agents act to satisfy some preferred mode of behaviour that leads to predictable states necessary for survival, i.e., states that are characteristic of that agent. In this chapter, we describe how active inference agents, equipped with preference learning, can exhibit these distinct behavioural modes – influenced by environment dynamics – to aptly trade-off between preference satisfaction and exploration. We validate this in a modified OpenAI Gym FrozenLake environment (without any extrinsic signal) with and without volatility under a fixed model of the environment. In a static (i.e., without volatility) environment, preference-learning agents accumulate confident (Bayesian) beliefs about their behaviour and act to satisfy them. In contrast, volatile dynamics led to preference uncertainty and exploratory behaviour. This demonstrates that active inference agents, equipped with preference learning, have the appropriate machinery to (i) engage in adaptive behaviour under appropriate levels of volatility, and (ii) learn context-dependent subjective preferences.

1 Introduction

Biological agents exhibit adaptive behaviour when engaging with their environment. These interactions are underwritten by extrinsic and intrinsic motivations that prescribe Bayes-optimal behaviour, in the sense of Bayesian design and active learning (Conant and Ross Ashby, 1970; Friston, 2019; Bruineberg et al., 2018b; Sajid et al., 2021b). Accordingly, agents have preferred modus operandi (or prior preferences) that influence their exchanges with the environment and may update under volatile conditions (Ashby, 1961; Schrodinger et al., 1992; Friston, 2013). These preferences can persist despite the absence of explicit signals from the environment (or sometimes in spite of it).

Briefly, preferences are a subjective assessment of what the agent would like to experience (i.e., desired states or outcomes). These are inherently different from, and should not be conflated with, objective rewards found in reinforcement learning and expected utility theory (Kahneman and Tversky, 2013; Fleming and Sheu, 2002) (Table 1). This is because objective rewards are set from the perspective of the designer (or environment), where the agent is not equipped with the ability to learn subjective preferences (i.e., what outcomes it prefers). Formulations in terms of reward or utility are difficult to reconcile with biological learning, where agents can gradually acquire preferences, settle into their ecological niche, and update their preferences as environmental contingencies change. Note a fundamental difference between reward—outcomes that explain observed behaviour–and preferences, which are an attribute of an agent's beliefs. Furthermore, reward is a uni-dimensional construct, while preferences apply to every state or outcome entertained by the agents generative or world

IOP Conf. Series: Materials Science and Engineering 1261 (2022) 012020 doi:10.1088/1757-899X/1261/1/012020

Table 1. Glossary

Term	Definition
Generative process	This describes transitions among latent states in the environment that generate observations.
Generative model	A statistical model of the generative process. The generative model is specified as a joint probability distribution over actions, states, and observations. This joint probability is usually expressed in terms of a prior and likelihood. In our (discrete) case, the prior specifies action-dependent state transitions, while the likelihood specifies the probabilistic mapping from states to outcomes or observations
Objective rewards	Particular outcomes determined by the environment e.g., an extrinsic teaching signal, reward signal, et cetera. Rewards are part of the generative process.
Subjective preferences	Beliefs about outcomes that the agent anticipates, i.e., what the agent 'likes'. These are inherently different from rewards, which are outcomes. Preferences are probabilistic specifications of every state or outcome an agent expects, a priori, to encounter. Preferences are part of the generative model.
Preference satisfaction	Consistent realisation of behaviour leading to preferred outcomes, i.e., exploitation of the environment to realise preferences.
Conjugate priors	Conjugate priors are chosen to ensure posterior is in the same family of distributions as the prior. This can sometimes finesse the tractability of deriving Bayesian belief updates.

model. In other words, an agent's preferences are just its a priori beliefs about the likelihood of any state or outcome. A preferred state is one that, a priori, the agent considers itself likely to occupy.

From a technical perspective, these preferred states are simply the attracting set of a pullback attractor that the agent returns to, despite external perturbations. Our formulation is motivated by this generic formulation of behaviour, i.e., how biological preferences shape and agent's experiences. For example, imagine a continuously changing world due to recurrent flooding. An agent equipped with preference learning could ensure its survival by appropriate niche construction (i.e., building a dam) or moving away from the floodplain. In this example, surviving agents would implicitly learn an aversion to drowning.

To investigate how preferences shape adaptive behaviour (under a fixed world model), we turn to active inference (Friston et al., 2017; Da Costa et al., 2020; Sajid et al., 2021a) and preference learning (Bruineberg et al., 2018b; Sajid et al., 2021a,c). Our aim here is to illustrate a generalisation of reward learning, in which preferences are themselves learned: i.e., 'learning to prefer'. This may sound like an oxymoron; in the sense that reinforcement learning operationally defines rewards in terms of their ability to reinforce behaviour. However, in active inference, the imperative that underwrites optimal behaviour is simply to maximise the (marginal) likelihood of an agent's generative or world model—thereby ensuring the likelihood of continued exchange with the world; i.e., survival. This means that an active inference agent can learn behavioural outcomes that are sustainable by the environment. In this setting, preferred outcomes are learnable. However, preference learning will clearly depend on environmental volatility. This speaks to a trade-off between exploration and preference satisfaction (i.e., seeking out preferred outcomes) that is constrained by the predictability of the environment (Sajid et al., 2021c). This trade-off is the focus of what follows.

Briefly, active inference is a formal account of how self-organising agents' interface with their environment and maintain a particular steady-state over time (Friston, 2019; Da Costa et al., 2020). It stipulates a feedback loop between the agent, and its interactions with the environment (Bruineberg et al., 2018b). Explicitly, the agent's actions shape the environment and shift the agents perception that influences future behaviour. For this, active inference agents optimise two complementary free energy functionals across time: variational and expected free energy. The variational free energy measures the fit between the agent's generative model and observable outcomes sampled from a generative process (i.e., the environment). Here, the generative process refers to the structure of the environment that generated the agent's observations, and the generative model embodies the agent's expectations about the causes of those observations. Conversely, the expected free energy scores possible action trajectories in terms of their ability to reach preferred outcomes via maximisation of extrinsic and intrinsic values. This equips the agent with a formal way to assess different hypothesis about the types of behaviour that can be pursued.

In short, one could summarise active inference as the inversion of a generative model that includes the consequences of action. Because consequences postdate causes, this necessarily entails a generative model of agency that has temporal depth. Inversion or (active) inference under these models evinces a kind of planning—or control—as inference (Attias, 2003; Botvinick and Toussaint, 2012; Millidge, 2020) that is Bayes optimal, in the dual sense of optimal Bayesian design and decision-making. In other words, plans or policies are inferred on the basis of optimally reducing uncertainty about the causes of outcomes, while minimising Bayesian risk, in relation to prior preferences.

Preference learning can be incorporated by equipping the agent's generative model with the right set of conjugate priors that facilitate the learning of preferences over states across time (Sajid et al., 2021a,c). During planning, these prior preferences influence how policies (i.e., action sequences) are evaluated, and shape subsequent behaviour. The ability to learn preferences, contingent upon environment dynamics, introduces an additional level of dependence between the agent-environment interactions. In other words, active inference with preference learning (Bruineberg et al., 2018a; Sajid et al., 2021c) changes the epistemic dispositions of an agent. In this instance, the agent expects the environment to unfold in a particular kind of way and – through its actions – ensures that those expectations are realised. For example, if the agent has prior preferences for residing in mild temperatures, then it will actively choose to alter the state of the environment to satisfy these preferences e.g., installing an air conditioning unit in hot weather, etc.

Here, we describe how preference learning can be introduced using conjugate priors. For this, we build upon the (state) preference learning mechanism — introduced in (Sajid et al., 2021c) — to provide a complementary exposition of why preference learning, over particular latent (i.e., hidden) states, can influence the agent's behaviour given changes in environmental contingencies. Specifically, we illustrate how: i) behaviour changes systematically with increased environmental volatility, and ii) how learnt preferences underwrite these behaviour shifts. We demonstrate this in a modified OpenAI Gym FrozenLake environment (without any extrinsic teaching or reward signal), with and without volatility. Our simulations reveal that in static environments, the agent learns confident (i.e., precise) preferences and acts to satisfy them. However, preference uncertainty leads to exploratory behaviour in a more volatile setting despite no changes to the agent's generative model. This speaks to a trade-off between exploration and preference satisfaction (i.e. exploitation of desired states) that is

IOP Conf. Series: Materials Science and Engineering 1261 (2022) 012020 doi:10.1088/1757-899X/1261/1/012020

constrained by the predictability of the environment (Sajid et al., 2021c). We conclude with a brief discussion of the relevance of this formulation for characterising adaptive behaviour under evolving dynamics, and how preference learning may influence this behaviour.

2 Active Inference

Active inference describes how adaptive agents interact with their environments, described by some generative process. The theory rests on a generative model of the agent's environment that is fitted to sampled observations (o) from the generative process to reduce surprisal or the variational free energy (Friston et al., 2017):

$$F(\tau) = \mathbb{E}_{q(s|\pi)}\left[\log q(s_\tau|\pi) - \log p(o_\tau, s_\tau|\pi)\right], \tag{1}$$

where $q(\cdot)$ is an approximate density over latent or hidden states (s), given some policy (π). Variational free energy plays the role of an upper bound on the log (marginal) likelihood of outcomes or evidence for the generative model (Winn et al., 2005). In this sense, active inference has been described in terms of self-evidencing (Hohwy, 2016). Additionally, these agents engage with their environment by selecting actions that minimise their uncertainty i.e., expected free energy (Parr and Friston, 2019):

$$G(\pi, \tau) = \mathbb{E}_{p(o_\tau|s_\tau,\theta)q(s_\tau,\theta|\pi)}\left[\log q(s_\tau, \theta|\pi) - \log p(o_\tau, s_\tau, \theta)\right] . \tag{2}$$

This objective is obtained from Eq.1 by supplementing the expectation under the approximate posterior with the likelihood, resulting in the following predictive distribution $p(o_\tau|s_\tau, \theta)q(s_\tau, \theta|\pi)$. Here, the next action is selected by sampling from the distribution:

$$p(\pi) = \sigma\big(-G(\pi)\big) = \sigma\Big(-\sum_{\tau>t} G(\pi, \tau)\Big) , \tag{3}$$

where $\sigma(\cdot)$ is the softmax function, and $p(\pi)$ has a Gibb's distribution.

3 Preference learning

Active inference agents do not have access to the generative process but can alter the environmental state by selecting particular actions that minimise expected free energy. To equip the agent with the ability to learn preferences, we extend the expected free energy functional to incorporate preferences over latent states by introducing conjugate priors over prior beliefs (i.e., hyper-priors) (Friston et al., 2017; Sajid et al., 2021c):

$$\begin{aligned} G(\pi, \tau) = & -\mathbb{E}_{\tilde{q}}\left[\log p(o_\tau|s_\tau)\right] & \text{(4a)}\\ & + \mathbb{E}_{\tilde{q}}\left[\log q(s_\tau|\pi) - \log p(s|C)\right] & \text{(4b)}\\ & + \mathbb{E}_{\tilde{q}}\left[\log q(\theta|s_\tau, \pi) - \log p(\theta|s_\tau, o_\tau)\right] . & \text{(4c)} \end{aligned}$$

where, $\tilde{q} = q_\phi(o_\tau, s_\tau, \theta|\pi) = q(\theta|\pi)q(s_\tau|, \pi)q_\phi(o_\tau|s_\tau, \pi)$, and $p(s|C)$ is the probability of a particular state given (learnt) prior preferences (C). Practically, because our distribution of interest is Categorical, we specify the Dirichlet distribution as the conjugate prior. This is defined as, $P(s) \sim Cat(D)$:

$$P(D^i|d^i) = Dir(d^i) \Rightarrow \begin{cases} \mathbb{E}_{P(D^i|d^i)}\left[D^i_{ij}\right] = \frac{d^i_{ij}}{\sum_k d^i_{kj}} \\ \mathbb{E}_{P(D^i|d^i)}\left[log(D^i_{ij})\right] = F(d^i_{ij}) - F(\sum_k d^i_{kj}) \end{cases} \quad (5)$$

where, F is the digamma function, $d \sim \mathbb{R}^+$ and same parameterisation holds for $P(r) \sim Cat(C)$. The posteriors for the Dirichlet hyper–parameters are evaluated by updating the prior using the following rule $\boldsymbol{d_{i,j}} + \alpha * s_{i,j}$ where $s_{i,j}$ are the observations for that particular category and α the learning rate. This allows the agent to accumulate contingencies (i.e., pseudocounts) and learn what it prefers.

Using this, a preference learning mechanism can be instantiated; introduced as PEPPER in Sajid et al. (2021c). Briefly, PEPPER stipulates a circular agent-environment dependence via a double loop preference learning. Here, the first loop is across time-steps during a particular interaction with the environment (i.e., a single episode). The second loop, evolving at a slower timescale, is across episodes and entails updating prior preferences using the information collected during each interaction. In subsequent time-steps, the updated preferences are used to select the next action via their influence on the expected free energy—and interactions with the environment used to update preferences. In turn, these preferences are used to select actions during the next episode, and so on. This formulation of preference learning highlights the circular causality between the agent and its environment. Specifically, the environment shapes the agent's behaviour via preference learning and the learnt preferences influence the environment via the agent's action on the environment. See Sajid et al. (2021c) for technical details.

4 Simulations

We used a modified OpenAI Gym's FrozenLake environment to evaluate how preference learning influences behaviour: in particular, the trade-off between exploration and preference satisfaction, under different environment dynamics.

4.1 FrozenLake environment

In the original FrozenLake task, the agent has to navigate a grid world comprised of frozen, hole and goal tiles, using 4 actions (left, right, down or up). Upon reaching the goal tile, the agent receives a reward of 10 or a penalty -0.25 if it moves to the hole. In our simulations, we included a sub-goal tile but removed the extrinsic reward signal (Fig.1). That is, although the agent can differentiate between tile categories – parameterised by its generative model – it does not receive reward from the environment: in active inference, reward or reinforcement signals are replaced by preferences for particular states or outcomes. These learnable preferences were encoded by an initial uniform (i.e., prior) Dirichlet concentration parameters encoding preferences over states, at the onset of each experiment. Effectively, this means that agents learn to prefer the states that they can reliably solicit through their behaviour.

To simulate different environmental dynamics, we introduced environmental volatility by switching the FrozenLake configuration every K-th step (Table 2). Additionally, the agent's starting position was varied across each episode. Note, this is different to the original FrozenLake environment, where the agent always starts the episode from the top right corner of the grid. This provided an appropriate test-bed to evaluate how adaptive behaviour was a natural consequence of shifts in prior preferences while engaging with volatile and non-volatile environments.

IOP Conf. Series: Materials Science and Engineering 1261 (2022) 012020 doi:10.1088/1757-899X/1261/1/012020

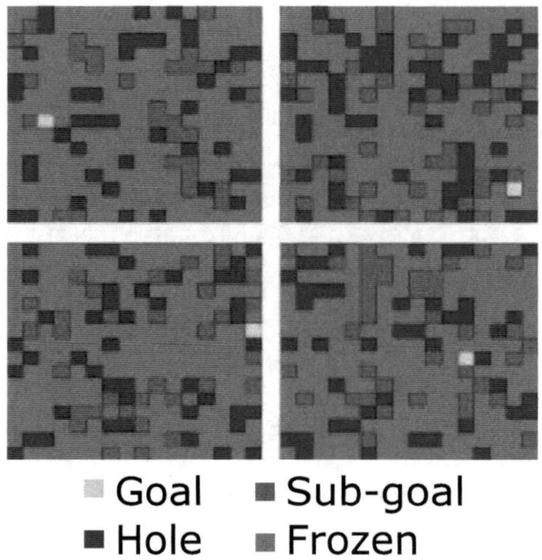

Fig. 1. The four panels show examples of the different OpenAI Gym FrozenLake 16x16 environments used for our simulations. Here, light green represents the goal tile, bottle green the sub-goal, blue the hole and red the frozen tiles.

Table 2. Training and Preference learning parameters

Parameter	Training	Preference learning
Planning Horizon	15 steps	15 steps
Episode Length	50 steps	50 steps
Reset Every	5 steps	1, 10, 20, 40, 50 steps
No. Episodes	50 episodes	50 episodes
No. State Categories	64 categories	64 categories
No. State Dimensions	50 dimensions	50 dimensions
No. Agents	·	10 agent

We assessed behaviour with and without volatility. Furthermore, the trained generative model network weights, optimised Eq.1 using ADAM (Kingma and Ba, 2014), were frozen during our experiments[1]. Thus, adaptive behaviour is a direct consequence of preference learning that induces differences in expected free energy estimation – and ensuing policy selection.

[1] Our agent was implemented as an extension to Dreamer V2 (Hafner et al., 2020) public implementation. Specifically, Dreamer's generative model training loop was used with an open-loop policy. However, the actor learning part of Dreamer was not incorporated, and the generative model was trained using Plan2Explore Sekar et al. (2020). The exact implementation details are available Sajid et al. (2021c).

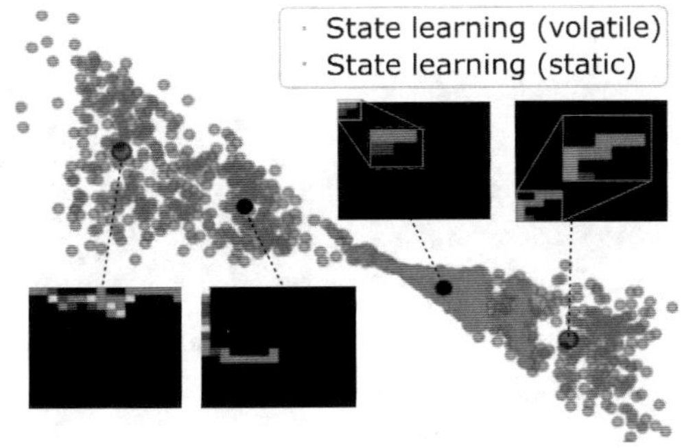

Fig. 2. Visualisation of the posterior latent states (estimated using $q_\phi(s_t|.)$) during state preference learning over episodes. The states have been projected onto their first two principal components, and the black circles represent the k-mean centroid of resulting clusters. Here, the accompanying graphics present a representative agent trajectory with tiles visited from the cluster highlighted in red.

4.2 Preference learning

Two types of behaviours were observed: exploratory and satisfaction of preferences (Fig.2). Here, preference satisfaction entailed restricted movement within a small section of the grid with gradual accumulation of prior preferences. Figure 3 presents examples of learnt states preferences across static and volatile environment. These accumulated preferences speak to the self-evidencing nature of our agents (Hohwy, 2016). That is as the agent sees similar observations across episodes it grows increasingly confident (via increased precision over the prior preference) that these are the outcomes it prefers. Conversely, we observe exploratory behaviour under the volatile setting with gradual preference accumulation of previously unseen states (Fig.3).

4.3 Exploration and preference satisfaction trade-off

We used Hausdorff distance (Blumberg, 1920), to evaluate the agent's adaptive behaviour, i.e., ability to trade-off between exploration and preference satisfaction (Fig.4). Hausdorff distance measures the maximum distance between the agents position in a particular trajectory with the nearest position taken in another trajectory. Thus, high Hausdorff distance denotes increased exploration. This is because trajectories observed across episodes differ from one other and a low distance entails consistent repetition of trajectories across episodes. Using this metric, we observed an inverted u-shaped association, between volatility in the environment and preference satisfaction for preference learning over the states. Here, 50% volatility shifted the agent behaviour from preference satisfaction to exploration, when faced with high uncertainty and an implicit inability to predict the future. Accordingly, agents in this setting pursued long paths from the initial location, and had the highest Hausdorff distance. (Fig.2).

Under complete volatility, the agents behaviour shifted back to satisfying its preferences. Interestingly, this is undergirded by bi-modal preference learning. In other words, regardless of how the environment changed, agents moved directly to a particular location given the initial position. We illustrate exemplar trajectories in Fig.5. This ability to disregard random, noisy information – with continuous environmental changes – highlights preference

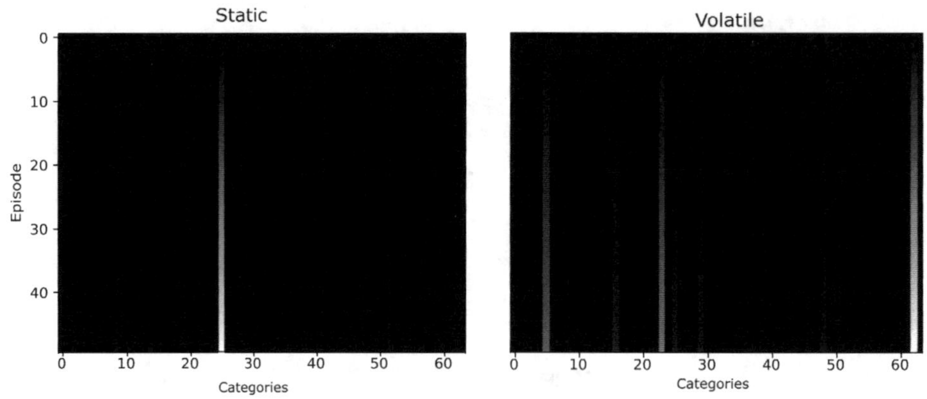

Fig. 3. An example of learnt state preferences for a single agent in a static and highly volatile setting. Here, 64 state categories are presented on the x-axis and episodes on the y-axis. The first panel is for preferences learnt under a static setting, and the second for preferences learnt under a volatile setting. The scale goes from white (i.e., high Dirichlet concentration) to black (i.e., low Dirichlet concentration), and grey indicates gradations between these.

satisfaction motivations beyond exploration under state preference learning. Conversely, the active inference agent without preference learning capacity, exhibited a shift in behaviour as the environment became volatile. This exploratory drive reflects the imperative to resolve state uncertainty i.e., the mutual information between the agent's beliefs before and after making a new observation.

5 Concluding remarks

Active inference agents, equipped with preference learning, prefer modes of behaviour that lead to predictable states, necessary for survival. In other words, these agents have certain kinds of expectation about the states of affair and act to realise them. Importantly, in a changing context, they can learn context-dependent subjective preferences. Here, these subjective preferences constitute what is preferred by the agent, instead of objective preferences, i.e., what is considered optimal by the environment 'designer'. This introduces an agent-centred formulation that allows the agent to construct its niche when interacting with the environment (Bruineberg et al., 2018b; Constant et al., 2018). We illustrate this through simulations, where preference learning agents develop bi-modal preferences when faced with extreme uncertainty but act to satisfy preferences in predictable settings.

Related work Reinforcement learning (RL) is another plausible framework for building adaptive agents. However, it relies on an extrinsic signal to reinforce modes of behaviour that are difficult to pre-define in real environments. Without these extrinsic signals, the agent must rely on intrinsic motivation to exhibit adaptive, exploratory behaviour (Ryan and Deci, 2000; Singh et al., 2004, 2005). Popular methods focus on exploration based on information gain (Houthooft et al., 2016; Still and Precup, 2011), prediction error (Achiam and Sastry,

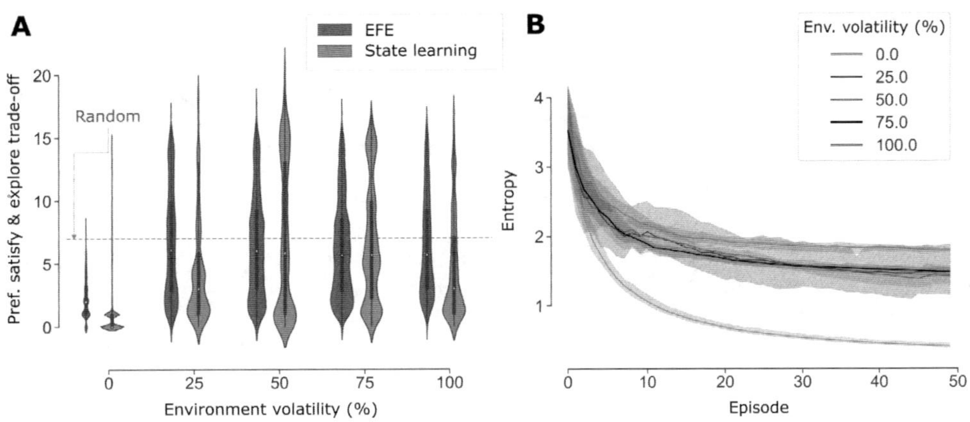

Fig. 4. A: The violin plot presents the preference satisfaction and exploration trade-off measured using Hausdorff distance (Blumberg, 1920) at different levels of volatility in the environment. The x-axis denotes environment volatility at constant map (0%), change in map every 40 steps (25%), 20 steps (50%), 10 steps (75%) and every step (100%). The y-axis denotes the Hausdorff distance. Here, red reports the agent optimising the standard expected free energy (EFE), and green for preference learning. **B:** The line plot depicts the entropy of (approximate) posterior beliefs over states, across varying levels of volatility in the environment. The x-axis represents the episodes, and the y-axis entropy (in natural units). Here, the dark lines represent the mean (across 10 seeds), and shaded area the 95% confidence interval. The pink line is for 0%, blue for 25%, green for 50%, black for 75% and red for 100% volatility in the environment.

IOP Conf. Series: Materials Science and Engineering 1261 (2022) 012020 doi:10.1088/1757-899X/1261/1/012020

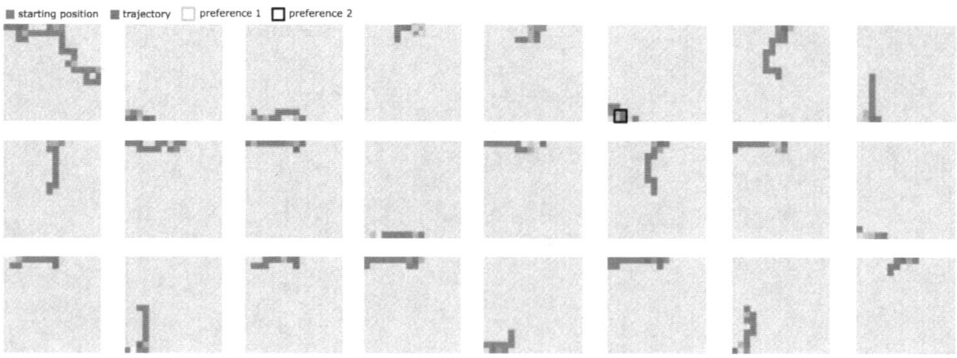

Fig. 5. Representative example of trajectories observed during preference learning in volatile settings. Each figure illustrates an agent's trajectory for a particular episode. Here, purple is the starting position, pink the trajectory, cyan square denotes the first learnt preference and dark blue denotes the second learnt preference.

2017; Pathak et al., 2017; Stadie et al., 2015), novelty search (Lehman and Stanley, 2008; Shyam et al., 2019), curiosity (Schmidhuber, 1991, 2007), entropy (Haarnoja et al., 2018; Lee et al., 2019), empowerment (Klyubin et al., 2005; Mohamed and Rezende, 2015) or control as inference Levine (2018); Todorov (2008).

Recently, the focus has shifted to self-supervised reward-free learning (Jin et al., 2020; Wang et al., 2020), but the ambition of such approaches is to learn representations that can be exploited in different (downstream) tasks. Conversely, our approach relies on agents learning subjective preferences that shape interactions with the environment. Thus, any observed exploration is simply an emergent behaviour of the agent, and not a contrived procedure for improving future task performance. Conceptually, preference learning active inference agents are most similar to open-ended learning algorithms Schmidhuber (2010); Standish (2003); Stanley et al. (2017), where agents are responsible for never-ending learning opportunities.

Limitations and future directions By equipping the agent with the capacity to learn subjective preferences, we forego control over the agent's behaviour. That is, by removing the ability to manipulate the agent preferences – regarding what is considered appropriate behaviour – we have removed the only clear communication channel that can be used to control agent behaviour and/or define task goals. However, given the flexibility of the active inference formulation, precise preferences – under direct supervision – could be installed as required. Practically, this simply involves equipping an agent with precise preferences over states (or outcomes) with suitable Dirichlet parameters (i.e., concentration parameters). To create agents that have confident beliefs in their preferences—and that are therefore more resistant to learning new preferences—the initial (prior) Dirichlet parameters can be set to high values. Effectively, this means that agents have to experience many more outcomes before they come to learn these are preferred or characteristic outcomes for the agent in question. Conversely, starting with very low Dirichlet parameters creates an impressionable agent who is predisposed to learning its preferences. Our simulations revealed that rich category spaces were required for learning preferences that establish distinct behavioural strategies; specifically, within a volatile setting. Thus, future research could evaluate the parameterisation of latent states suitable for equipping the agent with the capacity to learn apt preferences.

Preference learning of the sort considered in this work, emerges from minimising variational free energy across time, when agents are equipped with hyper-priors over their preferences. When parameterised using a conjugate (Dirichlet) hyper-prior learning corresponds to accumulating the relative frequency of different outcomes after enacting policies. The functional form of the ensuing variational updates is the same as commonly used learning rules, known as associative or Hebbian plasticity (Friston et al., 2016). Here, synaptic efficacy is reinforced by the simultaneous firing of pre-and post-synaptic neurons e.g., as more frozen tiles are observed, more evidence is accumulated in the synaptic connection to support the hypothesis that these are the sort of preferentially observed tiles. Future work could investigate how different learning strategies influence preference accumulation, and ultimately agent behaviour. Importantly, the current preference learning strategy did not employ 'reset' or removal of redundant preferences (i.e., Dirichlet concentration parameters): having accrued a preference for frozen tiles the agent could not remove these preferences even if the environment changed. The resetting of Dirichlet hyper-priors was evaluated in (Sajid et al., 2021c) and produces consistently exploratory agents that respond to epistemic (uncertainty reducing) affordances. The interesting question now is whether one can equip generative models with a further hierarchical level to produce agents that optimally reset their prior preferences, when moving to a new environment or context. Future research can investigate how the timescale, at which redundant preferences are removed, influences adaptive behaviour.

Acknowledgements

NS acknowledges funding from the Medical Research Council, UK (MR/S502522/1). PT is supported by the UK EPSRC CDT in Autonomous Intelligent Machines and Systems (grant reference EP/L015897/1). KJF is funded by the Wellcome Trust (Ref: 203147/Z/16/Z and 205103/Z/16/Z).

Bibliography

Joshua Achiam and S. Sastry. Surprise-based intrinsic motivation for deep reinforcement learning. *ArXiv*, abs/1703.01732, 2017.

W Ross Ashby. *An introduction to cybernetics*. Chapman & Hall Ltd, 1961.

Hagai Attias. Planning by probabilistic inference. In *International Workshop on Artificial Intelligence and Statistics*, pages 9–16. PMLR, 2003.

Henry Blumberg. Hausdorff's grundzüge der mengenlehre. *Bulletin of the American Mathematical Society*, 27(3):116–129, 1920.

Matthew Botvinick and Marc Toussaint. Planning as inference. *Trends in cognitive sciences*, 16(10):485–488, 2012.

Jelle Bruineberg, Julian Kiverstein, and Erik Rietveld. The anticipating brain is not a scientist: the free-energy principle from an ecological-enactive perspective. *Synthese*, 195 (6):2417–2444, 2018a.

Jelle Bruineberg, Erik Rietveld, Thomas Parr, Leendert van Maanen, and Karl J Friston. Free-energy minimization in joint agent-environment systems: A niche construction perspective. *Journal of theoretical biology*, 455:161–178, 2018b.

Roger C Conant and W Ross Ashby. Every good regulator of a system must be a model of that system. *International journal of systems science*, 1(2):89–97, 1970.

Axel Constant, Maxwell JD Ramstead, Samuel PL Veissiere, John O Campbell, and Karl J Friston. A variational approach to niche construction. *Journal of the Royal Society Interface*, 15(141):20170685, 2018.

Lancelot Da Costa, Thomas Parr, Noor Sajid, Sebastijan Veselic, Victorita Neacsu, and Karl Friston. Active inference on discrete state-spaces: a synthesis. *Journal of Mathematical Psychology*, 99:102447, 2020.

WH Fleming and SJ Sheu. Risk-sensitive control and an optimal investment model ii. *The Annals of Applied Probability*, 12(2):730–767, 2002.

Karl Friston. Life as we know it. *Journal of the Royal Society Interface*, 10(86):20130475, 2013.

Karl Friston. A free energy principle for a particular physics. *arXiv preprint arXiv:1906.10184*, 2019.

Karl Friston, Thomas FitzGerald, Francesco Rigoli, Philipp Schwartenbeck, Giovanni Pezzulo, et al. Active inference and learning. *Neuroscience & Biobehavioral Reviews*, 68:862–879, 2016.

Karl Friston, Thomas FitzGerald, Francesco Rigoli, Philipp Schwartenbeck, and Giovanni Pezzulo. Active inference: a process theory. *Neural computation*, 29(1):1–49, 2017.

Tuomas Haarnoja, Aurick Zhou, Pieter Abbeel, and Sergey Levine. Soft actor-critic: Off-policy maximum entropy deep reinforcement learning with a stochastic actor. In *ICML*, 2018.

Danijar Hafner, Timothy Lillicrap, Mohammad Norouzi, and Jimmy Ba. Mastering atari with discrete world models. *arXiv preprint arXiv:2010.02193*, 2020.

Jakob Hohwy. The self-evidencing brain. *Noûs*, 50(2):259–285, 2016.

Rein Houthooft, Xi Chen, Yan Duan, John Schulman, Filip De Turck, and Pieter Abbeel. Vime: Variational information maximizing exploration. In *NIPS*, 2016.

Chi Jin, Akshay Krishnamurthy, Max Simchowitz, and Tiancheng Yu. Reward-free exploration for reinforcement learning. In *International Conference on Machine Learning*, pages 4870–4879. PMLR, 2020.

Daniel Kahneman and Amos Tversky. Prospect theory: An analysis of decision under risk. In *Handbook of the fundamentals of financial decision making: Part I*, pages 99–127. World Scientific, 2013.

Diederik P Kingma and Jimmy Ba. Adam: A method for stochastic optimization. *arXiv preprint arXiv:1412.6980*, 2014.

Alexander S Klyubin, Daniel Polani, and Chrystopher L Nehaniv. Empowerment: A universal agent-centric measure of control. In *2005 IEEE Congress on Evolutionary Computation*, volume 1, pages 128–135. IEEE, 2005.

Lisa Lee, Benjamin Eysenbach, Emilio Parisotto, Eric P. Xing, Sergey Levine, and Ruslan Salakhutdinov. Efficient exploration via state marginal matching. *ArXiv*, abs/1906.05274, 2019.

J. Lehman and K. Stanley. Exploiting open-endedness to solve problems through the search for novelty. In *ALIFE*, 2008.

Sergey Levine. Reinforcement learning and control as probabilistic inference: Tutorial and review. *arXiv preprint arXiv:1805.00909*, 2018.

Beren Millidge. Deep active inference as variational policy gradients. *Journal of Mathematical Psychology*, 96:102348, 2020.

Shakir Mohamed and Danilo Jimenez Rezende. Variational information maximisation for intrinsically motivated reinforcement learning. In *NIPS*, 2015.

Thomas Parr and Karl J Friston. Generalised free energy and active inference. *Biological Cybernetics*, 113(5-6):495–513, 2019.

Deepak Pathak, Pulkit Agrawal, Alexei A. Efros, and Trevor Darrell. Curiosity-driven exploration by self-supervised prediction. *2017 IEEE Conference on Computer Vision and Pattern Recognition Workshops (CVPRW)*, pages 488–489, 2017.

Richard M Ryan and Edward L Deci. Intrinsic and extrinsic motivations: Classic definitions and new directions. *Contemporary educational psychology*, 25(1):54–67, 2000.

Noor Sajid, Philip J Ball, Thomas Parr, and Karl J Friston. Active inference: demystified and compared. *Neural Computation*, 33(3):674–712, 2021a.

Noor Sajid, Lancelot Da Costa, Thomas Parr, and Karl Friston. Active inference, bayesian optimal design, and expected utility. *arXiv e-prints*, pages arXiv–2110, 2021b.

Noor Sajid, P. Tigas, A. Zakharov, Z. Fountas, and Karl J. Friston. Exploration and preference satisfaction trade-off in reward-free learning. *ArXiv*, abs/2106.04316, 2021c.

Jürgen Schmidhuber. A possibility for implementing curiosity and boredom in model-building neural controllers. 1991.

Jürgen Schmidhuber. Simple algorithmic principles of discovery, subjective beauty, selective attention, curiosity & creativity. In *Discovery Science*, 2007.

Jürgen Schmidhuber. Formal theory of creativity, fun, and intrinsic motivation (1990–2010). *IEEE Transactions on Autonomous Mental Development*, 2(3):230–247, 2010.

Roger Schrodinger, Erwin Schrödinger, and Erwin Schr Dinger. *What is life?: With mind and matter and autobiographical sketches.* Cambridge University Press, 1992.

Ramanan Sekar, Oleh Rybkin, Kostas Daniilidis, Pieter Abbeel, Danijar Hafner, and Deepak Pathak. Planning to explore via self-supervised world models. In *International Conference on Machine Learning*, pages 8583–8592. PMLR, 2020.

Pranav Shyam, Wojciech Jaskowski, and Faustino Gomez. Model-based active exploration. In *ICML*, 2019.

Satinder Singh, A. Barto, and N. Chentanez. Intrinsically motivated reinforcement learning. In *NIPS*, 2004.

Satinder Singh, Andrew G Barto, and Nuttapong Chentanez. Intrinsically motivated reinforcement learning. Technical report, MASSACHUSETTS UNIV AMHERST DEPT OF COMPUTER SCIENCE, 2005.

Bradly C. Stadie, Sergey Levine, and P. Abbeel. Incentivizing exploration in reinforcement learning with deep predictive models. *NIPS*, 2015.

Russell K Standish. Open-ended artificial evolution. *International Journal of Computational Intelligence and Applications*, 3(02):167–175, 2003.

Kenneth O Stanley, Joel Lehman, and Lisa Soros. Open-endedness: The last grand challenge you've never heard of. *While open-endedness could be a force for discovering intelligence, it could also be a component of AI itself*, 2017.

Susanne Still and Doina Precup. An information-theoretic approach to curiosity-driven reinforcement learning. *Theory in Biosciences*, 131:139–148, 2011.

Emanuel Todorov. General duality between optimal control and estimation. In *2008 47th IEEE Conference on Decision and Control*, pages 4286–4292. IEEE, 2008.

Ruosong Wang, Simon S Du, Lin F Yang, and Ruslan Salakhutdinov. On reward-free reinforcement learning with linear function approximation. *arXiv preprint arXiv:2006.11274*, 2020.

John Winn, Christopher M Bishop, and Tommi Jaakkola. Variational message passing. *Journal of Machine Learning Research*, 6(4), 2005.

IOP Conf. Series: Materials Science and Engineering 1261 (2022) 012021 doi:10.1088/1757-899X/1261/1/012021

Embodiment and metaphorising in the learning of mathematics

Jorge Soto-Andrade[1,2], Daniela Díaz-Rojas[2,3], Amaranta Valdés-Zorrilla[1,2]

[1]Mathematics Department, Faculty of Science, University of Chile, Santiago, Chile.
[2]Centre for Advanced Study in Education, University of Chile, Santiago, Chile.
[3]Department of Education, University of Oxford, UK
E-mail: sotoandrade@uchile.cl

Abstract. In this paper we undertake a critical reading of the Ouroboric (circular) relationship between the guiding metaphors of artificial intelligence and those of natural human intelligence, which has evolved from a vicious circle, triggered by the metaphorising of cognition as information processing, to a virtuous one, fostered by the advent of the embodied and enactive turn in cognitive science. We describe then our own experimental and theoretical approach to mathematical thinking and learning, where metaphorization plays a key role, besides embodiment and enaction. We comment on some concrete examples of mathematical thinking with different types of learners, in combinatorics, arithmetic, geometry and probability, where idiosyncratic metaphorization emerges and enacting and embodiment make a dramatic difference. We outline finally some significant challenges to contemporary AI, cognitive sciences, and mathematics proper, suggested by our examples. These challenges involve metaphorising as the tip of the iceberg in human cognition, metaphorising in intelligent systems, Ouroboric circularity, mathematical collective improvisation as an analogue of musical improvisation, group creativity and human swarm intelligence.

1. Introduction

In our view, an Ouroboric[1] relationship has arisen between the guiding metaphors of artificial intelligence (AI) and those of "natural" human intelligence. Since AI aims at emulating and even surpassing human intelligence [34], an understanding of the latter, or better, of human *cognition*, is required.

For a long time, however, the prevailing metaphors for human cognition have been "the brain as a computer" and "cognition as information processing". The notion of computation was not seen as a metaphor but as part of a literal description of cognitive activity [52], [23]. These metaphors pertain rather to AI, though. This generated a sort of vicious circle in the first decades of AI.

More recently, however, we have witnessed an "embodied turn" in cognitive sciences and the emergence of an enactivist approach to cognitive sciences, as launched by Varela [81]. The arising of a virtuous circle became then possible, where new paradigms in cognitive sciences, like 4E (Embodied, Enactive, Embedded, Extended) Cognition and Ecological Cognition [24] informed AI and posed great challenges for the design of intelligent systems, as foreseen by Brooks [4] and acknowledged by Varela, Thompson and Rosch [81].

[1] The Ouroboros, the snake eating its own tail, is an outstanding metaphor of circularity, self-reference, and organisational closure in living systems [39], [67] and in cognition as enaction [81].

In this paper we focus then on embodied, enactive, and also metaphorical aspects of human cognition, drawing on our "experimental epistemological approach" to the supposedly most abstract variety of human thinking, to wit mathematical thinking. Notice that we refer to mathematical thinking and not to mathematical reasoning, proving, deducting or calculating. Indeed, together with some outstanding mathematicians, like Manin [37], Thom [71], Thurston [74][2], we deem that the axiomatic-deductive genre in mathematics is just *one* possible genre among many others [64].

Unfortunately, metaphorical, embodied, enactive aspects of mathematical thinking have been often neglected in traditional mathematics education, where - in our view - "robotic" teaching and learning prevail. In our country, we call metaphorically "donut factory teaching" this sort of standardized and pre-packaged massive teaching, where students are supposed to assimilate definitions and techniques and then exercise. In retrospect, this fits well with the first guiding metaphors for AI [34].

1.1. AI and natural intelligence: a vicious circle.

Good Old Fashioned Artificial Intelligence (GOFAI) adhered to the metaphors :

"perception = representation" & "cognition = computation";

Moreover, it identified thinking with reasoning and with logical symbol processing [47]. As it often happens, these questionable metaphors became "second nature" metaphors, i. e. unacknowledged and invisible ones. In the words of Pylyshyn [52, p. 435]:

> "For me the notion of computation ... is not a metaphor but part of a literal description of cognitive activity. This is not to say that there are not also metaphorical uses of computer concepts. But it seems to me that computation, and all that it entails regarding rule-governed transformations on intentionally interpreted symbolic expressions, applies just as literally to mental activity as it does to the activity of digital computers. Such a term is in no sense a literal description of the operation of electronic computers that has been metaphorically transported to the primary subject of mind"

GOFAI also claimed that "every aspect of learning or any other feature of intelligence can in principle be so precisely described that a machine can be made to simulate it" [41]. It accommodated well with a representationalist view of cognition, since it was taken for granted that to be able to reason one needs to construct an *abstract representation* of the object of interest, as in digital computing [47]. The metaphor of information became prevailing and ubiquitous, in a way quite analogous to the metaphor of phlogiston in early thermodynamics. The software-hardware dichotomy mirrored the Cartesian mind-body dichotomy.

Notice that this dichotomy may be seen already in the legend of mythical forerunner of an intelligent robot, to wit the XVI century *Golem*, in Jewish Prague, whose hardware was the sticky clay of the Moldau river and whose software was provided by the Hebrew word EMET (Truth) written on its forehead. The Golem was initially programmed to defend the Jewish community of that time in Prague [34]. Recall that the term *robot* was also coined in Prague.

So, a vicious circle of mutually justifying arguments in cognitive science and in AI arose [47]. AI at that time aimed at the implementation of mechanized logico-mathematical reasoning systems, because human intelligence was metaphorically understood in that way [34].

1.2. AI and natural intelligence: a virtuous circle?

In the last decades new paradigms have emerged in the cognitive sciences though, which counter representationalism and cognitivism. Particularly, the enactive paradigm and more generally Four E's (Embodied, Extended, Embedded and Enactive) Cognition [24]. This paradigm is based on an active and dynamic approach of the interactions between brain, body and environment [15], [24], [81], [82] and encompasses the hitherto neglected embodied, enactive, extended, distributed, socially and culturally situated dimensions of human cognition. In this vein, we should also mention "cognition as

[2] See also Fields Medallist Terence Tao's report on his lived mathematical experience in https://mathoverflow.net/questions/38639/thinking-and-explaining

material engagement" [36], [48]. The "enactive programme" in cognitive science launched by Varela [81] and its enactivist developments, emphasising also lived subjectivity [72], [73], [18] deserve special attention in this regard [23]. We describe them in more detail below.

This paradigm shift in cognitive sciences has dramatically affected the underlying metaphors for AI, which became also embodied and embedded, e.g. [4] - [6], [15], [82]. In this way, a virtuous circle arose, where, since the 1980s, AI has expanded into a broader study of the interaction between the body, brain, and environment, and of how intelligence emerges from such interaction. Notice that the authors of The Embodied Mind even hailed the emergence of Brooks's behaviour-based robotics as a "fully enactive approach to AI" [81, p. 212]. As a consequence of this embodied turn in cognitive sciences, nowadays AI also addresses the study of intelligent action in agents other than organisms or robots, thus metaphorising corporations, social systems, and networked embedded devices as intelligent and adaptive systems, acting, surviving and thriving in highly uncertain and unpredictable environments. A new vision emerges then of biologically based cognition and socially, spatially, and temporally distributed modes of supra-individual cognition [34], [48].

1.3. Aphorismic overview of our stance on mathematical thinking:

Núñez [45] asks how can an embodied view of the mind give an account of an abstract, idealized, precise, sophisticated and powerful domain of ideas, like mathematics, where direct bodily experience with the subject matter is not possible. Notice that this way of seeing mathematics suggests that GOFAI could successfully simulate mathematical thinking. After all, it has been taken for granted for a long time that to be able to reason one needs to construct an *abstract representation* of the object of interest, as in digital computing [47]. Núñez [45] makes a case for the embodied nature of mathematical ideas though, via *conceptual metaphor theory* (CMT), as presented in [33]. He argues then that if mathematics is embodied in nature, then any conceptual system would also be embodied.

We summarise below our own enactivist stance on mathematical thinking, which is not far from his, because we also argue that we need both embodiment and metaphorization to account for the construction of mathematical concepts.

Regarding intelligence, we agree with Smith and Gasser [62] in that "intelligence emerges in the interaction of an agent with an environment and as a result of sensorimotor activity".

In our view, mathematics is not utopic and uchronic: It is situated and embodied. Mathematical concept construction is multimodal and carried out via metaphorisation and enaction. We claim that mathematical abstraction is a pious lie: indeed, we metaphorise concretely - often unconsciously - when constructing "abstract" concepts or dealing with them.

Our radically enactivist theoretical stance on teaching and learning is well rendered by the provocative aphorism: "If you want somebody to learn something, the worst you can do is to try to teach him/her". This aphorism is appreciated by pioneers of mathematics education like Brousseau [8], [9] or Mason [38], but baffles a traditional educator, who looks upon learning as an *outcome* of teaching, unwittingly adhering to the aforementioned "donut factory" metaphor for school.

In our view, learning is neither following a path given in advance [83], nor climbing a staircase [43]; it is rather a random walk on a cognitive and epistemological topography [65], [66], [69]. This topography is not given in advance, it may change in the course of time and be modified by the unfolding of the random walk itself (an Ouroboric relationship). So, we see learning as a rhizomatic process [19].

A somewhat fundamentalist claim, now widespread in mathematics education, is that *problem solving* is the royal road to mathematics learning. We like to say that "this is not false, but it is not true either", because also *problem posing* matters. Indeed, from an enactivist viewpoint, problems are not "out there" waiting for us to solve them. They are co-constructed by us in *situation*, through an interactive circular process [81], [50].

We also dissent from the prevailing "gas fitter metaphor" for problem solving: "We look into our toolbox and choose the appropriate tool to solve our problem". From an enactivist perspective, the relevant tools are rather forged in the course of a random collective exploration of a situation, where questions or problems emerge [64], [50].

IOP Conf. Series: Materials Science and Engineering 1261 (2022) 012021 doi:10.1088/1757-899X/1261/1/012021

2. Our theoretical background regarding (natural) mathematical thinking and learning

We focus now on three theories in cognitive sciences and mathematical education which are especially relevant to the study of mathematical thinking and intelligence, particularly for the construction and learning of mathematical concepts, to wit:

- the theory of metaphor [49], [22], [33], [60], [61], [64], [65];
- the theory of enaction [81], [55], and
- the experimental epistemology of mathematics (theory of (a)-didactic situations) [8], [9], [11].

We could describe them tersely in metaphorical terms as follows (a tough act to follow for AI, we deem).

- Metaphorising: *"looking" at something and "seeing" something else*…i. e. *"hallucinating"* in the sense of Seth [57], [58].
- Enaction: "*Caminante, son tus huellas el camino, y nada más; caminante, no hay camino, se hace camino al andar*" *[Wanderer, your footsteps are the path, nothing else; there is no path, you lay down a path in walking]* [78].
- Experimental epistemology of mathematics (theory of a-didactic situations): *Mathematical concepts and processes emerge in a situation instead of being parachuted from Olympus* [8], [9], [11], [65].

Notice that Brousseau's experimental epistemology of mathematics may be seen as a forerunner (already in 1965 [8]) of an enactivist didactics of mathematics.

In our view, these naturally evolved cognitive capacities of our species: enaction and metaphorization, are most relevant to understanding how experience shapes embodied learning in contexts of mathematics education, wherein, traditionally, learners must cope with disembodied algorithms and abstract concepts [50], [51], [65], [21].

These theories intertwine in complex ways. Metaphorising may be used as a meta language to describe or fathom other theories, in particular metaphorising itself. This is just another avatar of the ubiquitous Ouroboros lurking in the field of cognition and intelligence. Recall in this regard that the first (Chilean) 1973 edition of Maturana and Varela's book [39] introducing autopoietic systems, pictured an Ouroboros on its cover, although the notion of self-reference is not explicitly mentioned there [67]. The theory of autopoiesis has however deep connections to the idea of self-reference. In fact, one of its creators, Varela, spent almost a decade searching for a suitable mathematical framework to formalise the notions behind these connections [75], [76]. [80], [68].

We describe below the main tenets of the theories of metaphor, enaction, and a-didactic situations.

2.1. Metaphorizing in cognitive science and math education

Notice that, because of our enactivist stance, we prefer to use the *verb* "metaphorising", an action carried out by a human being, instead of the *noun* "metaphor," a concept standing "out there" independently of an observer.

Regarding the ubiquitous role of metaphorising in human cognition, we would neither say nowadays that metaphorizing is just "transporting to one thing, the name that designates another" citing Aristotle [54, p. 23], nor that it is a *trope*, a figure of speech. We prefer the metaphorical description: "metaphorizing is looking at something and seeing something else," so that, in some sense, metaphorizing is *hallucinating* [57], [58], a key biological process.

Indeed, the sensorimotor structures that make it possible depend on the skeletal structures, muscles and the nervous system to enable movement [24]. Gould and Vrba [29] refer, on their side, to exaptation, i.e., the adaptive re-instrumentalization of existing organs and faculties to higher-order routines. Other researchers, like Dehaene [31], relate metaphorizing to the recruiting of neural mechanisms developed in the course of evolution to tackle new situations (e.g., subtracting as tracking an object moving backwards).

It is acknowledged nowadays that metaphorising serves as the often unknowing foundation for human thought [27], [60], [61], since our ordinary conceptual system, in terms of which we both think

IOP Conf. Series: Materials Science and Engineering 1261 (2022) 012021 doi:10.1088/1757-899X/1261/1/012021

and act, is fundamentally metaphorical in nature [32]. Metaphors are not just rhetorical devices but powerful and legitimate cognitive tools that help us in building or grasping new concepts, as well as in solving problems in an efficient and friendly way.

Evidence of this far-ranging role of metaphorising may be found in the pioneering work of Abrahamson et al. [1], Chiu [14], English [22], Lakoff and Núñez [33], Presmeg [49], Sfard [60], [61], Núñez [46], McGowen & Tall [42]. See also Díaz-Rojas et al. [21] and Soto-Andrade [64], [65].

Lakoff and Núñez [33] highlight the intensive (and often unconscious) use we make in our thinking, of *conceptual metaphors* that appear—metaphorically—as inference-preserving mappings from a more concrete or down-to-earth 'source domain' into a more abstract or opaque 'target domain', enabling us to fathom the latter in terms of the former. A classic example of a conceptual metaphor is "a teacher is a gardener," which maps gardening to education, allowing us to transfer some gardening wisdom to education (a less fathomable domain than gardening). For instance: a tree needs a favourable soil and water to grow; its growth cannot be accelerated ad libitum...

Let us remark, *en passant*, that *representations* go the other way around with respect to *metaphors*. Indeed, we *re-present* something given beforehand, typically to explain it, but we usually *metaphorise* to try to fathom something unknown or not yet constructed.

Our approach to the learning of mathematics emphasises the *poietic* (from the Greek *poiesis* = creation) role of metaphorising, which brings concepts into existence. Indeed, conceptual metaphors often help us to construct new concepts. They are not just *poetical*, like "the queen of the night" (the moon), or *rhetorical*, like "we are fighting against a powerful and merciless enemy" (the coronavirus, as seen by many).

For instance, we *create* the concept of probability when, while watching the random walk of a "fair frog" jumping equally likely right or left on a row of stones in a pond, we "hallucinate" [57], [58], seeing it *splitting* into two equal halves that go right and left, and so on! This "metaphoric sleight of hand" turns a random process into a deterministic one, and we just need to keep track of the frog's splitting into pieces to answer "impossible questions" like: "Where will the frog be after n jumps?". The probability of finding the frog at a given stone after n jumps is *constructed* as the *portion of the frog* landing there after n splittings [65].

So, in the framework of the conceptual metaphor theory (CMT) of Lakoff and Núñez [33], we have here a mapping from an existing source domain (sharing or fission processes) to a target domain which *does not exist yet* (stochastic processes). Thus, poietic metaphorising can be better metaphorised by an autopoietic Ouroboros [39] than by an arrow as in CMT, as suggested by Sfard [59]. In this sense, our approach to poietic metaphorising extends CMT as presented in [33] and exemplified in [45]. The ubiquity of metaphorising in mathematics education should not be underestimated. Besides bringing into existence mathematical concepts or objects and helping learners to fathom them, unconscious metaphorising often dramatically shapes the way teachers teach, for instance. A foremost example is afforded by the ubiquitous *second nature* metaphor 'teaching is transmitting knowledge'. Unperceived here is the *acquisition metaphor* [61], [64], which sees learning as acquiring an accumulated commodity. This is criticised in Plutarch's metaphor: 'A mind is a fire to be kindled, not a vessel to be filled' [61].

Paraphrasing Bachelard [3], who advocated epistemological vigilance, we suggest nowadays to practise metaphorical vigilance, i.e., the art of noticing [38] our unconscious or implicit metaphors, that shape our way of interacting with the world (even to construct it) and particularly our approach to teaching and learning.

Núñez [45], drawing on CMT, makes a case for mathematics, a foremost example of a complex abstract and formalised conceptual system, being embodied in nature. He argues that then *any* abstract conceptual system would also be embodied.

He remarks, on the other hand, that definitions and axioms often neither formalize nor generalize adequately human everyday concepts. Indeed, when we do mathematics, notably in the axiomatic way, we usually focus on something, and we ignore something else; so sometimes we throw away the baby with the bath's water. He exemplifies this with the classical formal definition of continuity of a function (the "epsilon-delta yoga" taught in Calculus 1) which is unfriendly for most learners and allows for teratological examples of continuity, including functions that oscillate wildly but with amplitude decreasing to 0 or the coastline of the fractal Koch Island.

We are led then, Núñez argues, to acknowledge the limitations of the axiomatic deductive genre in mathematics, which after all is just one possible genre among many, as Manin [37] points out.

This is however a never-ending story: mathematicians have also invented non-standard analysis and *hyper-real numbers* [44] and also *surreal numbers* [16] as a rather successful attempt to recover the infinitesimal baby thrown away with the axiomatic bath's water.

Last but not least, metaphorising plays also a key epistemological role. Its poietic power applies not just to concepts but also to whole theories. We have claimed elsewhere [64] that, metaphorically speaking, a theory is just the 'unfolding' of a *poietic* metaphor (a process, that may be laborious and technical, though). A paradigmatic example is the 'tree of life' metaphor in Darwin's theory of evolution.

We will use below metaphorising as a meta-language to describe other theories relevant to us, like enactivism or the theory of a-didactic situations, in terms of their generating metaphors, something more helpful to fathoming how they arose than just describing them a posteriori. This applies also to metaphor itself, in a circular way. This fits well with Ricoeur's saying that "There is no non-metaphorical standpoint from which one could look upon metaphor" [75]. Recall by way of example, Bruner's [13, p. 48], metaphorical description: "Metaphors are crutches to help us to get up the abstract mountain, but once up, we throw them away (even hide them)".

2.2. *Enaction and enactivism*

In recent years, embodied and enactive approaches to cognition have emerged, which counter cognitivism and highlight the active and dynamic interactions between brain, body and environment [81], [24], [18], [70]. This embodied enactive turn in cognitive science entailed a radical change from the traditional and representationalist way of metaphorizing our cognition, supported by cognitivism.

A landmark in this path was set by Varela, when he introduced what he called the *enactive approach* in cognitive science [77] - [79], [81]. In his own words:

> "The world is not something that is given to us but something we engage in by moving, touching, breathing, and eating. This is what I call cognition as enaction since enaction implies this bringing forth by concrete handling." [79, p. 8]

Varela fittingly cited Antonio Machado's well-known poem [35, p. 142], [78, p. 63]:

> "Caminante, son tus huellas el camino, y nada más; caminante, no hay camino, se hace camino al andar [Wanderer, your footsteps are the path, nothing else; wanderer, there is no path, you lay down a path in walking]";

an unfolding metaphor for enaction, indeed. Cognition as enaction was also metaphorized by Varela by the well-known "Drawing Hands" lithograph by Escher, where each hand draws the other into existence [77, p. 310].

Notice to what extent the "laying a path in walking" metaphor challenges the "second-nature" metaphor in education, which sees learning, especially mathematics learning, as following a well-marked path given in advance, as in the "learning trajectories" of Wittmann [83], or as "climbing a staircase to heaven" [43]. In the enactivist view, on the contrary, the curriculum is a fluid, enactive and unpredictable process emerging from the situated dialogical relation between students and teacher relation [17].

Key aspects of enaction are: perceptually guided action, embodiment and structural coupling through recurrent sensorimotor patterns [81], [53]. In an aphorism: 'All doing is knowing, and all knowing is doing' [40, p. 26].

Notice that the term "enaction" in mathematics education may be traced back to Bruner [12], who was following the traces of Dewey's [20] "learning by doing", when he described the *enactive, iconic and symbolic* modes of representation. Bruner's enaction, which means essentially acting out or doing, is however far less radical than Varela's, in that it does not challenge the notion of a reality 'out there', that we represent to ourselves more or less successfully. Dewey, however, emphasised the role of sensorimotor coordination in perception, acknowledging that movement is primary and sensation is secondary [26].

In what follows, to avoid confusion between Bruner's and Varela's enaction, we use the now prevalent terms *enactivism* and *enactivist* to refer to Varela's anti-representationalist 'enactive program', which sees cognition as enaction. For instance, an *enactivist approach* to problem solving or to mathematics education. On the other hand, unless otherwise explicitly stated, 'enact', 'enacting' and 'enactive' are to be understood in the sense of everyday language and Dewey [20] and Bruner [12], i.e., as synonyms of 'acting out' or 'acted out', in an embodied way. So 'enacting a metaphor' just means 'to act it out', with your body, as in Gallagher and Lindgren [26], where they refer to 'enactive metaphors' as opposed to what they call 'sitting metaphors'. We use the term 'enactive metaphorising' below in this sense.

Varela's enactive programme has provided remarkable new insights on learning, problem solving and problem posing [55], [50], [69]. Indeed, learning is now seen as arising from the interaction of the learner's structure and the environment, which plays at most the role of a 'trigger'. It is not a *result* or an *outcome* of teaching. Teaching and learning co-construct each other.

Traditionally, *problem solving* involves problems given beforehand, lying 'out there' in the world, waiting to be solved, independently of us. The prevailing usual *gas fitter metaphor* applies then, where solvers look into their toolboxes of predefined strategies and choose the appropriate one for solving the problem at hand.

In the enactivist perspective, because of our structural coupling with the world [81], we bring forth emergent problematic situations instead. This is what Varela calls *problem posing*. Moreover, mathematical strategies emerge continually in the interaction of solver and problematic situation [50].

This perspective is implemented in our courses, by not giving the students tasks to carry out, but instead, just presenting (adidactic) *situations* to them. What specific contents we tackle and how we work them out is strongly dependent on the students' reactions, on their previous learning history and on the ensuing classroom dynamics. We often take advantage of the salient aspects of the current situation the students are immersed in, e.g. the pandemics, so "laying down a course in walking".

In an enactivist didactics of mathematics, the teacher is an enactivist practitioner acting in situation and learning is an emergent, situated and embodied process [51]. For a survey of enactivism in mathematics education, see [28].

2.3. *Experimental epistemology of mathematics: a-didactic situations and didactic contract*

The term *experimental epistemology of mathematics* was coined by Brousseau already in 1963 (Brousseau, oral communication, July 14th, 2021) to describe his pioneering approach to the learning of mathematics, suggestively embodied in [8] and later developed in his theory of didactic situations [9], [11]. The latter may be seen as an unfolding of the "emergence metaphor" for mathematical content. Indeed, mathematical concepts or procedures we intend to teach, instead of being parachuted from Olympus as in traditional and abusive teaching, should *emerge* in a suitable challenging situation the learner is enmeshed in, as the only means to "save her/his life". This type of situation is called a *didactic situation*, because of the avowed didactical intent of the teacher who set it up. It becomes an *a-didactic* situation when the teacher steps back, to let the learners interact (enactively) on their own with the situation, unable to fathom beforehand the teacher's didactical design or the mathematical content she is aiming at.

Notice that Brousseau's theory of a-didactic situations may be seen as a forerunner of enactivist approaches to mathematics education. Remarkably Brousseau coined himself the term "experimental epistemology (of mathematics)" to refer to his theoretical approach, in the 60's, at the same time that Maturana and Varela [42] were setting up their laboratory with the same name, "experimental epistemology", in the context of biology of cognition, at the University of Chile.

Later, Brousseau [9], aiming at accounting for the actions and reactions of the partners involved in a didactic situation, also introduced the metaphor of a "didactic contract" as an interpretative embodiment of their mutual expectations, beliefs and commitments [10]. This sort of contract is necessarily tacit and unspoken; its effects may be nevertheless perverse and dramatic. Indeed, the prevailing didactic contract in our classrooms usually thwarts idiosyncratic metaphorising and enacting among the learners.

3. Illustrative examples of embodied, enactive and metaphorical thinking in the classroom.

We discuss below some examples, in combinatorics, arithmetic, geometry and probability, drawn from our teaching at the University of Chile, which illustrate important aspects of the implementation of our different embodied, enactive and metaphorical, theoretical perspectives and their intertwining.

Learners involved were prospective secondary school math teachers, undergraduate math students, humanistic first year students and in-service primary school teachers. They usually worked in (random) small groups of 3 to 4, and they were observed by a teacher and assistants, in the spirit of ethnomethodology [30].

3.1. The staircase climbing problem: enactive combinatorics.

As an example of the application of Machado's metaphor [78] in problem posing and solving, consider the *Staircase climbing problem: In how many ways can you climb a staircase?*

Students usually make sense of this problem by settling first for climbing up to two steps at a time, and notice then that the number F_n of possible ways of climbing an *n*-step staircase becomes the number of ways in which *n* decomposes as a sum of 1's and 2's, where order matters! Then they guess, after experimenting with small n's, the Fibonacci recursion formula : $F_{n+1} = F_n + F_{n-1}$, but they do not see how to prove it or why it is true. However, if we prompt them to (physically) enact the problem and "lay down a staircase in climbing", most of them immediately "see" why the recursion formula holds. Indeed, when they lift their foot to begin climbing, they realise they have to decide whether to take *one or two* steps. So the set of possible ways to climb the staircase splits in two, as in the recursion formula!

3.2. Angles in polygons and stars: enactive geometry [69].

One radically enactivist way in which we approached this topic (which includes the classical "sum of inner angles and exterior angles" problems, which are usually handled via Greek angular yoga) was to just show the students or participants in a workshop, the weird irregular seven-pointed star of Figure 1 and keep silent, waiting for their reactions to arise. If a student asks: "But, which is the question?" then our answer is: "That is the question!"

Participants and students were at first puzzled and intrigued. Some said "Splash!". Especially in service teachers and mathematics educators tried to make sense of the star by *decomposing* it somehow, into triangles. The reaction of *deforming* the given weird star to a friendlier, regular one, did not appear spontaneously. After some prompting: Do you like this star? What would you like to do with it? they recalled the magician's star, or the ubiquitous regular star in our flag.

Most students did not dare to change anything, because of the prevailing didactic contract: the star was not an object to be explored. Primary school teachers noticed the way the star was drawn, like the usual regular five-pointed star, approaching the problem in an enactive way in the sense of Bruner [12], see Fig. 2. They recalled that for the regular five pointed star, they could calculate easily the values of the inner acute angles, but not so with irregular stars. However, by deforming enactively their drawing of the star they realised that stretching one arm of the star decreases the corresponding inner acute angle and increases the others. So, the conjecture emerged that maybe the sum of all inner acute angles was *preserved* under deformation, a clever insight indeed! Notice that this way of thinking: looking for invariants under deformation of a system, is unfortunately not much fostered in our math classrooms.

Our approach was radically enactivist in that no specific problem or task was given to our subjects to solve, but only a situational seed or germ for them to explore and make sense of. They were enactive in Bruner's sense when they drew the stars themselves and deformed it. This enaction allowed them to bodily feel the compensation between increasing and decreasing inner angles, when varying their free hand drawing.

Nevertheless, a more enactive approach to the angle sum problem is possible, inspired by the "laying down a path in walking" metaphor. We have indeed described elsewhere [65] how the value of the sum of the *exterior* angles of a polygon may be "seen", with no calculation, just by "laying down a polygon in walking". When you go around the polygon, inflecting your rectilinear trajectory at each vertex as needed, you are summing up all exterior angles!

In our case, students were able to "lay down a star in walking"! And so they gleaned immediately the value of the sum of its inner acute angles. They noticed that the cleverest walking, among various possibilities, is to follow the path that we lay down when drawing the star, starting as indicated by the arrows in Figure 2. To add up the *inner* angles at the points of the star, we should be able to "span" or "sweep" them somehow, while walking across the star. If metaphorising is already unleashed among the students, many avatars of the walker may emerge. Among them, a metaphorical hummingbird flying *back and forth* along our star.

The bird begins its flight as indicated by the red arrow in Fig. 2, but when arriving to the next vertex, it sweeps (i.e. spans) the corresponding inner angle with its *tail* and then flies *backwards* along the next edge. Then it sweeps the inner angle at the next vertex with its beak and flies forwards, and so on. So it will sweep the inner angles: tail-beak-tail-beak-tail-beak-tail, to end up at the initial vertex again, but looking in the opposite direction. Moreover, during its flight, it always turned its beak clockwise at each vertex. So *our flying hummingbird teaches us that the sum of all inner acute angles is just half a turn.* A playful enactment arises, where one student enacts the hummingbird's flight, wearing a fake beak and tail, and another keeps track of the rotation of the beak or tail. In one of our workshops, a couple of primary school teachers discovered this idea on their own, walking across the star holding a duster.

We see that in this case an enactivist approach to a "concrete" geometric problem triggers all the same significant idiosyncratic metaphorising among the learners, which is also enactive (in Bruner's sense).

We may also notice that the first link of the activated metaphorical chain is a metaphorization of the polygon as a closed circuit (composed of straight segments) which you walk around!

Fig. 1 An irregular seven-pointed star

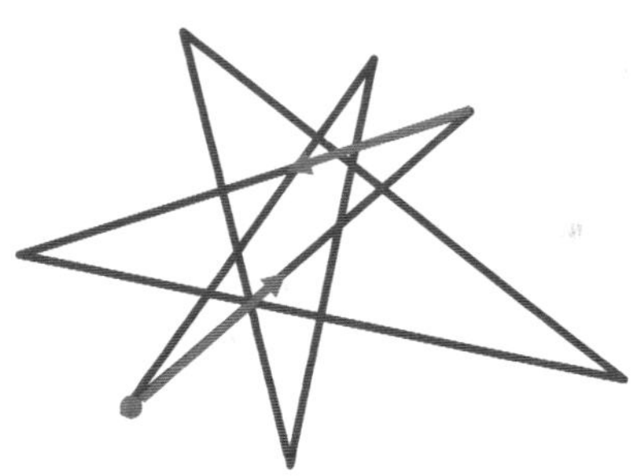

Fig. 2 The hummingbird's flight across the seven-pointed star

3.3. Various avatars of the frog's random walk: An enactive metaphorical construction of the concept of probability [66].

We claim that random walks constitute a royal road to probability, because they are ubiquitous, they constitute models for sundry probabilistic problems, which can be easily enacted and simulated, and when approached metaphorically, they allow learners to construct along the way the "abstract" notion of probability [63], [66].

A typical simple example is the case of a frog random jumping asymmetrically between just two stones A and B in a pond (with a non-zero probability of remaining stationary), which may be seen as a metaphor (or a model) of an unmerciful market struggle between two yoghurt producers A and B, which month after month seduce a part of each other's consumers according to a fixed pattern.

The random walk is a metaphor for the market struggle, or the other way around, according to whether the involved learner is more acquainted with jumping frogs or with market struggles.

A more involved example appears in our courses, when we told the students about Filomena, a frog which jumps randomly but symmetrically (like heads or tails) to one of its two next neighbours each time), on a row of 6 stones (tagged 0 to 5), starting at stone 3. Stones 0 and 5, however, are in fact two lurking alligators, called Anibal and Artemio, camouflaged as stones. Filomena will be instantly swallowed if it ever lands on the head of either alligator. Naturally, the students wonder about Filomena's fate...

We also told these students about the classical "ruin problem": two players *A* and *B*, having a certain amount of euros each, repeatedly flip a fair coin to decide who wins. After each flip, the winner receives one euro from the loser. No credit is available, so the game ends when any player runs out of money (is "ruined"). In our case, *A* would have an initial "fortune" of three euros and *B* only two (metaphorically, A's ruin means that Filomena was eaten by Anibal). Sundry questions arise, like:

How likely is that in the long run A, or B, becomes ruined?

How likely is that the game goes on forever?

Students tackling the ruin problem might move metaphorically to the frog's walk, and then in turn, to a splitting process or to a water draining process. Notice here that metaphorising involves a cascade of metaphors, so the skill to move between them becomes significant for the learners.

We investigated students' reaction when being proposed both, the frog's random walk and the ruin problem (with the same data) in a (pre-pandemic) test. We found that students who have been exposed to a metaphor thwarting teaching, had trouble in recognising both problems as the "same". Some even solved one right, and the other wrong! Less than 5% of them, noticed that they were given the same problem twice, but they did not dare to say so; another effect of the prevailing didactic contract, we deem.

On the "mathematical side", it is remarkable that this elementary example allows the students to re-discover Fibonacci numbers (as numerators of the fractions giving the probabilities of finding the frog at the different stones or in one of the two alligator bellies), to easily calculate infinite ("normalised") sums involving Fibonacci numbers, to imagine hydraulic metaphors (a water draining process) for easily calculating the required probabilities and also to connect metaphorically with thermodynamical systems in equilibrium and harmonic functions on discrete spaces.

3.4. The third ball drawn: enactive probability [65].

From an urn containing 2 red balls and 3 blue balls, a ball is drawn without replacement, thrice.

¿How likely is it that the third ball drawn is red?

Students usually calculate dutifully, using possibility trees, assigning probabilities and multiplying and adding fractions, to end up discovering that the probability of every drawn ball to be red, is the same: 2/5. They do not see why, though.

We suggest to them to enact the problem: to actually draw the balls from the urn (pebbles from a bag, in our implementation). Then, they have to decide what to do with the pebbles drawn. Usually, they do not throw them away or swallow them, but they align them carefully one by one, as they draw them from the bag. So doing, some students "see" immediately why the probabilities of "red" remain the same. Most of them, though, see that just for the first and last pebble drawn. To convince the others, one student had the idea of arranging the pebbles drawn otherwise on the table: in a circle! Then all agreed.

We see in this example how physically enacting problem may trigger new ideas and insights. It is not a matter of just thinking about the problem in your head, as some students (and teachers) believe.

4. Discussion

Our examples show the crucial role that enaction and metaphorisation play in mathematical thinking, particularly in the construction of "abstract" notions and in the insights that allow the learners to give sensible answers, even to "impossible questions" (like the one about the fate of a random walker).

IOP Conf. Series: Materials Science and Engineering 1261 (2022) 012021 doi:10.1088/1757-899X/1261/1/012021

We notice also that metaphorizing and enacting are ubiquitous emergents in our a-didactic situations, even when we are not aware of their presence, as it often happens as a consequence of a previous or current thwarting didactic contract in the classroom.

Our examples suggest – as an antidote to the prevailing didactic contracts - to pay attention to open up spaces where the emergence of (idiosyncratic) metaphorising is welcomed as a legitimate possibility of mathematical thinking, which deserves to be explored, especially in a collaborative and interactive setting. An enactivist approach in this context emphasises "learning by doing" and problem-posing instead of problem-solving. For this, we need to propose a-didactic situations where students have the freedom to interact with each other, re-construct and de-construct their understanding, while the teacher steps back and leaves room for them to co-construct their questions and problems.

We share with Maturana and Varela [40] the insight that cognition is a biological process, which is not a privilege of humankind (thought to be the "Apex of Creation"). What is in fact a privilege of humankind is the capacity we have of telling ourselves stories and denying the biological character of cognition, metaphorising it as information processing, manipulating abstract representations, or the like.

In Varela's sense, we are always 'enacting' a world, most of the time unconsciously. So, we cannot choose to enact or not to enact (in his sense); enaction is just the way we cognise as living beings. We may nevertheless entertain the 'representationalist illusion' (another privilege of humankind) that we are perceiving and representing an objective reality 'out there'. Also, we can choose to enact (in the everyday sense of the term, of bodily acting out) a given metaphor or situation or not, for instance. Paradoxically, we are definitely able to teach in a way that ignores enaction (in Varela's sense) and does not allow for enacting (as bodily acting out): a non-enactivist stance that paves the way for cognitively abusing the learners. Our enactivist approach to education, distilled in the 'lying down a path in walking' metaphor for cognition and learning, leads us on the contrary to foster metaphorising and enacting among the learners. Still, in metaphorical terms, it is very likely that what we currently describe as "metaphorising" in our species is just "the tip of the iceberg" regarding human cognition.

5. Open ends and interdisciplinary challenges

From our discussion above, it follows that several challenges arise for embodied AI. For instance:

Could an intelligent system scan its recorded previous experience and recognise analogy? In other words, emulate the "eye for resemblance" which Aristotle evoked when he said long time ago:

*The greatest thing by far is to be a master of metaphor. It is the one thing that cannot be learned from others; it is also a sign of genius, since a good metaphor implies an eye for resemblanc*e [2], p. 21:1459a.

An interesting related question is to which extent other species (e.g. octopuses) metaphorise? Or even social systems, including social networks?

The fact that metaphorisation may be just "the tip of the iceberg" in human cognition (particularly in regard to mathematical "abstract" thinking) poses also big questions for the design of intelligent systems. Recall in this context Seth's [58] insight of "cognition as controlled hallucinations"

We could imagine combining scans of previous history, analogy recognition, random walking (as in foraging) and iterative Bayesian construction of objects relevant for an artificial cognitive agent.

In our view, Brooks' insights regarding embodied artificial intelligence [4] - [6] and also regarding a new kind of mathematics being needed to address questions about the nature of life, cognition and intelligence [7], remain valid and pertinent today.

Enactivist approaches to mathematical problem solving may be more accessible to AI design, since nowadays there is increasing agreement in mathematics education on the relevance of random exploration in problem solving, especially in collaborative settings. A propitious setting for this kind of exploration is to have small random working groups, where the learners tackling a problem or exploring an open situation, engage in a "mathematical improvisation" quite analogous to a musical improvisation, where a key factor is that the "listen" to each other. A first step here could be to design artificial systems able to engage in musical improvisation.

Moreover, the fact that learners working in groups are more creative than working in isolation (the group is more creative than the sum of the individual creativities of its members), suggests that "swarm intelligence" may also unfold in suitable settings in the classroom, which foster interconnexion amongst the learners. Artificial Swarm Intelligence appears to be an interesting research direction to explore, either aiming at designing "artificial ants", for instance or setting up swarms of networked humans. In fact, recent progress has been made in the case of "human swarms" or "hive minds" [56] but not so much in the context of mathematics education yet, as far as we know.

Last but not least, the question remains open as to a new kind of mathematics being needed to properly address questions about the nature of life, cognition and intelligence [7], [75] - [77]. A mathematics of circularity and self-reference, an Ouroboric mathematics in other words, is most wanted. So, in a full circle, we come back to line 1...

Acknowledgements
The authors gratefully acknowledge funding from PIA-CONICYT Basal Funds for Centers of Excellence Project FB0003, DAAD Project 573 35022 D and University of Chile FIDoP Project 2019_19 FCS. All authors have contributed equally to this work.

References

[1] Abrahamson, D., Gutiérrez, J. F., & Baddorf, A. K. 2012. Try to see it my way: The discursive function of idiosyncratic mathematical metaphor. *Mathematical Thinking and Learning*, ***14***, 55–80.
[2] Aristotle. 1984. Poetics (trans: Bywater I). In J. Barnes (Ed.), *The Complete Works of Aristotle: The Revised Oxford Translation*. Princeton University Press.
[3] Bachelard, G. 1938. *La formation de l'esprit scientifique. Contribution à une psychanalyse de la connaissance objective*. Vrin.
[4] Brooks, R. A. 1991. Intelligence without representation. *Artificial Intelligence*, ***47***, 139–159.**,**
[5] Brooks, R. A. 1995. Intelligence without reason. In L. Steels & R. Brooks (Eds.), *The Artificial Life Route to Artificial Intelligence: Building Embodied, Situated Agents*. Routledge.
[6] Brooks, R. A. 1997. From earwigs to humans. *Robotics and Autonomous Systems*, ***20***, 291–304.
[7] Brooks, R. 2001. The relationship between matter and life. *Nature*, ***409***, 409–411.
[8] Brousseau, G. 1965. *Les Mathématiques du cours préparatoire*. Dunod.
[9] Brousseau, G. 2002. Theory of didactical situations in mathematics. Kluwer.
[10] Brousseau, G., Sarrazy, B., & Novotna, J. 2014. Didactic contract in mathematics education. In *Encyclopedia of Mathematics Education* (Lerman S, pp. 153–159). Springer.
[11] Brousseau, G., & Warfield, V. 2014. Didactic situations in mathematics education. In *Lerman S. (Ed.). Encyclopedia of Mathematics Education* (pp. 163–170). Springer.
[12] Bruner, J. 1966. *Toward a theory of instruction*. Harvard University Press.
[13] Bruner, J. 1986. *Actual minds, possible worlds*. Harvard University Press.
[14] Chiu, M. 2000. Metaphorical reasoning: Origins, uses, development and interactions in mathematics. *Educational Journal*, ***28 (1)***, 13–46.
[15] Clark, A. 1997. The dynamical challenge. *Cognitive Science*, ***21***, 461–481.
[16] Conway J 1976 *On Numbers and Games*, Academic Press
[17] Dall'Acqua, L. 2010. Learning Path and Assessment Criteria in the Conception and Development of an e-course based on the PENTHA ID Model. *Lecture Notes in Engineering and Computer Science*, *2186*.
[18] De Jaegher, H., & Di Paolo, E. 2007. Participatory sense-making: An enactive approach to social cognition. *Phenomenology and the Cognitive Sciences*, ***6***, 485–507.
[19] Deleuze, G., & Guattari, F. 1980. *Mille Platteaux*. Éditions de Minuit.
[20] Dewey, J. 1997. *How we think*. Dover Publications. (Original work published 1910)
[21] Díaz-Rojas, D., Soto-Andrade, J., & Videla-Reyes, R. 2021. Enactive Metaphorizing in the Mathematical Experience. *Constructivist Foundations*, ***16***, 265–274.

[22] English, L. D. 1997. *Mathematical reasoning: Analogies, metaphors, and images*. L. Erlbaum Associates.
[23] Froese, T. 2007. On the Role of AI in the Ongoing Paradigm Shift within the Cognitive Sciences. In M. Lungarella, F. Iida, J. Bongard, & R. Pfeifer (Eds.), *50 Years of Artificial Intelligence* (pp. 63–75). Springer.
[24] Fuchs, T. 2017. *Ecology of the Brain: The phenomenology and biology of the embodied mind*. Oxford University Press.
[25] Gallagher, S. 2005. *How the body shapes the mind*. Oxford University Press.
[26] Gallagher, S., & Lindgren, R. 2015. Enactive metaphors: Learning through full-body engagement. *Educational Psychology Review*, ***27***, 391–404.
[27] Gibbs, R. W. (Ed.). 2008. *The Cambridge handbook of metaphor and thought*. Cambridge University Press.
[28] Goodchild, S. 2014. Enactivist theories. In S. Lerman (Ed.), *Encyclopedia of Mathematics Education* (pp. 209–214). Springer.
[29] Gould, S. J., & Vrba, E. S. 1982. Exaptation-a missing term in the science of form. *Paleobiology*, ***8***, 4–15.
[30] Ingram, J., & Elliott, V. 2020. *Research methods for classroom discourse*. Bloomsbury Academic.
[31] Knops, A., Thirion, B., Hubbard, E. M., Michel, V., & Dehaene, S. 2009. Recruitment of an area involved in eye movements during mental arithmetic. *Science*, ***324***, 1583–1585.
[32] Lakoff, G., & Johnson, M. 2003. *Metaphors we live by*. University of Chicago Press.
[33] Lakoff, G., & Núñez, R. E. 2000. *Where mathematics comes from: How the embodied mind brings mathematics into being* (Nachdr.). Basic Books.
[34] Lungarella, M., Iida, F., Bongard, J., & Pfeifer, R. (Eds.). 2007b. *50 Years of Artificial Intelligence: Festschrift*, LNAI 4850. Springer.
[35] Machado, A. 1988. *Selected poems*. Harvard University Press.
[36] Malafouris, L. 2013. *How things shape the mind: A theory of material engagement*. MIT Press.
[37] Manin, Y. I. 2007. *Mathematics as metaphor*. Amer. Math. Society.
[38] Mason, J. 2002. *Researching Your Own Practice: The Discipline of Noticing*. Routledge.
[39] Maturana, H. R., & Varela, F. J. 1980. *Autopoiesis and Cognition: The Realization of the Living*. Springer.
[40] Maturana, H. R., & Varela, F. J. 1992. *The tree of knowledge: The biological roots of human understanding* (Rev. ed). Shambhala.
[41] McCarthy, J., Minsky, M. L., Rochester, N., & Shannon, C. E. 2006. A Proposal for the Dartmouth Summer Research Project on Artificial Intelligence, August 31, 1955. *AI Magazine*, ***27***, 12–12.
[42] McGowen, M.A., & Tall, D. 2010. Metaphor or Met-Before? The effects of previous experience on practice and theory of learning mathematics. *Journal of Mathematical Behavior*, ***29***, 169–179
[43] NAEYC/NCMT. 2002. *Learning paths and teaching strategies in early mathematics*. USA: National Association for the Education of Young Children.
[44] Nelson E 1977 Internal set theory: A new approach to nonstandard analysis. *Bull. Amer. Math. Soc*. **83(6)**. 1165-1198.
[45] Núñez, R. 2004. *Do real numbers really move? Language, thought, and gesture: The embodied cognitive foundations of mathematics* In F. Iida & R. Pfeifer, Eds.). Springer.
[46] Núñez, R. 2008. A Fresh Look at the Foundations of Mathematics: Gesture and the Psychological Reality of Conceptual Metaphor. In A. Cienki, & C. Müller (Eds.), *Gesture and Metaphor* (pp. 93-114). John Benjamins.
[47] Penny, S. 2017. *Making sense: Cognition, computing, art, and embodiment*. MIT Press.
[48] Penny, S. 2021. Sensorimotor debilities in digital cultures. *AI & SOCIETY*. https://doi.org/10.1007/s00146-021-01186-0
[49] Presmeg, N. C. 1997. Reasoning With Metaphors and Metonymies in Mathematics Learning. In L. D. English (Ed.), *Mathematical Reasoning: Analogies, Metaphors, and Images*. Routledge & CRC Press.

[50] Proulx, J., & Maheux, J.-F. 2017. From problem solving to problem posing, and from strategies to laying down a path in solving: Taking Varela's ideas to mathematics education research. *Constructivist Foundations*, ***13***, 160–167.
[51] Proulx, J., & Simmt, E. 2015. *Enactivism and mathematics education: Sources, meanings, and research*. Information Age Publishing.
[52] Pylyshyn, Z. W. 1979. The rate of "mental rotation" of images: A test of a holistic analogue hypothesis. *Memory & Cognition*, ***7***, 19–28.
[53] Reid, D. A., & Mgombelo, J. 2015. Survey of key concepts in enactivist theory and methodology. *ZDM*, ***47***, 171–183.
[54] Ricoeur, P. 1975. *La Métaphore Vive*. Éditions du Seuil.
[55] Riegler, A., & Vörös, S. 2017. Missing the woods for the trees: Neglected aspects of Francisco Varela's work. *Constructivist Foundations*, ***13***, 1–190.
[56] Rosenberg, L., & Willcox, G. 2020. Artificial Swarm Intelligence. In Y. Bi, R. Bhatia, & S. Kapoor (Eds.), *Intelligent Systems and Applications* (pp. 1054–1070). Springer
[57] Seth, A. 2019. Our Inner Universes: The Neuroscience of Reality. *Scientific American*, *321*.
[58] Seth, A. 2021. *Being you: A new science of consciousness*. Faber & Faber.
[59] Sfard, A. 1997. Commentary: On metaphorical roots of conceptual growth. In L. D. English (Ed.), *Mathematical Reasoning: Analogies, Metaphors, and Images*. Routledge & CRC Press.
[60] Sfard, A. 2008. *Thinking as communicating: Human development, the growth of discourses, and mathematizing*. Cambridge University Press.
[61] Sfard, A. 2009. Metaphors in education. In *H. Daniels, H. Lauder & J. Porter (Eds.) Educational theories, cultures and learning: A critical perspective* (pp. 39–50). Routledge.
[62] Smith, L., & Gasser, M. 2005. The development of embodied cognition: Six lessons from babies. *Artificial Life*, ***11***, 13–29.
[63] Soto-Andrade, J. 2013. Metaphoric Random Walks: A Royal Road to Stochastic Thinking. In B. Ubuz, C. Haser, & M. A. Mariotti (Eds.) Proc. *CERME8* (pp. 890–900) ERME & METU.
[64] Soto-Andrade, J. 2014. Metaphors in Mathematics Education. In S Lerman (Ed.), *Encyclopedia of Mathematics Education* (pp. 447–453). Springer.
[65] Soto-Andrade, J. 2018. Enactive metaphorising in the learning of mathematics. In G. Kaiser, H. Forgasz, M. Graven, A. Kuzniak, E. Simmt, & B. Xu (Eds.), *Invited Lectures from the 13th International Congress on Mathematical Education* (pp. 619–638). Springer.
[66] Soto-Andrade, J., Diaz-Rojas, D., & Reyes-Santander, P. 2018. Random walks in the didactics of probability: Enactive metaphoric learning sprouts. In C. Batanero & E. J. Chernoff (Eds.), *Teaching and learning stochastics: Advances in probability education research* (pp. 125–143). Springer.
[67] Soto-Andrade, J., Jaramillo-Riveri, S., Gutiérrez, C., & Letelier, J. 2011. Ouroboros avatars: A mathematical exploration of self-reference and metabolic closure. In T. Lenaerts, M. Giacobini, H. Bersini, P. Bourgine, M. Dorigo, & R. Doursat (Eds.), *Advances in Artificial Life ECAL 2011: Proceedings of the Eleventh European Conference on the Synthesis and Simulation of Living Systems* (pp. 763–770). The MIT Press.
[68] Soto-Andrade, J., & Varela, F. J. 1984. Self-reference and fixed points: A discussion and an extension of Lawvere's Theorem. *Acta Applicandae Mathematicae*, ***2***, 1–19.
[69] Soto-Andrade, J., & Yáñez-Aburto, A. 2019. Acknowledging the Ouroboros: An enactivist and metaphoric approach to problem solving. In P. Felmer, P. Koichu, & B. Liljedahl (Eds.) *Problem Solving in Mathematics Instruction and Teacher Professional Development* (pp. 67–85). Springer.
[70] Stewart, J., Gapenne, O., & Di Paolo, E. A. D. 2010. *Enaction: Toward a new paradigm for cognitive science*. Bradford.
[71] Thom, R. 1994. Spectre bord d'un centre obscur. In M. Porte (Ed.), *Passion des formes* (pp. 13–24). ENS Éditions Fontenay-Saint Cloud.
[72] Thompson, E. 2005. Sensorimotor subjectivity and the enactive approach to experience. *Phenomenology and the Cognitive Sciences*, ***4***, 407–427.

[73] Thompson, E. 2007. *Mind in life: Biology, phenomenology, and the sciences of mind*. Belknap Press of Harvard University Press.
[74] Thurston, W. 1994. On Proof and Progress in Mathematics. *Bulletin of the American Mathematical Society*, ***30 (2)***, 161–177.
[75] Varela, F. 1975. A calculus for self-reference, *Int. J. Gen. Systems, 2*:5-24
[76] Varela, F. J. 1979. Principles of Biological Autonomy. North Holland
[77] Varela, F. 1984. The creative circle: Sketches on the natural history of circularity. In P. Watzlawick (Ed.), *The invented reality: How do we know what we believe we know? : Contributions to constructivism* (1st ed., pp. 309–323). W. W. Norton.
[78] Varela, F. J. 1987. Laying down a path in walking: A biologist's look at a new biology. *Cybernetic*, ***2***, 6–15.
[79] Varela, F. J. 1999. *Ethical Know-How: Action, Wisdom, and Cognition*. Stanford University Press.
[80] Varela, F. J., & Goguen, J. A. 1978. The arithmetic of closure. *Journal of Cybernetics*, ***8***, 291–324.
[81] Varela, F. J., Thompson, E. & Rosch, E. 1991. *The embodied mind: Cognitive science and human experience*. MIT Press.
[82] Wheeler, M. 2005. *Reconstructing the cognitive world: The next step*. Bradford.
[83] Wittmann, E. 2013. Fundamental Ideas of Mathematics as the Natural Source of Teaching and Learning. *Quaderni Di Ricerca in Didattica (Mathematics)*, ***23*** Supplemento 1. 98–110.

The deep time perspective: How insights from millennia of evolving and extinct animals can inform robot design decision making

Aja Carter* and Sonia Roberts

Electrical and Systems Engineering
University of Pennsylvania
3401 Grays Ferry Ave
Bldg 6176 Fl 3

E-mail: caja@seas.upenn.edu

Abstract. Animals on the visible scale have been experimenting with body form and function in enumerable enviroments for the last 540 million years. Almost all of the variation in the history of life is now gone, leaving only a tiny fraction of what is possible alive in modern animals. Recently biological and engineering techniques have made it possible to robustly answer questions only the fossil record can provide, such as the history and original functions of certain behaviors. Robotics has already begun to champion inspiration from biology, but only from the small variation seen in extant taxa. In this chapter we introduce paleontological study of a now famous source of bioinspiration: The modern African Cheetah. We discuss additional forms of high-speed pursuit predators from the lineage that demonstrate an evolutionary experiment in manuverability on uneven terrain. We also discuss how the evolution of sensing and perception does not always follow or work in concert with the evolution of the musculoskeletal system. We end by summarizing the process of our analysis so that our methods can be repeated for other animals that inspire engineers. We hope this brief primer introduces paleobiology to roboticists and demonstrates the need for biologically inspired robotics to engage with paleobiological processes, to the benefit of both robotics in developing new technology and paleobiology in answering long-standing questions about relationships between extinct forms and their functions.

1. Introduction

Roboticists are often drawn to nature for inspiration. Nature is creative, offering pre-built solutions to many different kinds of engineering problems. Designing components of a robot's body, sensors, and control simultaneously to accomplish a task (co-design) is extremely challenging because the performance of each component depends so heavily on the parameters of the other components. By looking to biology, we can find clues to combinations of body structures, sensors, and controls that may at least provide a useful starting point, provided that we know for what task and in what environment that animal is adapted. Animals have many structures and body plans that inspire roboticists: The many gaits that a horse uses at different speeds [66], the incredible ability of cockroaches to navigate obstacles larger than their hip height [48], and the Cheetah's springy spine [51, 53] and tail [46] that allow it to run and make turns at astonishing speeds. While each of these taxa – and many more – are deeps wells of inspiration, all of them are products of their evolutionary history.

Animals throughout Earth's history have evolved at specific times as well as specific places. The features of their bodies that roboticists find so inspiring perform certain biological functions in the animals' environments, but the development of these features is constrained by the animal's evolutionary history (i.e., phylogeny), how the structure is made (i.e., fabrication), and its biologial function [8]. These biological structures also have some effectiveness in their environment. When changes in an environment occur (e.g., desertification, mountain range formation), the relationship between the animal's form and effective functionality may also change. Through geologic time, drastic changes to environments have occured [50]. These environmental changes induce selective pressures on animals that either adapt, or fail to adapt and go extinct [38]. While what causes an animal to go extinct may not be of interest to bio-inspired engineers, the behavior and effectiveness of extinct forms may yet serve as an untapped well for biological inspiration.

Investigating the response of biological systems to changes in the environment over deep time is exactly what paleobiology does [8]. For example, a change to the skeletal structure of an animal that lives on an open plain may be an adaptation to the open plain environment, or it may be a result of other selective pressures unrelated to locomotion [45]. Only by considering biological structures through geologic time can one generate and test hypotheses surrounding the selective pressure and evaluate their effectiveness, use, or sufficiency in a given environment. For roboticists, who do not have the constraints of evolutionary history and have different fabrication constraints than animals do, animals that happened to go extinct may serve as biological inspiration just as well as animals that happen to be alive now. What matters is the selection pressures that led to the evolution of that structure, which we can only see by examining evolutionary processes on an extremely large time scale. This *deep time perspective* is necessary for a truly biologically inspired robotics.

"Biological inspiration" does not need to stop at either inspiration from specific structures or body plans, or even the process of evolution itself, as in evolutionary robotics [19, 1, 11, 9, 41]. Unlike robots evolving under artificial selection, animals do not evolve outside of a phylogenetic framework. Through time, small changes in biological structure are better suited for an environment and in turn selected for in subsequent generations. These small changes are constrained by the history of the animal [8]. However, in an animal's evolutionary history, novel traits that substantially increase behavioral performance or allow a taxon to interact with the environment in a completely new way may also occasionally evolve. We refer to these drastic changes as *key innovations* [29, 63]. By understanding the evolutionary process rather than mimicking it, we can trace lineages of animals backward and forward through evolutionary time and identify opportunities for biological inspiration that come from several animals simultaneously rather than one animal.

In this chapter, we will discuss the evolutionary history of an animal that has provided great inspiration to roboticists: The Cheetah. We end the chapter by describing our analysis procedure in detail so that it can be replicated for other animals. We aim to convince the reader that it is important for roboticists interested in taking inspiration from nature to use a deep time perspective, considering animals throughout evolutionary history and not just the snapshot of time we currently live in. As roboticists, we can take inspiration from the process of evolution and not just its artefacts.

2. Natural experiments on maneuverability and speed in ancient Cheetah species

The high-speed African Cheetah (*Acinoynx jubatus*) that has captured the interest of engineers and inspired robotic design is just one living member of the Puma group of cats (Pumas, Cheetahs, and the Jaguarundi) and the genus *Acinonyx* [58, 60]. Three extinct relatives we discuss here are *Acinoynx pardineneis*, *Miracinonyx trumani*, and *Miracinonyx inexpectus* [59, 58, 60]. The closest extinct relative of the modern Cheetah, *Acinonyx pardineneis* (1.9

Figure 1. Comparison between the modern African Cheetah (foreground) and its most recent extinct ancestor, the Giant Cheetah (background)

mya), shares many skeletal similarities to the modern African Cheetah but was twice as large compared to modern Cheetahs, giving the animal its colloquial name: The "Giant Cheetah" [26, 20, 52]. In addition to size, *A. pardineneis* had features of the skull and dentition more akin to Pumas than the modern Cheetah [20, 60]. The giant Cheetah's locomotor apparatuses were similar to the modern African Cheetah, suggesting the extinct species was physically capable of a high-speed pursuit [60, 58]. The mosaic of features – larger overall size, less specialized dentition, elongated limbs, but a lack of synthesis on postcranial material of the Giant Cheetah – has left paleobiologists hypothesizing that the Giant Cheetah had a similar ecology to its modern relative.

Miracinonyx had more nuanced differences in the locomotor apparatus. Both species of the "North American Cheetah" (in quotes because they are not true Cheetahs), *Miracionyx trumani* and *Miracinonyx inexpectus*, have several critical skeletal structures that are similar to modern African Cheetahs [59, 60]. These characters, such as elongated vertebrae in their lower backs, the morphology of their forelimbs, and their general body size, suggested a close phylogenetic relationship to the modern African Cheetah [59]. The similarity in locomotor traits between the "North American Cheetah" *Miracinonyix* and the true African Cheetah *Acinoynx* suggested similar high-speed ambush predation methods [59]. However, recent studies have demonstrated that *Miracinonyx* is not a close relative of the Cheetah, but rather a much more ancient ancestor of the Puma clade that invaded North America [4]. As the North American Cheetahs *Miracinonyx* is not a direct relative of the true African Cheetah, but the body morphologies are similar to the African Cheetah, paleobiologists consider *Miracionyx* an example of ecomorphological convergence [4, 60]. This one-to-one mapping of morphology to ecology (i.e., behaviors) occurs when selective pressures are strong enough to override ancestral morphology [67]. Species that conserve ecologically converge morphologically; that is, species which behave similarly in similar environments develop similar morphologies over evolutionary time. Even though the North American *Miracinonyx* and the modern African Cheetah generally have locomotor similarities, some unique locomotor morphological features suggest variations on high-speed predation in *Miracinonyx*.

Additional morphological investigations have demonstrated that the North American *M. trumani* had more robust limb elements and an additional rotational degree of freedom in the wrist and ankle joints [18, 60]. This means that *M. trumani* could climb and grasp onto prey, unlike the modern African Cheetah [18]. Though this has not been demonstrated explicitly in the "North American Cheetah," some paleobiologists have also hypothesized this internal degree of freedom led to more maneuverable behaviors on uneven terrain in a separate cursorial taxon [65]. The morphology of the sacrum, the triangular bone at the base of the spine connecting to the tail, suggests that the North American *M. trumani* could rotate its tail for balance in a similar manner to the modern African Cheetah, whereas the North American *M. inexpectatus* could not [43]. The calcaneum, the main bone in-lever for plantar flexion in the hindlimb (extending the foot), is longer in *M. trumani* than in *M. inexpectatus*, suggesting that *M. trumani* is more adapted to running than its ancestor, but not more than the African Cheetah [18, 43]. Within four taxa, the modern African Cheetah, the extinct giant African Cheetah, and the two species of "North American Cheetah," the fossil record has recorded an in-group experiment surrounding the degree of adaptation to running, size, degrees of freedom in the forelimb, and reorientation using a tail during high-speed pursuit. While it is always difficult to point out why a structure exists (indeed by Darwinian evolution, it can be present simply because it is not maladaptive enough not to be selected for), differences in biological structure and means to develop and test hypotheses as to why changes in a structure have occurred are the hallmarks of paleobiological study.

Table 1. A natural experiment in Cheetah and Cheetah-like species. Note that this table is a simplification of the complex morphology in each animal. The degrees of freedom listed for the forelimbs and tails of the animals are their dominant degrees of freedom, not the only possible degrees of freedom.

Species	Forelimb DoF	Tail DoF	Hunting location	Possible functionality
M. trumani	Rotational Wrist extension	Rotational	Forest clearings [59]	Uneven terrain Maneuverability Speed
M. inexpactatus	Rotational	No rotation	Forest clearings [59]	Uneven terrain Maneuverability
A. jubatus	Wrist extension	Rotational	Flat savannahs	Maneuverability Speed
A. pardineneis	Wrist extension	Rotational	Flat savannahs [26]	Maneuverability Speed

M. trumani, M.inexpectus, A. pardineneis, and *A. jubatus* represent an experiment conducted by evolutionary pressures. Within these four animals, there are animals that have a rotational degree of freedom in the forelimb and the tail (*M. trumani*), that have a rotational degree of freedom in the forelimb but not the tail (*M. inexpactatus*), that have no internal degrees of freedom in the forelimb but a very mobile tail (*A. jubatus*), and that are identical to the modern Cheetah *A. jubatus* but twice as large (*A. pardineneis*). See Table 1 and Figure for a summary. It is not enough to demonstrate that there are differences present, and indeed many studies cited in

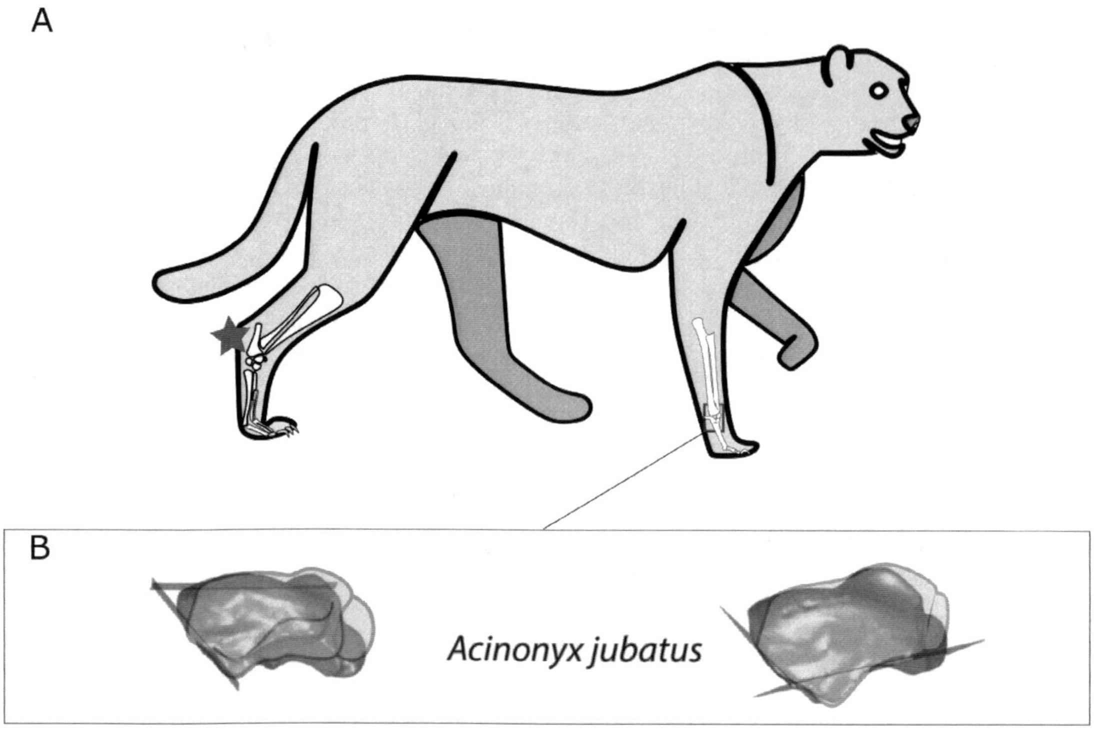

Figure 2. Cheetah outline for anatomical context (A) and subtle differences in the scapholunar bone (B) between the North American *M. trumani* and the modern African Cheetah *A. jubatus*. For additional information on the calcneum (star) see [43]. The box in the forelimb highlights the carpal bones, one of which is the scapholunar bone. Differences in surface area and angle of artiulation with additional bones (radius left, trapezium right) give *M. trumani* (in light blue) the ability rotate its wrists, representing an additional degree of freedom in comparison with the true African Cheetah (in gray) [18]. Modified from [18] with permission from Wiley and Sons.

this work go beyond presenting differences. Many paleobiological studies demonstrate what these differences may mean for the animal, using classic techniques in paleobiology (e.g., geometric morphometrics, comparative phylogenetic analyses, joint reconstruction) and borrowing some from engineering fields (e.g., finite element analyses).

A collaboration between paleobiologists and engineers could be fruitful in investigating these variations on degrees of freedom and behavior. Indeed there have been some successful collaborations between paleobiologists and engineers [10, 39, 42]. Essential questions asked and answered in paleobiological studies are what these morphological differences changed in terms of locomotion; what, if anything, these changes afforded the animal; and how they evolved in response to which selective pressures (e.g., environmental change, change in the behavior of their prey). Likewise, engineers interested in creating fast and maneuverable robots may be able to benefit more from understanding these natural experiments than from simply looking to the one species of Cheetah that is currently alive. For example, to build a robot that is maneuverable and fast on uneven terrain, it may be more useful to add a rotational degree of freedom to the forelimbs than an abduction-adduction degree of freedom.

IOP Conf. Series: Materials Science and Engineering 1261 (2022) 012022 doi:10.1088/1757-899X/1261/1/012022

3. The key innovation that gave us the modern Cheetah

3.1. Understanding key innovations through analogy: Direct-drive motors and legged robots

Recall that a key innovation is a novel trait which substantially improves a taxon's performance in some behavior or makes it possible to interact with the environment in a new way. An example that may make this concept accessible to roboticists is the introduction of direct-drive motors to the field of legged robotics.

Legged robots have historically relied on the high torques produced by heavily geared motors [22, 6, 49, 25, 5]. However, these heavy gearboxes affect the sensitivity of the motors to position errors at the end effector due to the inertia of the gearbox [34]. In contrast, direct-drive motors – that is, electric motors with no gearboxes – have been in use for decades in robot arms [2], which do not have to carry large amounts of weight and therefore have lower torque requirements. These arms have benefited greatly from the sensitivity or "backdriveability" of direct-drive motors, making it possible to detect forces at the end effector using position sensors on the motors [33, 64].

The MIT Cheetah was the first legged robot to use custom direct-drive motors which had been specially designed to have larger torque outputs [2, 34] in a highly dynamic robot [51]. Since then, the Cheetah [7, 44, 17] and other quadrupedal robots with direct-drive motors [34, 31] have consistently impressed with dynamic locomotion behaviors that use the extremely fast and precise sensing capabilities of the direct-drive legs [64, 30]. For example, one small direct-drive robot was able to jump up to a door handle, feel for and then push the door handle with one foot, and then push the door open with its other foot – all during ballistic flight [62, 61, 54].

High-torque direct-drive motors can be seen as a key innovation for legged robots. By improving the speed and precision with which the robots could sense forces on the end effectors, it became possible for these robots to perform acrobatic feats that had previously been impossible.

3.2. The Cheetah's key innovation was a sensing organ

A diet consisting only of meat invokes a very high selection pressure on the musculoskeletal system of the cat family, particularly high-speed ambush behaviors, as seen in the Cheetah [27, 47]. Holliday et al. [27] demonstrated that in hypercarnivores such as Cheetahs, there is a high degree of phenotypic integration: That is, their skeletal elements have a strong covariance because all elements of the limbs and skull are involved with hunting prey. They also note that the Cheetah has specialized skeletal forms, likely due to its speed [27]. A recent study on bone-crunching ancient hypercarnivores also indicated that the Cheetah skull morphology is unique, likely due to necessary constraints [56]. Indeed, the skeletal features required for running and highspeed ambush predation have been present in the Cheetah-puma line for much of their evolutionary history.

Even with the skeletal morphologies to run at high speeds, and one could hypothesize the soft tissue requirements like lung capacity, could ancient Cheetahs run as quickly as modern Cheetahs? Without sensing abilities similar to the modern Cheetah, it was unlikely. A recent study described a key innovation not seen in the musculoskeletal system of the Cheetah [23]. This was a sensing element not previously considered: The inner ear. Grohe et al. [23] demonstrated that the otolith organ, which is responsible for maintaining head posture and detecting angular and linear accelerations and decelerations in the modern African Cheetah, is larger and different in morphology than its more recent ancestor, the Giant Cheetah [23]. This difference in auditory sensing ability suggests that the skeletal morphology for high-speed ambush behaviors evolved long before sensing capabilities [23]. While the Giant Cheetah could likely run fast, and indeed a predator only needs to be a modicum faster than what it is chasing, it is only with the modern Cheetah and subsequent sensing abilities that the modern Cheetah can run at truly incredible speeds [23, 58].

This enhanced auditory sensing in modern Cheetahs suggests two things. First, it is an example of a key innovation which likely allowed the modern Cheetah to move up an adaptive peak to a degree that had not previously been possible. And second, it is an example of how evolution generally acts on one of the triple of behavior, musculoskeletal body plan, or sensing at a time. Without a careful understanding of how evolutionary processes do and do not work, co-design of sensory, control, and morphological components may not scale. There is already some evidence that attention to the deep time perspective of evolution can pay off. For example, Cheney et al. [11] achieved success in co-optimizing morphological and sensorimotor aspects of embodied agents by the use of a clever trick to encourage a similar "stepwise" co-design process to that which appears to arise naturally from evolution. An agent that has recently changed its morphology experiences a temporary relaxation of selection pressures so that the sensorimotor control can adapt to the new body shape and how it interacts with the environment.

4. Understanding natural selection to improve bio-inspired robot evolution

Living animals are diverse in form and behavior. They have proliferated in innumerable environments, but they are only a minuscule subset along the continuum of visible life that began 540 million years ago. All animals evolve in specific times and places and either adapt to their environments or go extinct. However, while extinction is forever, it does *not* mean inferior. Some extinct animals could enact novel behaviors not present in modern taxa, as the fossil record is the only source of most variation in biological systems [37, 21]. Biologists and bio-inspired engineers already have a firmly established and fruitful relationship. However, when modern biological techniques and questions are meaningfully applied not only to modern living animals but also to the fossil record, the results are unique and can only aid in finding solutions to bio-inspired engineering concerns [28]. The deep time perspective can help roboticists and paleobiologists to investigate paleobiological questions together that have deep implications for both fields.

4.1. Biological questions with a deep time perspective

A key aspect of paleobiology is understanding the variation of morphological traits as they change through time. Such inquires include: phylogenetic burdens that constrain morphological novelty, selective pressures on key innovations that lead to specialization or adaptive radiations(large amount of speciation within a bauplan), and ecomorphological convergence. Convergence on a morphology in a particular ecology, especially across distantly related taxa [57], can indicate that a morphology is universally adaptive to an ecology [32]. One example would be paddle-like limbs in aquatic mammals, birds, and ancient reptiles. This can only be discovered through a deep time perspective. A focus of paleobiological inquiry is to resolve which trait is a key innovation and how that will affect subsequent populations and adaptability. Determining the adaptability of a key innovation scenario is strongest when form-function relationships from extant taxa is combined with data from the fossil [29]. A key question is whether a key innovation will lead to a specialist, such as the high-speed cheetah, or "adaptive radiation," which describes the phenomenon of a large amount of speciation with this new body plan or structure, creating a whole new group of species. In other words, how generally useful is this innovation for animals adapted to different ecologies?

4.2. Co-design is a testable hypothesis of biological systems, not a standard rule

Evolution is one of the few ways that controllers, morphological features, and sensors have been successfully developed together over time – an example of co-design [11]. An intuitive idea in paleobiology is that organisms are made of parts, some of which are more interrelated than others. Indeed, there are methods to investigate whether and how much structures evolve together [21, 40]. However, this idea is testable and quantifiable, not necessarily a clearly established rule in the body, a group of animals, or even time. We demonstrate in this paper

IOP Conf. Series: Materials Science and Engineering 1261 (2022) 012022 doi:10.1088/1757-899X/1261/1/012022

that Cheetahs had musculoskeletal forms associated with extremely high speeds. However, only the modern African Cheetah has sensing for maintaining gaze at high speeds [23]. The behavioral capability differences in the body in the Cheetah is a case where sensing came after changes in the body. This is not the only case of sensing capabilities not evolving strictly in concert with mechanical abilities. For example, 360 million years ago, when tetrapods first began making the transition to land, they had low-level aerial hearing capabilities before they had other musculoskeletal traits required for terrestrial locomotion [12]. Morphological changes in the musculoskeletal system have rates of change through deep time, and so do sensing capabilities; these rates are not necessarily the same [15, 16].

There are moments in life's evolutionary history in which sensory systems may have evolved simultaneously with musculoskeletal systems. For example, *Tiktaalik* had greater neck mobility and visual capabilities than its more fish-like ancestors; a combination of these traits (among others) permitted *Tiktaalik* to hunt at the water-to-air interface [36, 55]. However, we again urge embodied cognition roboticsts to recognize these events as testable hypotheses, not biological norms. We recommend [14, 13] and citations within for further reading. As a field, we have a relatively good understanding of the rate of change of auditory, orientation, and balance sensing systems in particular because these structures leave hard bony parts in the fossil record. A shortcoming of the fossil record can be the lack of preservation of transitional forms or elements that may change the rates of these morphological changes.

5. Conclusion and future directions

Without grabbing a shovel or perusing fossil collection in a museum and analyzing them with the trained eyes of a paleontologist, how can engineers use the ideas presented in this paper to tap into the fossil record for biological inspiration? The method we used in this paper can be summarized as follows. First, we identified an animal that often inspires engineers: The Cheetah, which has amazing speed and maneuverability. We then investigated the literature on this animal's evolutionary development to identify the key innovation that set it apart from similar animals. In some cases, such as the gecko's sticky feet which allow it to climb vertical surface with incredible speed [3], it might be appropriate for engineers to take inspiration directly from the key innovation that made it possible for geckos [35, 24]. For the Cheetah, the key innovation was a sensory organ which roboticists could mimic in any robot using a gyroscope or an accelerometer. However, the body plan that made it possible to make use of this key innovation – a quadruped with a spine and a tail – looks quite different from most existing legged robots. In this case, rather than focusing on the Cheetah's key innovation, it is possible to look at animals with similar musculoskeletal systems and identify natural experiments in design that may best serve the engineer's potential applications.

To do this, we traced back through evolutionary time, identifying animals with similar ecologies to the Cheetah before its key innovation. Since evolution proceeds by random chance, it is an accident that the true African Cheetah was the species to develop this sensory organ. We therefore expect that any of these other animals should theoretically have been able to run and maneuver as well as the true African Cheetah if we were able to drop an otolith organ into their ear. However, each of these animals had a slightly different ecology and operated in a slightly different environment.

This makes it possible to identify the ecology that best matches the potential use cases for engineers. In this case, engineers have identified that wheels are energetically efficient for vehicles, but legs are useful for navigating uneven ground. There are no "Cheetahs" with wheels, so "Cheetahs" must run with legs even on flat ground. However, there was a "Cheetah" that ran with legs on uneven ground. Comparing the differences between the "Cheetah" adapted for uneven ground with the "Cheetahs" adapted for flat ground suggests that the rotational degree of freedom in the forelimbs might be more useful for maneuverability on uneven ground than the adduction-abduction degree of freedom that we often see in quadrupedal robots.

While this is a new method that is not guaranteed to produce insights that will improve the design of any biologically inspired robot, we believe that this level of analysis could be repeated for any animal that engineers find inspiring. Paleontology offers the tools to perform this analysis. For ideas of other inspiring animals that may benefit from this type of analysis, we suggest investigating flight in bats and pterosaurs [37] or looking to review papers with examples of co-design throughout evolutionary history such as [13, 14, 37].

Acknowledgments

We gratefully acknowledge The University of Pennsylvania Vice Provost Fellowship for Academic Diversity. We are also grateful for support from NSF NRI-2.0 grant #1734355 and NSF INSPIRE grant #1514882.

References

[1] Alattas, R. J., Patel, S., and Sobh, T. M. Evolutionary modular robotics: Survey and analysis. *Journal of Intelligent & Robotic Systems 95*, 3 (2019), 815–828.

[2] Asada, H., and Youcef-Toumi, K. *Direct-drive robots: theory and practice.* MIT press, 1987.

[3] Autumn, K., Liang, Y. A., Hsieh, S. T., Zesch, W., Chan, W. P., Kenny, T. W., Fearing, R., and Full, R. J. Adhesive force of a single gecko foot-hair. *Nature 405*, 6787 (2000), 681–685.

[4] Barnett, R., Barnes, I., Phillips, M. J., Martin, L. D., Harington, C. R., Leonard, J. A., and Cooper, A. Evolution of the extinct sabretooths and the american cheetah-like cat. *Current Biology 15*, 15 (2005), R589–R590.

[5] Bartsch, S., Birnschein, T., Cordes, F., Kuehn, D., Kampmann, P., Hilljegerdes, J., Planthaber, S., Roemmermann, M., and Kirchner, F. Spaceclimber: Development of a six-legged climbing robot for space exploration. In *ISR 2010 (41st International Symposium on Robotics) and ROBOTIK 2010 (6th German Conference on Robotics)* (2010), VDE, pp. 1–8.

[6] Bartsch, S., Manz, M., Kampmann, P., Dettmann, A., Hanff, H., Langosz, M., von Szadkowski, K., Hilljegerdes, J., Simnofske, M., Kloss, P., et al. Development and control of the multi-legged robot mantis. In *Proceedings of ISR 2016: 47st International Symposium on Robotics* (2016), VDE, pp. 1–8.

[7] Bledt, G., Powell, M. J., Katz, B., Di Carlo, J., Wensing, P. M., and Kim, S. Mit cheetah 3: Design and control of a robust, dynamic quadruped robot. In *2018 IEEE/RSJ International Conference on Intelligent Robots and Systems (IROS)* (2018), IEEE, pp. 2245–2252.

[8] Briggs, D. E. Seilacher, konstruktions-morphologie, morphodynamics, and the evolution of form. *Journal of Experimental Zoology Part B: Molecular and Developmental Evolution 328*, 3 (2017), 197–206.

[9] Brodbeck, L., Hauser, S., and Iida, F. Morphological evolution of physical robots through model-free phenotype development. *PloS one 10*, 6 (2015), e0128444.

[10] Brooks, A., MacCready, P., LISSAMAN, P., and MORGAN, W. Development of a wing-flapping flying replica of the largest pterosaur. In *21st Joint Propulsion Conference* (1985), p. 1446.

[11] Cheney, N., Bongard, J., SunSpiral, V., and Lipson, H. Scalable co-optimization of morphology and control in embodied machines. *Journal of The Royal Society Interface 15*, 143 (2018), 20170937.

[12] Christensen, C. B., Christensen-Dalsgaard, J., and Madsen, P. T. Hearing of the african lungfish (protopterus annectens) suggests underwater pressure detection and rudimentary aerial hearing in early tetrapods. *Journal of Experimental Biology 218*, 3 (2015), 381–387.

[13] Cisek, P. Evolution of behavioural control from chordates to primates. *Philosophical Transactions of the Royal Society B 377*, 1844 (2022), 20200522.

[14] Cisek, P., and Hayden, B. Y. Neuroscience needs evolution, 2022.

[15] Clack, J. The stapes of acanthostega gunnari and the role of the stapes in early tetrapods. In *The evolutionary biology of hearing.* Springer, 1992, pp. 405–420.

[16] Clack, J. A. Patterns and processes in the early evolution of the tetrapod ear. *Journal of neurobiology 53*, 2 (2002), 251–264.

[17] Di Carlo, J., Wensing, P. M., Katz, B., Bledt, G., and Kim, S. Dynamic locomotion in the mit cheetah 3 through convex model-predictive control. In *2018 IEEE/RSJ international conference on intelligent robots and systems (IROS)* (2018), IEEE, pp. 1–9.

[18] Dunn, R. H., Cooper, C., Lemert, J., Mironov, N., and Meachen, J. A. Locomotor correlates of the scapholunar of living and extinct carnivorans. *Journal of morphology 280*, 8 (2019), 1197–1206.

[19] Eiben, A. E., Hart, E., Timmis, J., Tyrrell, A. M., and Winfield, A. F. Towards autonomous robot evolution. In *Software Engineering for Robotics*. Springer, 2021, pp. 29–51.

[20] Geraads, D. How old is the cheetah skull shape? the case of acinonyx pardinensis (mammalia, felidae). *Geobios 47*, 1-2 (2014), 39–44.

[21] Goswami, A., Binder, W. J., Meachen, J., and O'Keefe, F. R. The fossil record of phenotypic integration and modularity: a deep-time perspective on developmental and evolutionary dynamics. *Proceedings of the National Academy of Sciences 112*, 16 (2015), 4891–4896.

[22] Grimminger, F., Meduri, A., Khadiv, M., Viereck, J., Wüthrich, M., Naveau, M., Berenz, V., Heim, S., Widmaier, F., Flayols, T., et al. An open torque-controlled modular robot architecture for legged locomotion research. *IEEE Robotics and Automation Letters 5*, 2 (2020), 3650–3657.

[23] Grohé, C., Lee, B., and Flynn, J. J. Recent inner ear specialization for high-speed hunting in cheetahs. *Scientific Reports 8*, 1 (2018), 1–8.

[24] Hawkes, E. W., Eason, E. V., Christensen, D. L., and Cutkosky, M. R. Human climbing with efficiently scaled gecko-inspired dry adhesives. *Journal of The Royal Society Interface 12*, 102 (2015), 20140675.

[25] Haynes, G. C., Pusey, J., Knopf, R., Johnson, A. M., and Koditschek, D. E. Laboratory on legs: an architecture for adjustable morphology with legged robots.

[26] Hemmer, H., Kahlke, R.-D., and Vekua, A. K. The cheetah acinonyx pardinensis (croizet et jobert, 1828) sl at the hominin site of dmanisi (georgia)–a potential prime meat supplier in early pleistocene ecosystems. *Quaternary Science Reviews 30*, 19-20 (2011), 2703–2714.

[27] Holliday, J. A., and Steppan, S. J. Evolution of hypercarnivory: the effect of specialization on morphological and taxonomic diversity. *Paleobiology 30*, 1 (2004), 108–128.

[28] Hsieh, S., Schassburger, A., and Plotnick, R. E. The modern and fossil record of farming behavior. *Paleobiology 45*, 3 (2019), 395–404.

[29] Hunter, J. P. Key innovations and the ecology of macroevolution. *Trends in ecology & evolution 13*, 1 (1998), 31–36.

[30] Hyun, D. J., Seok, S., Lee, J., and Kim, S. High speed trot-running: Implementation of a hierarchical controller using proprioceptive impedance control on the mit cheetah. *The International Journal of Robotics Research 33*, 11 (2014), 1417–1445.

[31] Kau, N., Schultz, A., Ferrante, N., and Slade, P. Stanford doggo: An open-source, quasi-direct-drive quadruped. In *2019 International conference on robotics and automation (ICRA)* (2019), IEEE, pp. 6309–6315.

[32] Kelley, N. P., and Pyenson, N. D. Evolutionary innovation and ecology in marine tetrapods from the triassic to the anthropocene. *Science 348*, 6232 (2015), aaa3716.

[33] Kenneally, G., Chen, W.-H., and Koditschek, D. E. Actuator transparency and the energetic cost of proprioception. In *International Symposium on Experimental Robotics* (2018), Springer, p. 485–495.

[34] Kenneally, G. D., De, A., and Koditschek, D. E. Design principles for a family of direct-drive legged robots. *IEEE Robotics and Automation Letters 1*, 2 (2016), 900–907.

[35] Kim, S., Spenko, M., Trujillo, S., Heyneman, B., Santos, D., and Cutkosky, M. R. Smooth vertical surface climbing with directional adhesion. *IEEE Transactions on robotics 24*, 1 (2008), 65–74.

[36] MacIver, M. A., and Finlay, B. L. The neuroecology of the water-to-land transition and the evolution of the vertebrate brain. *Philosophical Transactions of the Royal Society B 377*, 1844 (2022), 20200523.

[37] Martin-Silverstone, E., Habib, M. B., and Hone, D. W. Volant fossil vertebrates: potential for bioinspired flight technology. *Trends in Ecology & Evolution 35*, 7 (2020), 618–629.

[38] McGhee, G. R. *The geometry of evolution: adaptive landscapes and theoretical morphospaces.* Cambridge University Press, 2006.

[39] McInroe, B., Astley, H. C., Gong, C., Kawano, S. M., Schiebel, P. E., Rieser, J. M., Choset, H., Blob, R. W., and Goldman, D. I. Tail use improves performance on soft substrates in models of early vertebrate land locomotors. *Science 353*, 6295 (2016), 154–158.

[40] Michaud, M., Veron, G., and Fabre, A.-C. Phenotypic integration in feliform carnivores: Covariation patterns and disparity in hypercarnivores versus generalists. *Evolution 74*, 12 (2020), 2681–2702.

[41] Nolfi, S., and Floreano, D. *Evolutionary robotics: The biology, intelligence, and technology of self-organizing machines.* MIT press, 2000.

[42] Nyakatura, J. A., Melo, K., Horvat, T., Karakasiliotis, K., Allen, V. R., Andikfar, A., Andrada, E., Arnold, P., Lauströer, J., Hutchinson, J. R., et al. Reverse-engineering the locomotion of a stem amniote. *Nature 565*, 7739 (2019), 351–355.

[43] Panciroli, E., Janis, C., Stockdale, M., and Martín-Serra, A. Correlates between calcaneal morphology and locomotion in extant and extinct carnivorous mammals. *Journal of Morphology 278*, 10 (2017), 1333–1353.

[44] Park, H.-W., Wensing, P. M., and Kim, S. High-speed bounding with the mit cheetah 2: Control design and experiments. *The International Journal of Robotics Research 36*, 2 (2017), 167–192.

[45] Parker, A. K., McHorse, B. K., and Pierce, S. E. Niche modeling reveals lack of broad-scale habitat partitioning in extinct horses of north america. *Palaeogeography, palaeoclimatology, palaeoecology 511* (2018), 103–118.

[46] Patel, A., and Braae, M. Rapid acceleration and braking: Inspirations from the cheetah's tail. In *2014 IEEE International Conference on Robotics and Automation (ICRA)* (2014), IEEE, pp. 793–799.

[47] Randau, M., and Goswami, A. Shape covariation (or the lack thereof) between vertebrae and other skeletal traits in felids: the whole is not always greater than the sum of parts. *Evolutionary biology 45*, 2 (2018), 196–210.

[48] Saranli, U., Buehler, M., and Koditschek, D. E. RHex: A simple and highly mobile hexapod robot. *The International Journal of Robotics Research 20*, 7 (2001), 616–631.

[49] Schneider, A., Paskarbeit, J., Schilling, M., and Schmitz, J. Hector, a bio-inspired and compliant hexapod robot. In *Conference on Biomimetic and Biohybrid Systems* (2014), Springer, pp. 427–429.

[50] Schweitzer, M. H., Schroeter, E. R., and Czajka, C. D. *Dinosaurs: How We Know What We Know.* CRC Press, 2020.

[51] Seok, S., Wang, A., Chuah, M. Y., Otten, D., Lang, J., and Kim, S. Design principles for highly efficient quadrupeds and implementation on the mit cheetah robot. In *2013 IEEE International Conference on Robotics and Automation* (2013), pp. 3307–3312.

[52] Spassov, N. Acinonyx pardinensis (croizet et jobert) remains from the middle villafranchian locality of varshets (bulgaria) and the plio-pleistocene history of the cheetahs in eurasia. *Estud. Geol 67* (2011), 245–253.

[53] Tang, Y., Chi, Y., Sun, J., Huang, T.-H., Maghsoudi, O. H., Spence, A., Zhao, J., Su, H., and Yin, J. Leveraging elastic instabilities for amplified performance: Spine-inspired high-speed and high-force soft robots. *Science advances 6*, 19 (2020), eaaz6912.

[54] Topping, T. T., Vasilopoulos, V., De, A., and Koditschek, Daniel, E. Composition of templates for transitional pedipulation behaviors. p. (in press).

[55] Trestman, M. The cambrian explosion and the origins of embodied cognition. *Biological Theory 8*, 1 (2013), 80–92.

[56] Tseng, Z. J. Testing adaptive hypotheses of convergence with functional landscapes: a case study of bone-cracking hypercarnivores. *PLOS one 8*, 5 (2013), e65305.

[57] Tseng, Z. J., and Wang, X. Do convergent ecomorphs evolve through convergent morphological pathways? cranial shape evolution in fossil hyaenids and borophagine canids (carnivora, mammalia). *Paleobiology 37*, 3 (2011), 470–489.

[58] Turner, A., and Howell, F. C. *The big cats and their fossil relatives: an illustrated guide to their evolution and natural history.* Columbia University Press, 1997.

[59] Van Valkenburgh, B., Grady, F., and Kurtén, B. The plio-pleistocene cheetah-like cat miracinonyx inexpectatus of north america. *Journal of Vertebrate Paleontology 10*, 4 (1990), 434–454.

[60] Van Valkenburgh, B., Pang, B., Cherin, M., and Rook, L. The cheetah: Evolutionary history and paleoecology. *Biodiversity of the World—Cheetahs: Biology and Conservation. 1st ed. San Diego: Elsevier* (2018), 25–32.

[61] Vasilopoulos, V., Topping, T. T., Vega-Brown, W., Roy, N., and Koditschek, D. E. Sensor-based reactive execution of symbolic rearrangement plans by a legged mobile manipulator. In *2018 IEEE/RSJ International Conference on Intelligent Robots and Systems (IROS)* (2018), IEEE, pp. 3298–3305.

[62] Vasilopoulos, V., Vega-Brown, W., Arslan, O., Roy, N., and Koditschek, D. E. Sensor-based reactive symbolic planning in partially known environments. In *2018 IEEE International Conference on Robotics and Automation (ICRA)* (2018), IEEE, pp. 1–5.

[63] Wainwright, P. C., and Price, S. A. The impact of organismal innovation on functional and ecological diversification. *Integrative and Comparative Biology 56*, 3 (2016), 479–488.

[64] Wensing, P. M., Wang, A., Seok, S., Otten, D., Lang, J., and Kim, S. Proprioceptive actuator design in the mit cheetah: Impact mitigation and high-bandwidth physical interaction for dynamic legged robots. *Ieee transactions on robotics 33*, 3 (2017), 509–522.

[65] Wood, A. R., Bebej, R. M., Manz, C. L., Begun, D. L., and Gingerich, P. D. Postcranial functional morphology of hyracotherium (equidae, perissodactyla) and locomotion in the earliest horses. *Journal of Mammalian Evolution 18*, 1 (2011), 1–32.

[66] Xi, W., Yesilevskiy, Y., and Remy, C. D. Selecting gaits for economical locomotion of legged robots. *The International Journal of Robotics Research 35*, 9 (2016), 1140–1154.

[67] Zelditch, M. L., Ye, J., Mitchell, J. S., and Swiderski, D. L. Rare ecomorphological convergence on a complex adaptive landscape: body size and diet mediate evolution of jaw shape in squirrels (sciuridae). *Evolution 71*, 3 (2017), 633–649.

IOP Conf. Series: Materials Science and Engineering 1261 (2022) 012023 doi:10.1088/1757-899X/1261/1/012023

What Could Models of Superorganismal Cognition Offer to Embodied AI?

Edmund R. Hunt

Department of Engineering Mathematics, University of Bristol, Ada Lovelace Building, University Walk, Bristol, BS8 1TW, UK

E-mail: edmund.hunt@bristol.ac.uk

Abstract. Superorganisms such as ant or honeybee colonies exhibit extraordinary collective intelligence, such as an ability to identify and choose the best available nest site in an uncertain world. This collective cognition is inextricably reliant on the embodiment of individual agents, specifically their movement through space. We have recently developed models of superorganismal cognition based on a compelling analogy with techniques in Bayesian statistics, which are likewise aimed at grappling with the uncertainty and incompleteness of real data sources. These models foreground some potential lessons for the design of embodied artificial intelligences, such as robot swarms. For example, the spatial distribution of independently judging agents can convey valuable information, relaxing expectations that regular inter-agent ('inter-neuronal') communication is necessary for cognition, which points to the potential of minimal field swarm robotics. Meanwhile, the importance of individual heterogeneity to effective and resilient collective cognition in biology suggests great potential in this area for engineering.

1. Introduction

Many animals live in groups and make common actions together, such as choosing foraging patches or resettling to a new home. Such collective action has been described as 'collective cognition' [1], and the associated collective information processing is particularly remarkable in social insects such as ants or honeybees [2]. This is because their close-knit societies resemble single organisms, or 'superorganisms' [3], and their cognition in certain respects resembles a single brain [4, 5]. Such collective 'brains' have recently been described as 'liquid brains', where the individual 'neurons' have mobility and transient connections with each other, rather than the persistent architecture of interactions seen in 'solid brains' [6]. Indeed, the spatial movement of individuals within superorganisms is undoubtedly key to their cognition, and by association, so is their embodiment. This might be something of a statement of the obvious, but working in artificial intelligence, one may tend to under-appreciate the information that is intrinsically conveyed by an embodied agent's decision to be in one location rather than another. The field of 'movement ecology' has emerged in the scientific push to understand animal movement within the broader context of internal animal states, traits, constraints and interactions among themselves and with their environment [7]. In our own effort to make clear the inseparable relationship between movement, environment and superorganismal cognition, we have recently developed a new perspective that highlights the close analogy between colony behaviour and certain Bayesian statistical methods. These models may indicate some future directions for embodied artificial intelligence.

IOP Conf. Series: Materials Science and Engineering 1261 (2022) 012023 doi:10.1088/1757-899X/1261/1/012023

2. Models of Superorganismal Cognition

2.1. Movement for Foraging

The acquisition of food resources is a key problem for any living organism. For superorganisms like ant or honeybee colonies, there is the added challenge of acquiring food not only for oneself, but for the group. In this case, one would expect the evolution of a colony-level foraging strategy that maximizes colony fitness (i.e. long-term reproductive success). In the 'ergonomic' stage of a colony's life cycle, the focus is on exponential growth, where workers collect food to be 'reinvested' by the egg-laying queen into more and more workers. This has parallels with betting, where one can reinvest one's winnings into bigger bets. The optimal strategy there is a 'Kelly', or probability-matching, strategy whereby one makes bigger bets on more likely outcomes (e.g. to put a bigger bet on the favourite in the horse race). If one sees the environment as having a variety of probabilities of food payoff in different locations, then one would expect the colony to 'allocate' (via self-organization) a higher proportion of worker ants to food patches with more reliable yields (probability matching behaviour has recently been identified in honeybees [8]). In an unknown environment, an individual is initially unaware of the distribution of such probabilities, but the distribution can be estimated via Markov chain Monte Carlo (MCMC) methods such as the Metropolis-Hastings algorithm; in this case the 'walker' is the animal itself [9]. MCMC walking naturally results in an individual spending proportionally more time in regions of higher probability (probability matching), following a 'burn in' exploration period. Coordination via pheromones could speed up the dispersion of walkers across the space [10]. The colony's performance is maximised when the distribution of ants across the environment, as represented by a probability distribution, best matches the patch payoff probability distribution, which can be quantified using the Kullback-Leibler divergence. Thus, we see the relevance of information theory measures in the evolution of (embodied) movement strategies, beyond the application of information theory to e.g. sensory-motor coordination [11], and could be seen a macro-level expression of the 'principle of information self-structuring' [12]. Such a MCMC movement strategy is inherently relevant to superorganisms because probability matching as a strategy is only realisable with a 'bank' of organisms, permitting bet-hedging; it would not be possible or desirable for a single organism.

2.2. Movement for Nest Site Selection

Apart from finding food, obtaining a good place to live is of fundamental importance to most animals, and this certainly the case for superorganisms like colonies of honeybees or ants. The ant *Temnothorax albipennis* is very good at choosing between different nest site options, a type of 'best-of-n' problem [13]. This is despite the fact that individual ants do not need to make direct comparison of the options in order for the colony as a whole to end up at the best available site [14, 15]. This is because if a number of ants simultaneously explore the options, and accept/reject depending on an individual quality threshold, a larger number will tend to accumulate at higher quality sites (Figure 1).

We have developed a model of this nest-finding process [16] as being analogous to Approximate Bayesian Computation (see Sunnåker et al. for an introduction to ABC [17]). One might call the ants' behaviour *Spatial* Approximate Bayesian Computation ('SABC', see Algorithm 1.). Essentially, the individual acceptance or rejection of nest sites can be compared to the accept/reject of simulated parameter values in ABC, except in SABC the accept/reject is enacted by the physical (embodied) movement of an agent to a different location. Just as after the accept/reject of a large number of parameter samples, one obtains a posterior distribution for that parameter, here we argue that the distribution of ants over the environment can be seen as a posterior estimate of the probability that each location is likely to be the best available nest site. For the colony to make the best choice, then, it simply needs to identify the highest probability location: where most of its own worker scouts have decided to put themselves.

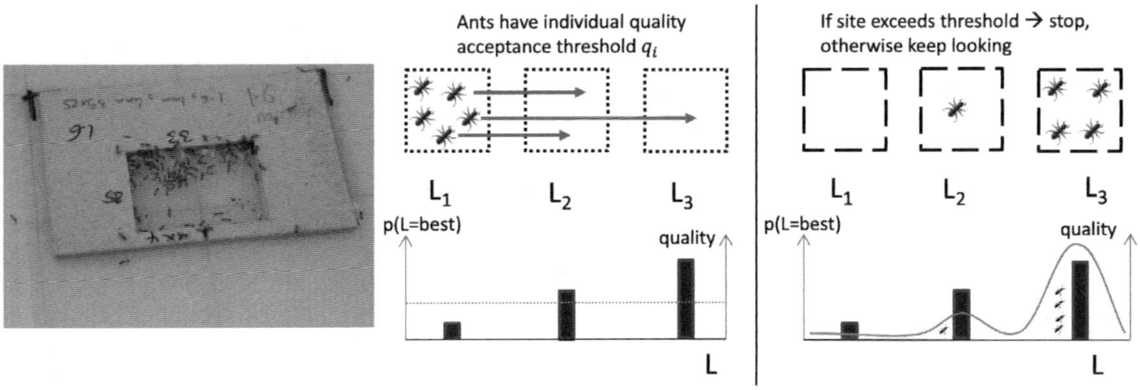

Figure 1. Simple model of superorganismal nest site selection. Left: real ants (size ~2mm) leaving an artificial nest site that has been destroyed (roof removed). Right: Simulated non-interacting ants inspect nest sites and compare quality observations to an internal threshold. If this is exceeded, they stop searching, otherwise they keep looking for alternatives. Their macroscopic distribution approximates the posterior distribution of location qualities. Four out of five ants at location 3 corresponds to an 80% probability it is the best one.

Additional mechanisms to allow inter-agent interaction can be included, such as the one-to-one leading of naive workers to high quality sites ('tandem running'), which we compare to particle filtering (preferentially sampling important parts of the distribution) [16].

Algorithm 1: Pseudocode for simple SABC nest finding method.

```
Result: probabilities p_L each location is best
initialize: N ants, true location L qualities Q_L, individual ant priors re. location
  qualities μ_iL, σ_iL, collective decision threshold p_quorum
while ∀p_L < p_quorum do
    for ant i=1:N do
        rand. draw from L priors q_iL ~ N(μ_iL, σ²_iL)
        move to L with highest draw
        make quality observation
        update q_iL for chosen L using Kalman filter
    end
    calculate each p_L = Σ_L(ants at location L)/N
end
```

3. Relevance to Embodied Artificial Intelligence

The models of foraging [9, 10] and nest site selection [16] foreground some interesting insights into the nature of embodied intelligence, which are also relevant to the design of embodied artificial intelligences, especially robot swarms. Here are three ideas.

First, at least in the *Temnothorax* ant system we have examined, the spatial distribution of individuals conveys valuable information, beyond the knowledge of any single agent. This information is emergent from the independent action of many individuals across the environment, and is exploited in nature via a quorum-sensing mechanism, where agents judge the local agent density to estimate the number of nestmates in the same place [18]. In an artificial system, a user could exploit a similar self-organized approach, or would also have the option of simply observing the distribution of agents from the 'top down', assuming some kind of tracking or 'bird's-eye view' is possible. That is to say, the p_L in Algorithm 1 can be estimated/calculated by the agents within the system, or by an onlooker. As we develop cyber-physical systems, we should not lose sight of what the overall physical view of system components is telling us. The MCMC-based models of movement also point toward the applicability of information-theory-based insights at higher levels of (biological) self-organisation, at the interaction between colony movement strategies and their ecology. In the swarm robotics context, one indeed will be aware that that movement strategies for individual robots must be optimised for group-level performance. This is because swarm performance is emergent from individual interactions, sometimes purposeful cooperation. To develop individual controllers via artificial evolution, information-theory-based metrics comparing the swarm's spatial distribution to various user-defined targets could be employed as objective functions, in an exploratory fashion. This holistic focus on finding the 'shape' of high-performance swarms goes beyond thinking about information theory descriptions at the individual level, for instance in sensory-motor coordination.

Second, interaction between agents can be rather limited, or not needed at all, to nevertheless obtain useful information from their macroscopic distribution. Pointing in this direction, a recent study has confirmed the value of less intensive interactions in robot swarm decision-making [19]. For a user deploying truly minimal field swarm systems, for example biodegradable floating balls that change colour when in a pollution patch [20], the probability distribution of the most polluted patches could still become evident even if the 'robots' do not communicate. Nevertheless, one might still venture to describe this as some form of cognition, emergent from the encounter with the environment.

Third, biological research points toward the value of heterogeneity in behavioural thresholds for more accurate and resilient collective decisions [21, 22]. For instance, a small proportion (3-12%) of *Temnothorax* workers are persistent in exploring during a nest-site decision [23], which may correspond to high acceptance threshold 'picky' ants who keep looking for the very highest quality nest sites. This may ensure that the best option (global maximum) is usually discovered eventually. A distribution of behavioural traits in robot swarms could likewise be key to unlocking higher performance in tasks such as exploration and decision-making. Obtaining a suitable trait distribution is likely to rely on a mixture of pre-deployment (offline) optimisation in artificial evolution and laboratory trials, and also online flexibility to adapt to the actual environment. I have suggested that this flexibility can be understood as part of a general 'phenotypic plasticity' needed for successful embodied intelligence [24]. Neurons in a brain are not homogeneous, neither in their morphology nor their connectivity, and likewise to realise embodied artificial intelligence we should be more willing to experiment with agent heterogeneity, either by design or accident. This suggests that traditional engineering paradigms that seek to control variation and noise will need to be somewhat inverted. There is some growing awareness in e.g. the field of cyber security that deliberately engineered differences in components can make systems more resilient [25], but such insights have yet to permeate more broadly. Although these three ideas arise from trying to model the statistical computation of superorganismal cognition, they are not dependent on specific model details, and could be among general features of distributed embodied cognition.

References

[1] Couzin I D 2009 *Trends in Cognitive Sciences* **13** 36–43
[2] Detrain C, Deneubourg J L and Pasteels J M 1999 *Information Processing in Social Insects* (Springer Science & Business Media)
[3] Kennedy P, Baron G, Qiu B, Freitak D, Helanterä H, Hunt E, Manfredini F, O'Shea-Wheller T, Patalano S, Pull C, Sasaki T, Taylor D, Wyatt C and Sumner S 2017 *Trends in Ecology and Evolution* **32**

[4] Marshall J A R, Bogacz R, Dornhaus A, Planque R, Kovacs T and Franks N R 2009 *Journal of The Royal Society Interface* **6** 1065–1074
[5] Sasaki T and Pratt S C 2017 *Annual Review of Entomology* **63** 259–275
[6] Solé R, Moses M and Forrest S 2019 *Philosophical Transactions of the Royal Society B: Biological Sciences* **374**
[7] Nathan R, Getz W M, Revilla E, Holyoak M, Kadmon R, Saltz D and Smouse P E 2008 *Proceedings of the National Academy of Sciences of the United States of America* **105** 19052–19059
[8] MaBouDi H, Marshall J A R and Barron A B 2020 *Proceedings of the Royal Society B: Biological Sciences* **287** 20201525
[9] Baddeley R J, Franks N R and Hunt E R 2019 *Journal of The Royal Society Interface* **16** 20190162
[10] Hunt E R, Franks N R and Baddeley R J 2020 *Journal of The Royal Society Interface* **17** 20190848
[11] Sporns O and Pegors T K 2004 *Embodied Artificial Intelligence* (Springer) pp 74–85
[12] Lungarella M and Sporns O 2005 *Proceedings. The 4th International Conference on Development and Learning, 2005* pp 25–30
[13] Franks N R, Mallon E B, Bray H E, Hamilton M J and Mischler T C 2003 *Animal Behaviour* **65** 215–223
[14] Robinson E J H, Smith F D, Sullivan K M E and Franks N R 2009 *Proceedings of the Royal Society B: Biological Sciences* **276** 2635–2641
[15] Robinson E J, Franks N R, Ellis S, Okuda S and Marshall J A 2011 *PLoS ONE* **6**
[16] Hunt E R, Franks N R and Baddeley R J 2020 *Artificial Life Conference Proceedings* **32** 315–323
[17] Sunnåker M, Busetto A G, Numminen E, Corander J, Foll M and Dessimoz C 2013 *PLoS Computational Biology* **9**
[18] Pratt S C 2005 *Behavioral Ecology* **16** 488–496
[19] Talamali M S, Saha A, Marshall J A R and Reina A 2021 *Science Robotics* **6** eabf1416
[20] Hunt E R and Hauert S 2020 *Nature Machine Intelligence* **2** 420–422
[21] O'Shea-Wheller T A, Hunt E R and Sasaki T 2020 *Annals of the Entomological Society of America* **114** 562–574
[22] Hasegawa E, Mizumoto N, Kobayashi K, Dobata S, Yoshimura J, Watanabe S, Murakami Y and Matsuura K 2017 *Scientific Reports* **7** 1–11
[23] Valentini G, Masuda N, Shaffer Z, Hanson J R, Sasaki T, Walker S I, Pavlic T P and Pratt S C 2020 *Proceedings of the Royal Society B: Biological Sciences* **287** 20192950
[24] Hunt E R 2020 *Frontiers in Robotics and AI* **7** 23
[25] Eder K 2021 *Proceedings of the 1st International Workshop on Verification of Autonomous & Robotic Systems* VARS '21 (New York, NY, USA: Association for Computing Machinery)

IOP Conf. Series: Materials Science and Engineering 1261 (2022) 012024 doi:10.1088/1757-899X/1261/1/012024

Embodied Intelligence in Physical, Social and Technological Environments

Ilona Straub

Department of History and Sociology
University of Konstanz, Germany

ilona.straub@uni-konstanz.de

Abstract. The aim of this contribution is to highlight different aspects of embodied intelligence that add to a broader interdisciplinary perspective on this phenomenon. This article intends to bring together three forms of embodied intelligence from the fields of technology and social research, starting from robotics research that deals with the technical replication of organic bodies, to insights from the sociology of the body outlining the relevance of corporeality in social encounters, and finally to the forms of embodied intelligence in the connection of technical forms with organic bodies and their effects on the perception of self and others within the hybrid research field of "Bodies of Technologies". This tripartite division of embodied intelligence deals with the biophysiological properties of an organism in self-environment relationships (the subject of robotics research), the interaction of a being or system with other beings in self-environment relationships (the subject of the sociology of bodies), and finally, as a third, new level of embodied intelligence, the technically mediated experience of (virtual) self-environment relationships through the somatechnical fusion of bio- and socio-physical properties in human-machine hybrids is presented. Accordingly, the article will name the specificity of each form of embodied intelligence and trace the conditions and peculiarities in the gradual increase of complexity from simple to interactive and finally to virtual bodily intelligence.

1. Introduction

A basic prerequisite for organic life is the organism's relationship to its environment as well as to other living beings as part of its experienced world. In this context, living beings with their properties do not buzz through the world as immaterial life forms, but are equipped with bodies that enable them to harmonise their needs with the challenges of the world, according to their perceptual and sensory specifications.

The research field of embodied intelligence has set itself the task of getting to the bottom of precisely these properties of organisms and recreating them in all their nuances through artificial life forms. The research field of embodied intelligence is a relatively young field that made its first attempts in the 1990s [1] and has gradually established itself as an inter- and transdisciplinary branch of research between computer science, biology, engineering, philosophy, and neuroscience. Parallel to the development in the technical sciences, a "body turn" took place in the humanities and social sciences, starting from the prevailing linguistic paradigm, where the body increasingly became the focus of consideration of entities that physically interact with each other or form practices in dealing with objects. Thus, here too, the pure consideration of language and linguistic idiosyncrasies or the focus on cognitive mechanisms of social relations was brought back to the foundation of bodily materiality, and in sociology the field of sociology of the body was established. As in so many areas of the natural and social sciences, the results of the two fields, which are intensively worked on in their own fields of research, have not been merged, although this merging of findings would be a great gain for the disciplines involved. As Pfeifer & Iida state, it can therefore be assumed that a bringing together of the findings from the different disciplines would lead to both a more sophisticated theoretical level and a more advanced practical knowledge for the application of the concepts in embodied intelligence research.

"With the fundamental paradigm shift from a computational to an embodied perspective, the nature of research topics, the theoretical and technical issues, and the disciplines involved have changed dramatically, or in other words, the "landscape" has changed completely." (Pfeifer & Iida 2003:1)

In their contribution "Embodied Artificial Intelligence: Trends and Challenges," Pfeifer and Iida [2] point to a paradigm shift from emulating information processes of higher cognition in computer science to an embodied perspective on intelligence through imitation of organic beings. Similarly, this article highlights the implications of the sociology of the body for robotics research and new findings on bodily properties for the technical imitation of embodied beings in environmental relationships. Insights into the fit of embodied technologies to everyday, socially habituated life situations reveal, on the one hand, undiscovered bodily as well as corporeal capacities for theorizing corporeal actors and, on the other hand, contribute to recognizing somatechnical hybrids as part of the body, the self, the alter ego, and even as part of society, from which new social practices emerge.

A step towards merging concepts of robotics and social research is taken in this contribution by extracting the basic levels of embodied intelligence research from the various research findings at this point and systematically relating them to each other. Accordingly, an overview of the basic forms of embodied intelligence research is given, followed by a proposal for determining their characteristics up to a leveling of their degrees of abstraction. In this way, it becomes visible to what extent embodied intelligence increases in complexity from lower order to higher order intelligence analogously to the degrees of cognitive intelligence and even keeps pace with the degree of abstraction of symbol processing in cognitive systems.

In the following pages, three levels of embodied intelligence are considered: first, *physical embodied intelligence* as reflecting a self-environment relationship in robotics; second, *socially embodied intelligence* as reflecting a co-world relationship in social encounters; and third, *embodied intelligence with somatechnically integrated objects* in terms of new embodied qualities emerging in the hybrid relationship between physical human-technology interfaces through effects of controllable technical extensions on the self-environment relationship and the co-world relationship.

2. Embodied Intelligence in Self-Environment Relations

In order to tap the potentials of controllable human-robot constellations as technical body extensions for the discovery of hitherto unrecognized basic capabilities of embodied intelligence in human actors, it is first necessary to gain an impression of the physical properties of organisms acting in the world. Therefore, the conditions for a stable self-environment relationship, which is aimed at in robotics research as a basis for imitating living beings, will be reproduced first. The approaches discussed in the workshop illustrate the detailed development of embodied entities acting in variable, local environmental situations, from the selection of materials for the flexibility of shape change (morphology) as the emulation of simple physical intelligence, to the selection of relevant information for obstacle-free navigation in the terrain (locomotion), to the estimation of distances, velocities, or pressure compensation, e.g., as forms of advanced bodily estimations of local situations with embodied intelligence. Robotics research draws on results from a broad inter- or transdisciplinary field of biology, neuroscience, engineering, biomechanics, material science, and dynamical systems to emulate living beings (cf. Pfeifer & Iida 2003:10) in order to understand the physical properties of system-environment relationships in everyday, real-world situations: "The reason for the strongly transdisciplinary character of Embodied Artificial Intelligence is that intelligence, especially embodied intelligence, has to do with behaviour, with interaction in the real world, and since we are dealing with physical agents, many aspects and components are involved: Materials, morphology, sensors, actuators, energy supply, control, planning, cognition and perhaps even consciousness." (Iida 2003:V) [3].

To successfully develop animated machines based on biophysiology, robotics advocates the "physical intelligence" (PI) or "embodied intelligence" (EI) approach and focuses on biomimetic simulations of living beings. Researchers in embodied intelligence research primarily refer to Pfeifer and Bongard [4] for the definition of embodied intelligence, who name a connection between the morphology of the body and the environment in relation to the action to be performed and propose the emergence of bodily-based intelligence from movement, locomotion, and manipulation abilities. Sitti [5] characterizes intelligence in autonomous machines in general as "(...) ability to perceive (sense, interpret), control (decide, plan, predict, regulate), act (move, change, affect, coordinate) and learn (adapt, evolve, acquire experience, infer) continuously and automatically" (2021:1) and physical/bodily intelligence in particular as "(...) physically encoding sensing, actuation, control, memory, logic, computation, adaptation, learning and decision-making into the body of an agent." (ibid.).

While Sitti names gestalt morphology as "physical intelligence" (PI) and thus a condition for the formation of embodied intelligence (EI), Pfeifer and Bongard as well as Howard et al.'s [6] definition of "embodied intelligence" includes the effect of bodily information processing on cognitive functions of e.g. decision making: "Natural life strives in unstructured environments through a specific type of intelligence known as 'embodied cognition'. Intelligent behaviour arises from a close connection between an agent's body and brain and the environment, not just the brain." [6]. This definition attributes intelligent behavior in unstructured environments to the coupling of body, cognition, and environment, and contrasts with top-down approaches to symbol processing that focus on behavioral instructions starting from higher cognitive process architectures to accomplish tasks. Kramer-Bottiglio defines the emergence of embodied intelligence as a bottom-up learning process in interaction with the entity's body constitution using the concept of "morphological computing" as follows: "Embodied intelligence is the idea of embodying an ability to learn or apply knowledge, which might include the adaptive morphology of agents." [7].

This requires a close relationship of the system to the events in the environment. The main principle of robotics research for modeling AI is thus: *True intelligence requires bodily interaction of the system with the real world.* Robotics aims to understand and reproduce the fundamental physical conditions that enable an actor to orient itself spatiotemporally and to adapt its shape and behavior to material environments, as well as to autonomously perform tasks in unstructured and complex real-world environments. To this end, it follows both ontogenetic and phylogenetic approaches as applied in the subdisciplines of (evolutionary) developmental robotics, biorobotics, biomechanics, and morphogenesis. Together with developmental robotics, biorobotics or biomechanics strive to create viable organisms adapted to the environment. These branches study how certain behaviors and physical characteristics of real-world living creatures can be transferred to artificial agents. The research area includes studies on the distribution of tasks between materials of the body structure as well as on their morphology in the self-control of the organism in interaction with the environment. From this perspective, the basic behavior of organisms in self-environment relationships depends on the morphology, locomotion dynamics, self-organization by energy supply, and the forms of gestalt based on material properties of the organism that enable the engineered system to flexibly deal with complex environmental obstacles based on varying gestalt adaptations. Examples include obstacle avoidance through environmental mapping as well as autonomous responses, effortless navigation, orientation and path planning in rough terrain, or purposeful reaching for objects. A living organism surrounded by objects performs local interactions that are aligned with biological and physical laws of nature and can therefore be accurately measured and calculated for technical implementation. The challenge for robotics research in this area is to mimic a self-regulating and self-configuring entity at the level of low-threshold sensorimotor processes in order to achieve a basis for successful practical access to the world via flexible gestalt formations. If the robot is supposed to have the shape as well as the range of action of a human being, this requires a balance between the multilayered components of the sensorimotor control system, which in its totality of sensory stimuli must be regarded as a complex multimodal and multisensory system.

For example, an everyday task that is simple by our standards and that we perform automatically on the basis of habitualized action sequences, such as removing a water bottle from the refrigerator and then filling a glass with water, requires a complex interaction between multisensory data: These are required to detect the refrigerator as a target object, to move to the refrigerator and open it, to maintain balance while opening and closing the refrigerator, to estimate the pressure balance while opening the refrigerator door, to visually detect and locate the water bottle, to perform an arm movement in the direction of the desired object, tactile fine motor skills in grasping the object, pressure equalization and resistance in grasping the object according to the weight and texture of the material of the bottle, observing the environment to avoid hitting a) the bottle or b) one's body against a fixed or moving obstacle when lifting out and stepping back, and other small-step, complex actions to put the water in a glass. It is this multimodal body-based system that links, processes, and mediates sensory stimuli with motor signals that, in its nature and function for the whole organism in relation to the environment, can be cited as a crucial moment for a next higher level of multisensory embodied intelligence. The more sensors are involved in the information acquisition, the more complex becomes the consolidation and evaluation of the data to be processed in parallel as well as the algorithmic calculation of the system reaction in the form of an environmentally appropriate action.

To relieve the limited computational capacity and energy supply, intelligent behavior should already be incorporated in the design of the form as well as the material structure (PI): Thus, sustainable smart materials are embedded in the hardware (the body) of the robot. This includes robust exoskeletons against collisions with the environment as well as a multi-functional and adaptable body morphology that interacts with the environment and, for example, draws its energy from the environment based on stimulus-responsive and programmable smart materials (Sitti 2021:3). Further, at the level of physical intelligence that is embedded in the gestalt, a flexible system can respond to environmental stimuli, including changes in material properties such as shape, volume, strength, density, thermal conductivity, or adhesion, and subsequently derive its own self-localization based on the self-position, self-motion, and self-awareness (proprioception) of body parts in relation to environmental conditions and translate this into subsequent actions (cf. Sitti 2021:5). The initial motto of robotics, "True intelligence requires interaction with the real world," is joined by an increase in embodied intelligence through the linking of diverse multisensory information, which AI researchers assume is the source of the emergence of consciousness in a further step. The combination of the variety of sensory information from self-environmental events should finally lead to the generation of possible subjective qualities as "qualia" in the form of lived experiences, emotions or motivations in technical systems [8].

From the above discussion of the sensitivities of body-based intelligence that prove to be obstacles in the attempt to technically reproduce living organisms, the foundations of a) basal physical embodied intelligence, based on the materials of gestalt, and b) embodied intelligence, based on the ability of the system to operate in environments, can be named.

Embodied intelligence manifests itself on a basic physical level in successful goal-directed action through autonomous and flexible adaptation of the gestalt to possibly resistant or changing environmental conditions (a). Here, the assumption becomes clear that a being has embodied intelligence when it, on the basis of diverse sensory information, acts autonomously in relation to its environment and, in doing so, is able to differentiate and regulate itself as a multisensory proactive self from what is happening in the environment (b) [9]. Thus, the degree of embodied intelligence required here is at a fundamental level in the interplay of multiple sensorimotor processes for self-control of an embodied sensory unit with the material world and shall be termed as *physical embodied intelligence*.

As explained above, the adaptation of organic and technical actors to dynamic environmental conditions is in itself a highly complex endeavor involving the support of physical-material properties, the evaluation of multisensory information, the design and capabilities of successful navigation, etc. Once the environment is populated by other self-moving actors, it means that the system is confronted with other systems acting with unpredictable route, speed, intensity or direction in the environment. This greatly increases the complexity of multisensory path computations. Moreover, these self-moving actors may not only present additional obstacles in the environment, but also interact with the computing system. This circumstance creates additional interfaces and necessities for linking self-regulating systems as interactive action systems or as multi-agent systems capable of acting collectively and generating higher levels of embodied intelligence. The corresponding skills of embodied intelligence will be discussed in the next section on the embodied intelligence in co-world relations, making visible another component of subtle body-based knowledge.

3. Embodied Intelligence in Co-World Relations

As has become clear in the section above, the requirements for implementing embodied intelligence increase with the amount of sensory data, i.e., multimodal and multisensory data collection. Another component that points to an additional multi-layered form of embodied intelligence is social encounter. Based on rules for local system-environment interactions, embodied intelligence reaches its next level of complexity in co-environmental multi-agent systems or multi-party interactions. Here, the system shares the local environment with other systems with which it is able to perform relations or corporeal interactions. By focusing on physically co-present systems, which are themselves multisensory and can interact a) with the environment and b) with the system itself, additional further information for navigation as well as interaction activities of the system with the co-world arise, which include understanding of the social context and group dynamics for extended context perception and sense making. Through the social theoretical discussion of the physical-communicative conditions of social encounters, embodied intelligence will be addressed in the following section at a next-higher level of local physical environmental orientation. To this end, we will make comparisons with social beings living in the social world, elaborate the particularities of embodied intelligence in social worlds, and identify the bodily capacities that should be considered in the construction of socially acting technical beings. In doing so, the qualities of embodied intelligence in social beings are divided into the following: 1) subject-object differentiation, 2) perspective expansion to a common sense of world experience through a reciprocal expansion of the senses in interaction, and 3) a habituated sense for interaction.

3.1. Subject-Object Differentiation

While the replication of "physical embodied intelligence" focuses primarily on the analysis and reproduction of form, materiality, and successful navigation in the self-environment relationship of organisms, everyday experience shows that embodied intelligence comes into play not only in environmental actions, but also in social encounters and thus in co-worlds. If, for example, dogs meet each other on a dog run, they often approach each other, circle around each other or run together towards a goal. Further it does not remain with the four-legged friends to sniff each other or to bark to each other. In joint play, such as tugging on a stick, the rich detail of coordinated behavioral sequences becomes apparent, as does the coordinated action of an individual and the shifting of emphasis from movements to joint behavioral coordination. Structurally identical forms are observable in everyday interpersonal encounters between human such as moments of interaction and coordination, although the type of encounter differs from animal encounters [9]. Overall, the behavior between organisms that can relate to each other differs from moments of orientation of an actor to a purely material environment. For adequate acting in co-worldly relations, correspondingly further specific abilities of embodied intelligence are required than in the case of local interactions of embodied entities with the material world. Simply the skill to distinguish and recognize other potential social partners from mere objects in the environment, in order to then extend the self-environment reference to a self-co-world reference that includes the social partner, requires additional specific abilities of embodied intelligence through divergent discretionary powers of sense processing of perceptions. This distinction between a self, an external object, and another acting subject can be mentioned as an essential feature of embodied intelligence in the co-world relation.

The moment of (inter)bodily encounter is the basis for social processes in local co-worlds and thus crucial for the further analysis of embodied intelligence practiced in co-worlds. Lindemann calls this subject-object-differentiating moment of (interbodily) encounter with another bodily self in a shared situation "moment of touch" [10] and Breyer [11] speaks of "embodied empathy" when it comes to interpreting the other as a potential social partner in the "moment of touch" based on shape and bodily expressive behavior. The "moment of touch" refers less to the tactile touch of the actors, but rather to the initiating moment of contact, which, through the social and personal presence of an actor, evokes a form of "affectation" on a social-perceptual level. This "moment of touch" is crucial when the ego encounters a social actor in the local situation (alter ego), i.e., in shared presence, and distinguishes him as a social partner from objects. In doing so, he can perceive the alter ego unilaterally as a being that is able to separate itself from the environment (indirect touch) - e.g., a searching animal in the forest - or to initiate a reciprocal perceptual relationship (direct touch). The latter, for example, in that the alter ego appears as a figure beckoning to him. Here ego recognizes this as a bodily expressive movement directed at itself, by which ego is "directly touched" and which it can follow up with further actions based on a bodily expression directed at the alter ego [10]. This moment of direct touch is subliminal and bodily inherently mediated, thus it is not grounded in a reflexive-cognitive thematization for sensuously locating the appearing gestalt as a social being. It is rather a form of tangible certainty as a drive for spontaneous bodily actions oriented to alter ego. Accordingly, direct perception in social encounters analogous to the mutual perception of bodily selves and the resulting possibility to influence each other is a key feature of social encounters. This ability to differentiate between subject and object is an essential moment of embodied intelligence in socially acting beings, which has to be implemented in the body architecture of social robots. In the next section, we will show to what extent the requirements for the implementation of embodied intelligence change in the transition from environments to co-worlds with respect to perspectival diversity, perceptual synergies, and the resulting common sense once the actors act as the "intermediate body" of ego and alter ego.

3.2. Co-Sensoriality, Con-Sensoriality, and Common-Sensoriality

According to Merleau-Ponty, the primacy of the social encounter lies in the dual aspect of perceptible and perceptual physicality as the basis of expression [12]. For example, by seeing that alter ego sees me, by feeling that he touches me, I can assume that alter ego perceives me, as it were. So, in this situation I feel myself as touching and by my touching I also feel alter ego and I know further about his foreign and self-perception towards me. With the awareness of the mutual touch, mutual orientation as well as reference to each other, the "double aspectivity" of the bodily perception is given. Even more, Merleau-Ponty explains the "living body" (Leib) as the foundation of "intercorporality", as a bodily analogue to an intersubjectivity based on sensory-mental processes [12].

Understanding thus becomes bodily understanding via shared (knowledge of) bodily perceptual practices. Intercorporeality thus represents the bodily counterpart to intersubjectivity, which, in contrast to mental processes, derives its meaning from the praxeology of bodily co-world relations based on shared perceptions and their interpretations. The dual aspectivity of bodies creates the basic condition for a shared sense of the experienced material world and, together with the fusion into an intervening "inter-body," is a feature of embodied intelligence in social encounters. Especially this fused co-world experience has to be technically realized in the implementation of robots in interaction settings, which is why a detailed social-theoretical elaboration of the perceptual convergences necessary to achieve "inter-bodyness" follows at this point.

So how exactly does such a bodily fusion of shared perceptions into a shared body proceed? Meyer [13] addresses subtleties of shared bodily perceptions in moments of incorporation by co-sensoriality, con-sensoriality, and common-sensoriality, underscoring essential bodily experiences in co-worlds. With these nuances, differences from physical embodied intelligence and the complexity for replicating embodied intelligence in social bodily encounters in technical gestalts become apparent. Co- and con-sensoriality in social encounters illustrate a) *multiplicity of perspectives* through alternating perceptual positions of actors, b) *synaesthetic perception* as an amalgamation of perceptual expectations, knowledge, and practice, and lead to c) agreement on shared (co-)semantics of co-/con-sensorial incorporated perceptions in *common-sensoriality* [12, 13]. Because of their potential for differentiation from the contents of incorporeal experiences of co-worlds as an essential point of socially embodied intelligence, the three aspects are briefly presented below.

Meyer emphasizes the co-sensory nature of the participants in social encounters, whereby the multi-sensory impressions of the individual actors complement each other and the situation can be experienced from a "multiplicity of perspectives" [12, 13]. For example, the perspectives of several parties on a

IOP Conf. Series: Materials Science and Engineering 1261 (2022) 012024 doi:10.1088/1757-899X/1261/1/012024

landscape complement each other during a hike or during communal work such as screwing on an engine or the like. The different, spatial standpoints of the actors towards the material environment mutually reinforce each other through the perspectives of the alter egos and lead to sensory intersubjectivity via perceptions of "kinesthetic, equilibrioceptive and proprioceptive dimensions, (...) thermoception, chronoperception, and (...) the presence or absence of body-internal aerial, liquid, or solid 'objects'" responsible for "'moods' and 'atmospheres', or the perceived 'sense of agency' and 'self-efficacy'" (Meyer 2021:3). Shared perception extends the first-person perspective and multimodal perception of situatedness to include the multiperceptivity of sensory impressions brought to bear through the encounter structure with alter ego, leading to a supervenient form of shared world-perception. On the one hand, this concerns the perception of the environment as a co-world as well as the shared sense of social experience. Aware that the alter ego is thirsty or in pain, for example, the overriding sense of the participant and the mood of the action is directed toward remedying the deficiency based on bodily realizations (finding something to drink or taking a break to rest the leg). In the co-sensorial dimension of intercorporality, the integration and adoption of alter ego's perspective into an overall picture of the social situation succeeds through (bodily) perspective adoption with alter ego as well as the shared and complementary world perception.

In addition to co-sensoriality, Meyer refers to con-sensoriality as the fusion of local-situational perceptions about physical experiences that participate in the situational apprehension of the world. This con-sensoriality is a kind of "apperceptive addition" (orig. "apperzeptive Ergänzung" [14]) to the given sensory impressions of the environment, in which one adds "cross-sensorial stocks of knowledge" [13] to what is perceived. "(...) when I perceive an object, such as a house from the front, the back is involved in this perception not merely as a possible perception which I judge could be produced if I walked around the house, nor as a necessary implication of the concept 'house.' Instead, the back is experienced as actually co-present—concealed but suggested by the appearance of the front" [Meyer 2021:5).

Similarly, Piaget's object permanence in children [15] establishes an apperceptive completion of world perception through knowledge of the constant presence of objects. Here the child learns; that an object is still present even when it is out of sight, e.g. in the case of a nutshell game, where a ball - hidden under a nutshell - is moved. Even when an object is occluded, knowledge of the permanence of objects in the world in the presence and absence of perception, as well as in changing spatial positioning, plays a prominent role in planning or actively executing individual or communal movements, and in understanding absent facts. With reference to Merleau-Ponty, Meyer further refers to "synesthesia," in which a sensory stimulus in one sensory modality triggers an additional perception in at least one other sensory modality and, in our cases, con-sensorially links the perceptions [13]. This means that when we look at an object, e.g., a steel measuring blade, the sharpness of the blade, the coolness of the metal, or the softness of the wooden handle are simultaneously delivered to us as con-sensory perceptions and appear to guide our actions. At the same time, the motor knowledge and memory about the required hand posture, grip technique, grip intensity and lifting force of the weight in the hand is given, which is different from the motor requirements, when picking a daisy. "We see a piece of glass, and our hand immediately knows how it must grasp it to raise it. We see its hardness and brittleness, and the same holds for other properties of materials. (...) Thus, our whole body is engaged in perception, preparing for taking up opportunities for action." (Meyer 2021:5). The con-sensory connection is essential for the "being ready-to-hand" [13] and for the conceivable execution of motor actions in relation to the object. More than this, through the synesthesia of perception, a knowledge of the properties of the object is conveyed that goes beyond mere current perception. Without this con-sensorial linkage of direct perception and implicit knowledge of the character of things in the world, all action would require constant re-experience of objects and the way we act with them. "Perceptual synesthesia" along with "apperceptive addition" of covert perceptual moments, when applied to robots, is already required for the physical embodied intelligence of motoric awareness and navigation, but finds further complexity of embodied intelligence in social encounters. What do such con-sensorial experiences of the world look like in social encounters? Con-sensoriality in intermediate actions enables bodily actors to act as a unity, as is evident in craft coordination work in teams, synchronized speeds in walking, or movement synchronizations or

IOP Conf. Series: Materials Science and Engineering 1261 (2022) 012024 doi:10.1088/1757-899X/1261/1/012024

gestalt fusions in couple dancing. In interbody connections with alter ego, we generate the atmosphere of a setting, knowledge of complementary courses of action or experience sensibilities based on con-sensory categorizations that we synaesthetically grasp and from which the situation is experienced on the multidimensional body level.

Already the co-sensorial "multiplicity of perspectives", which is fundamental for jointly coordinated actions in social encounters, illustrates the shared basis of perception and interpretation of material environments for social partners. The same is true for the communal convergence of con-sensorial perceptual synesthesia as soon as we e.g. glimpse the surface structure of an object or can estimate both the distance and the direction of a detonating object based on sounds. Co-sensorial and con-sensorial aspects illustrate that the world we relate to, as well as the expectations we have toward the interpretation of world, provide objective access to the world via shared sensory perception. The agreement on the convergence of sensory experiences via the co-sensorial addition of viewpoints to the diversity of perspectives as well as the simultaneous con-sensorial implication of perceptual synesthesias of human actors leads to an access to the environment and to the world that can be evaluated as objective. Meyer, following Hannah Arendt [13], calls this interplay of multiple perceptual implications in social interaction common-sensoriality or common sense, which offers us the objectifying basis for shared, bodily understanding. Accordingly, the interplay of co-, con-, and common-sensoriality is fundamental to embodied knowledge, the successful technical replication of which could capture essential moments of environmental and co-environmental sensoriality and provide key moments for intercorporeal encounters in interaction settings.

3.3. Habitual Senses Of Perception In Interaction

Following the multi-perspectival perception synesthesia, the embodied intelligence in social encounters manifests itself additionally through actively interrelated movement sequences in non-verbal and para-verbal moments of interaction. This includes spontaneous as well as ritualized encounter sequences, which are bodily realized through the application of social behavior codes such as non-verbal gestures, facial expressions, posture, personal distance through proxemics, the allocation of personal and spatial territories or also meta-linguistic properties of utterances such as tone of voice, volume or intensity of vocal utterances [16]. But how is it possible for social encounters to take place in an orderly fashion without the actors having to constantly reinterpret the social situation? In this regard, Gugutzer refers to Bourdieu's "practical sense" [17] or "Spürsinn" [18] for a situation: both mean a form of bodily intuition in which actors have the inherent ability to make right decisions according to situational requirements. Through the sense of intuition, the agent orients himself to the circumstances of the situation and succeeds in recognizing practical contexts of meaning and options for action on a bodily level. This sense of intuition acts as a pre-reflexive, bodily-practical sense guiding action and manifests itself by means of repeated routinized movements in unquestioned, everyday practices and finally results in habitual bodily knowledge. In the direct encounter of actors who are about to initiate a relation to each other, an almost automatic alignment and adaptation to the gestalt of the other takes place together with offers of interaction via expressive gestalts. In this process, actors face each other in their posture or adapt, for example, the width, or intensity of their gesture production to the distance or possible visual or auditory restrictions of the counterpart [19]. The adaptation of the individual thereby aligns itself almost bodily-intuitively with the perceptual skill of the other. This can vary from person to person due to limitations e.g. of sight, sounds or mobility and is thus bound to the social situation and the cooperative action context with an ongoing reconfiguration of kinesthetic information, e.g. by variations in the width of the gesture, the location of the sight, the volume of the address, the intensity of the grasp, walking speed or the like.

With regard to encounters with the world, the sense of intuition or “Spürsinn” is imparted as part of an existing, but historically variable cultural social order in the socialization process and serves as an instructive practical execution, for example, in that a sensitivity for socially appropriate physical proximity is present in social encounters [20]. Sociocultural imprinting forms thought patterns and knowledge stocks as habitus [17]. These habitually internalized and incorporated - in this case embodied

- bodies of knowledge now shape perceptions of the social world as well as of other- and self-bodyliness and guide decisions about action. Even though habituated bodily behaviors are learned and non-biogenetically anchored practices, at a subliminal-intuitive level they are embodied knowledge practices that are free from mental influences in their execution and help minimize the uncertainties of embodied socio-interactive behavior. In habitual, practical execution, free of reflective processes, one thus selects which objects and aspects of the situation are relevant for bodily action or in which way one interacts with the social world [17]. Accordingly, the practical sense assists in the coordinated co-worldly encounter with alter ego and the intermediate corporeality that unfolds as a result. The bodily intelligence that becomes immanent in co-world encounters is revealed in the "Spürsinn" for the motor-bodily demands of a social situation.

In accordance with sections 3.1 – 3.3 above, embodied intelligence includes, in addition to the sensorimotor level with physical self-environment reference, a "sociomotoric level" with co-world reference, found at the levels of perception (a), (b) praxeology, and (c) interaction. This on a fundamentally bodily, i.e., pre-reflexive, pre-mental level with the functions of a) perceiving other beings and distinguishing them from objects, b) experiencing the co-world in which the encountering actors meet "interbodily" and unfold shared perceptual and world interpretations through common actions, as well as c) forming an action-instructive sense for interaction units. Therefore, the question, "How do robots succeed in a) distinguishing social actors from environmental situations, b) combining multisensory self-perceptions with the complementary co- and con-sensory perspective diversity of alter egos to form a common sense with the objective world shared by human actors, and c) developing a practical "Spürsinn" to communicate with each other intuitively?" is significant for the formation of a technical, integral intercorporeality. Clearly, the simulation of social intelligence requires a sociomimetic replication of actors' capabilities that goes beyond the biomimetic construction of environmental relations. To what extent such a sociomimetic co-world could be replicated via extended embodied intelligence, is presented in the following, based on the concept of *sociomotorics* together with the principle of *suggestion of movement*, as a sense for situation-appropriate sequences of movements.

3.4. Application of Socially Embodied Intelligence to Robotics Research

The development of sociomimetic robots requires the modeling of fundamental epistemic features of socially embodied intelligence, which include at least the skills of subject-object differentiation as well as multimodal, multisensory, and multiperspectival perceptual synergies, copresence experience, intercorporeality, and a sense for normative expressions in interaction. On the basis of these criteria of embodied intelligence, a convergence of co- and con-sensory experiences to a common co-worldly context of meaning can succeed, which extends self-referential actions of physical embodied intelligence by the experience of a synergetically co-operating intermediate body.

Social actors, then, are mobile beings that are difficult to predict; they direct their movements both according to environmental conditions and their own motives, and in a coordinated way according to other beings. Encounters with other social actors thus increase the necessary complexity of a social robot's expressive behavior in situations that are now transforming from environment to co-world, and require the adaptation of embodied technologies to the follow-up actions of variable, moving bodies that, moreover, exhibit collaborative-interactive behavior. In co-worlds, the robot must now be able to extend its physical interaction not only to static environmental constants alone, but also to human-related environments (and objects of cultural value therein). Kinetic actions such as navigation, orientation, path planning, and motion control of autonomous systems accordingly require additional trajectory prediction that incorporates actor-centered "social navigation". This includes the interpretation of bodily expressions and motion sequences like localization, the consideration of motion directions, dynamics and velocities, and many more. Consequently, sensorimotor skills, which have already played a prominent role for the self acting in an environment, expand in embodied intelligence in social encounters to a level that can be termed as "sociomotoric skills" [21]. In contrast to and in continuation of sensorimotor movement processes, which enable actors to navigate, act, and manipulate the environment as well as the surroundings and contribute to a delimiting experience of one's own body with its efficacy on surrounding objects, etc., the orientation and coordination of movements on and with other socially enabled living beings is referred to as *sociomotoric skills* at this point. With sociomotoric skills, the sensorimotor feedback of perceptions through one's own movement is maintained, with the experiential component expanded to include experiencing through and with others - based on a social-kinetic component. The sociomotoric processes occur during the transition from the self-environment relationship to the co-world relationship and are the motor-kinetic basis for interbody cooperative actions. Sociomotoric actions include both the movements of actors with and toward each other, as well as their distances and postures, and the resulting overall gestalt that appears to the observer of social relations as "communication gestalt" [21]. The term "communication gestalt" refers to the gestalt-like wholeness that results from the interaction of the actors and is visible to the observer as a fused gestalt unit. It is the externally observable counterpart to intercorporeality.

In social robotics, it is precisely the translation from sensorimotor to sociomotor that allows the complexity of adapting motion components to be multiplied in real-time interaction scenarios. The robot must be able to act sociomotorically to cooperate in multi-agent environments, and at a subtle level of embodied intelligence to shape and coordinate its movements, body expressions, and interactions with others. At this point, the intertwining of sociomotoric skills with intercorporeality, intuitive "Spürsinn", and the necessary multisensory experience of multiperspectivity and perceptual synesthesia becomes clear. The technical realization of this extended motor activity could be the key for a convergence of sensory, bodily and world experiences to interbody sensoriality and thus for the basis of a (second level of) *socially embodied intelligence.* However, these contributions to a sense of communal "incorporation" or "compresence" still seem technically difficult to realize and model at the current state of research. Thus, for the development of social robots, the question is whether and how the intuitive "Spürsinn" and the associated sociomotor schema can be transferred to technical actors.

As a solution for modeling social involvement, which at least offers an approach to the mediation of sociomotoric sense, we would like to refer here to *suggestion of movement* (orig. "Bewegungssuggestion" [22]), which is oriented to norm-guided behavior in social situations and could contribute as a behavioral resource to select prefabricated movements. "Suggestion of movement" enables bodily engagement with environments and co-worlds, through "imposition of a movement that is felt in one's own body" [18]. It offers bodily impulses for coping with practical challenges and bridges the gap between the visible, the audible, the tactile and the bodily sensible, thus contributing to an intuitive way of dealing with the world. It is unimportant whether the suggestion comes from alter ego's movement, physical-mechanical or from something unmoved. Suggestions of movement relate directly to the inherent motor schema and guide instinctive-spontaneous bodily actions. They thus close the gap between multisensory, multiperspectival, and synesthetic-complementary perceptions to situationally appropriate actions in the surrounding and co-world, in that, following the multilayered levels of perception, movement impulses become accessible as practical solutions to local situational demands.

For successful human-robot interaction, the translation of social norms into sociomotor suggestions of movement is essential for robotic systems in human environments. Interaction is expressed in minimal physical units of action that enable culturally and collectively borrowed interpretive schemes and meaning attributions, thus providing rule-driven guidance for imitation. In other words, the normative set of rules makes it possible to select from a multitude of possible options for action and movement that follow-up behavior which is oriented to already established movement sequences. Rules that are influenced by culturally and socially accepted behaviors such as body distances, postures, length and intensity of touch or eye contact must therefore be included as sociomotoric suggestions of movement in the construction of a socially conforming robot behavior. Should this succeed, a behavioral architecture based on sociomotoric data would emerge, which could be oriented to established practices and meaning attribution by other social partners and could be updated in interaction. Accordingly, technical architectures with intuitive "Spürsinn" are needed that translate the plurality of multiperspectival and synaesthetic levels of perception into a socially-compatible world view and anchor recurrent suggestions of motion into bodily routines, a sense of behavioral norms, and ultimately bodily knowledge.

Conferring to the above, the socially embodied intelligence includes not only the conditions of the self's adaptability to complex environmental events, but also the additional inclusion of social behaviors, such as unknown movement paths and variable movement patterns of animate beings in co-worlds. Accordingly, modeling social embodied intelligence requires a perceptual and movement architecture that multiplies the complexity of environmental relations at various levels. Starting from the experience of a "moment of touch" for subject-object differentiation, we could see in sections 3.1 – 3.3 that technical systems with co-environmental references require enhanced navigation, perception and interaction capabilities and contribute to enhanced bodily experiences. Perspective variety and multisensory perceptual synesthesia provide the basis for a convergence of sensory experiences and, moreover, for interbodily shared world references as well as practical, situation-appropriate behaviour. Furthermore, for immediate moments of interaction, the body-intuitive "practical sense" was highlighted as 1) a selection instance of movement alternatives as well as 2) knowledge of contexts of meaning and action, and 3) sensitivities to "body arrangements" - such as the selection of nonverbal gestures from a pool of sociomotor suggestions of movement - for interaction. Finally, the sociomotoric behavioral architecture of socially embodied intelligence is highly complex and bound to the perception and realization of social events through selection processes of standardized movement sequences in interaction situations that are supposed to be technically reproduced.

4. Embodied Intelligence in Relation to Embodied Technologies

In the workshop presentations, it was emphasized that in teleoperated embodied technologies, such as robots, there is no embodied intelligence at play in the execution of movements or adaptations to the environment. In this case, there would be no autonomy and thus no intelligence provided by the mechanical algorithm for the robot's self-control. The dimension of embodied intelligence, however, manifests itself in another domain: namely, in the physical ability of the controlling teleoperator to adapt to the robot's shape, location, as well as situational requirements, and to successfully navigate it to another, distant location to meet the demands of the material and social world there. Here we see that embodied intelligence not only concerns the ability to control one's own body, but also includes the adaptation of objects up to and including virtualized bodies into one's own "body schema" [23]. These controllable somatechnical objects can thereby a) serve as extensions of one's own body parts (e.g. exoskeletons) or refer in embodied intelligence to b) a high degree of plasticity for adaptable body forms (into one's own body schema e.g. in remotely controllable robotic avatars). This becomes visible in the sensory, motor, and situational, i.e., bodily, fit of the controller to mechanical devices that operate in remote locations as independent entities [24]. Examples include controlled robots in inaccessible environments on exoplanets (e.g. Mars Rovers), in deep-sea exploration or drone operations, and social interaction robots in service or care robotics.

As explained in the section above, object recognition as well as encounters with embodied actors (be they technical or organic entities) are based on a fundamental and inherent level of embodied intelligence given in organic bodies. Both manipulations in the environment and in encounters with other social entities can therefore serve as examples of forms of embodied intelligence. The distinction between a self, an external object, and another acting subject is an essential feature of the embodied intelligence of social beings. Moreover, the distinction between an object that has no living properties and neither belongs to the self nor represents another self, and the perception of objects as part of one's own or another's body (e.g., in the case of prostheses, exoskeletons, wearables, or the like) has important consequences for the further analysis of stages and applications of embodied intelligence (e.g., in human-machine hybrids). If one considers the ability of humans to bodily adapt to the specific properties of various objects that function as a kind of "technically medialized in-between corporeality" [25], be it mere objects, tools or machines, up to prostheses, a mutability of our body in dealing with objects unfamiliar to the body becomes clear. A walking stick or a prosthetic leg change both the motor dimensions of the body, adapted to the resistance of the support, and the way the subject experiences himself and the world on the one hand, and how he is perceived by others on the other [26, 27]. At this point, another quality of embodied intelligence emerges, revealing the consequences of body-additive technologies as

IOP Conf. Series: Materials Science and Engineering 1261 (2022) 012024 doi:10.1088/1757-899X/1261/1/012024

additional, new and undiscovered capabilities of *embodied intelligence via somatechnically integrated objects* as well as relevant effects on the subjective and social experience of the body.

This "plasticity" as openness of the body for material extensions also allows us to adapt to the physical, sensorimotor, sociomotor, situational, and personal characteristics of teleoperated robots when controlling them [24]. Controllable robot avatars serve as surrogates for the physical presence of the controller and are intended to provide the experience of acting as a "cybernetic being" in interaction scenarios through "telepresence" or "teleexistence." They enable the controller to place himself in remote environments and to realize precise manipulations and interactions in remote scenarios by means of wearable control units or interfaces via real-time remote control.

In addition to technical applications of individual body parts (e.g., hand, arm, or leg prostheses), advanced interfaces also enable the connection of one's own body with whole-body surrogates equipped with bidirectional feedback on visual, haptic, acoustic, and kinesthetic levels, thus contributing to multisensory impressions for the user. The effects on the user resulting from the symbiosis of the technical body surrogates reveal previously unknown dimensions of embodied intelligence with somatechnically integrated objects, which can be divided into three levels: Firstly, into the effects of the physical body experience, secondly, into the subjective self-experience during the adaptation of an object-like body part, or to the shape of a body surrogate and further as an effect on the external perception as being a unique robotic identity [26]. Accordingly, the third form of "embodied intelligence with somatechnically integrated objects" experienced with technical body augmentations goes beyond the "physical embodied intelligence" as well as the "socially embodied intelligence" in social encounters, in that the altered physicality through object-like body additives or whole-body surrogates influences self-perception, self-presentation, situational perception, as well as social perception by others.

4.1. Effects on Body, Self-Experience and Perception by Others

Technical and neuroscientific findings on the effects of prostheses on individual body parts (see [28] for the example of the "Rubber-Hand-Illusion") or of whole-body surrogates on body experience (see e.g. "Body Ownership Transfer" [29, 30, 31]) indicate motor adaptations and multisensory integration of the mediated "body representation" into the teleoperator's own "body schema" [26], leading to the impression of "body ownership" and agency [32]. In this context, users indicate that they perceive sensory stimuli, such as touching the robot avatar a) from material environment and b) from other people touching the robot surface, as proprioceptive impulses during control. These sensory impressions, together with the synchronous movements of the robotic avatar, give the controller the impression that the avatar is an extension of his/her own body or that the touches are on/with his/her own body [33]. Others report alternating personal experience of a second body detached from one's own body, through which a self-transforming ego forms into its own telepresent actor [34, 35].

Thus, immersion in a robotic avatar already occurs at the perceptual level through the experience of synchronous sensorimotor activity of the avatar with the movements of the remote controller and expands to immersion in and identification with the robotic figure at a further interactive level together with the mediation of social-communicative resonance of the interaction partner [36, 37]. In this process, the degree of incorporation gradually increases to the point of personal transposition into the robot figure and communicative self-representations of the controller from the perspective of its own robot identity, which differs from the original identity of the controller [38]. Thus, initial technoscientific and sociological studies point to a transposition into the robotic avatar on a physical-bodily, personal, and social level: A) On the physical-bodily level, the controller experiences an inner-bodily expansion of sensory proprioception both when controlling individual limbs (e.g., prostheses) and when transferring to an avatar as a whole body, leading to a partial symbiosis of the technical apparatus with one's own body [39]. B) On the personal and social level, the technically embodied figure and the communicative staging have an additional effect on the fit of a narrative identity representation towards the interaction partner, which is oriented towards the characteristics of the technical figure [37]. This leads to a "fusion" between embodied technology and the organic body and thus to an inner corporeality that transcends the material and phenomenal "being".

IOP Conf. Series: Materials Science and Engineering 1261 (2022) 012024 doi:10.1088/1757-899X/1261/1/012024

This "fusion" resembles the interbody resonance between interaction partners, however, in this case it concerns a body unit between technology and organic body. Exactly this newly experienced corporeality leads again to a new body identification of the controlling person and thus to the extension of the personal identity genesis [for this concept see 40] by the new body characteristics, which leads also in teleoperated interaction scenarios to a new identity quality of the controlling person (on the basis of individual and social experience). In this case, embodied intelligence manifests itself in the incorporation of material-technical additions into bodily experiences of self, world, and other, and thus into the identity processes that originally transcend the body. Human users thereby animate the material body through their embodied knowledge, which becomes transferable between bodies with objects. This embodied intelligence with somatechnically integrated objects enables them to perform social actions towards remote interaction partners and to mediate their own form of (intermediate) corporeality via the body surrogate. In this way, an entity independent of the controller emerges for the interacting users, which conveys its own social identity [37].

To what extent the interbodily fusion of actors in human-robot interaction with technical objects succeeds and to what extent co- and con-sensibilities, common sense and a sense for sociomotor movement suggestions between the controller, the controlled robot avatar and the interaction partner emerge and are transferable cannot be dealt with here and strongly requires further empirical investigation in order to learn more details about the specifics of social encounters between controller, controlled agents and their interaction partners.

4.2. Embodied Intelligence by Technological Substitution of Selfhood

Embodied intelligence through technological body substitutions makes material augmentations accessible on two dimensions: a) inward (internal), to one's own body (e.g. through prostheses) and b) outward (external), through controllable body surrogates (avatars or wearables). Both forms of augmentation alter the phenomenally subjective experience of both the augmented body's affiliation to one's own body and the user's experience of identity to one's very own bodily identity. These results show the plasticity as multivariability of the body to material body augmentations that show effects on the bodily identification of the self [24]. This plasticity for material body extensions suggests an adaptation of bodily, personal and social spheres to remote situations within the modes of experience of one's own body and points to a third form of embodied intelligence with somatechnically integrated objects, more progressive than the physical and socially embodied intelligence, which enables bodily resonance via material body extensions and thus virtual dislocation of body, person and social encounters.

Robotic avatars irritate the bodily experience of the individual by superimposing a second body, acting in spatially distant situations, on their own corporeality. The technical surrogate enables the personal actor to extract genuine bioorganic properties of (proprio- and exterio-) perception, kinesthesia as well as sensorimotor functions from his local, singular bodily environment. This allows him to perform actions according to his novel environment and co-world via the object-like embodied surrogate and to integrate these actions into his holistic bodily experience. Thus, embodied intelligence is manifested in the ability to incorporate "transbody technologies" into one's bodily structure, which is reflected in the adaptation of physiological and phenomenal experiences to the situatedness of the body surrogate.

This "immersion" in the robotic body demonstrates the plasticity of the corporeal self, which seems to be capable of accumulating subjective phenomena about distant bodily experiences by temporarily transposing itself into another body, self, situation, etc. The transference is associated with a felt presence of the self in the virtually transferred environment: Our ability to shift through other shapes enables us to experience "body transfer illusions" by technical means and to fully empathize with another body as well as to adapt to its own spatio-temporal conditions. This marks the experience of a new self-, time- and space-perception and shifts the experience of re-embodiment of the self, triggered by feeling with the other body, from a purely mental perspective-taking to an additional virtual bodily gestalt-taking and empathic-bodily dislocation. Somatechnologies dock with our unique embodied intelligence with their analog form, perceptual and expressive capacity and by mediating somatosensitive experiences. They enable us to break free from subject-centered constraints such as a single biography, an unchanging body physiology or singular social identities through limited attributions of self-identity via a single body and simple, monolocal world references.

Thus, embodied technologies (at least temporarily) dissolve the assumption of enduring core identities [35] not only at the personal level, but also perspectives limited to monolocal corporeality, revealing that the self instead appears dynamically as a constant response to multifaceted bodily-perceptible experiences in environmental and co-environmental relations. Physically immersive technologies thus abolish monolocal subject-object relations and reveal open boundaries to the multilocally shaped genesis of the bodily and identitary self. It becomes clear that the self is able to decenter itself from its own body and to enter into an interplay with materials, which enables an autonomous self-perception that deviates from one's own body. Thereby the experience expands from the present environment to an experience of the world, which stands apart from the spatio-temporal and bodily concordance of a local co-presence and can be dispersed in mediable "multiverses".

What is otherwise made possible in reflexive cognition and communicative use of symbols capable of abstracting from the present to the absent [14], can also be experienced on a bodily level through embodied intelligence with somatechnically integrated objects. Accordingly, the body is capable of symbolic abstraction, as it were, to have bodily experiences that are removed from the here and now, and to sense decentralized, multiversal local atmospheres through body surrogates elsewhere (for the term "atmosphere" see [41]). This experience of decentralized displacement of bodily life through internal and external techno-material augmentations characterizes humans as beings capable of multibodily, multilocal, and multimaterial world experience and world adaptation, as well as multiple, alternating forms of identity. The virtual ability to move to distant places and into other bodies is thus a high-level achievement of embodied intelligence that abstracts from the "here and now" and allows us to be present and bodily involved in distant places.

4.3. Concluding Remarks

Through technologies that transcend the boundaries of the phenomenal self and the distinctions between one's physical and virtually simulated body as well as the local and mediated environment, we experience the paradox of being one self and simultaneously in the guise of another [35]. Here, the specificity of the gestalt for self-perception and self-presentation of a personal identity becomes clear. It is characterized by new techno-physical impressions and induces the user to adopt a personal nature analogous to the gestalt's traits (see the concept of the "Proteus effect" in [42]).

The embodied intelligence thus becomes visible on another level. In addition to the environmental and co-environmental reference, distant virtual worlds become primary realities, and through multisensory adaptation to the technical gestalt, practices of simulating ways of acting and ways of experiencing the world change. By experiencing partial material gestalts as self-originating, the plasticity of intrastructural bodily representations becomes visible through the entanglement of corporeality and materiality through controllable somatechnologies [43, 44]. The malleability of bodily experiences and their transferability to material body surrogates reveal the limits and possibilities of body technologies and map undiscovered properties of embodied intelligence in human-machine hybrids, which include a diversity of self-concepts and the formation of a new, genuine identity for the technically embodied gestalt.

Consequently, it can be postulated that the physical, personal, environmental, and social shifts of experiences and perceptions toward an external, controllable body are extended features of a higher-level embodied intelligence with somatechnically integrated objects. The teleoperation of embodied technologies is thus an example of embodied actions, which consequently cannot be reduced to purely human activities, but are realized as hybrid human-technology actions [45].

For technically mediated self-environment relations, we thus assume that the combination of organic properties with technical body surrogates leads to extended experiences of one's own body, environments as well as co-environment relations. In these bodily-technical constellations, the almost full performance potential of organic bodies and thus of embodied intelligence is revealed in the sense that a (bodily-structurally and) functionally analogous, technical body is able to suspend the immediacy of

IOP Conf. Series: Materials Science and Engineering 1261 (2022) 012024 doi:10.1088/1757-899X/1261/1/012024

bodily and environmental contexts in favor of manifold, bodily self-experiences. Consequently, this means that the self with all its properties can be multiplied and that the levels of embodied intelligence of organic bodies can also be transferred to material body surrogates together with their relation to self, environment, and co-world.

In this view, the potential of bodily experiences is corresponding to expanded cognitive capacities, as evident in symbol theory in the transition from representational sign use to abstracting symbol use [14, 46]. Similarly, the experience of distant environments in technically mediated bodies exceeds field perception purely related to the body in the here and now (synonymous with representations) and transcends it through technical shifts into virtual, distanced body-surroundings (synonymous with symbolic thinking). This paper thus assumes that the embodied intelligence of the human being fully unfolds in parallel with the expansive forms of the body in (technically) expanded relations to distant environments and co-worlds, where it is further extended and endowed with new properties. To what extent this gradation of forms of embodied intelligence into object-worldly, social-worldly, and virtual displacement capacities is analogous to the subdivision from lower to higher levels of cognition [47], or what further peculiarities it holds, is to be examined in more detail elsewhere. This paper, however, already shows evidence for the increase of levels of embodied intelligence from a) *physical embodied intelligence* by simple rules in "local interactions" with objects (low-level sensorimotor processes) via b) *socially embodied intelligence* as more complex adaptations to social-specific behavior and bodily experience in "moments of touch" in social encounters as well as by means of rule-based normative expectations in social interactions up to c) high-level *embodied intelligence with somatechnically integrated objects* as advanced displacement ability in technical-material bodies, operating in distant, virtual worlds where they have real-world effects on projected self-concepts, environments, and co-worlds.

This investigation of forms and properties of embodied intelligence is intended to contribute to the differentiation of various situation-specific performance potentials that should be considered in the design of social robots as well as controllable robotic avatars. The differentiation is a first step towards the multi-layered and interwoven performance characteristics that a technical system should exhibit when using embodied intelligence. Thus, this study represents a first draft of a holistic concept of inter- and transdisciplinary research on embodied intelligence, which will certainly be complemented by further results in the coming years.

References

[1] Brooks, R.A., and Stein, L.A. (1993). Building brains for bodies. Memo 1439, Artificial Intelligence Lab, MIT, Cambridge, Mass.
[2] Pfeifer, R. & Iida, F. (2004). Embodied Artificial Intelligence: Trends and Challenges. In F. Iida (Ed.), Embodied artificial intelligence: International seminar, Dagstuhl Castle, Germany, July 7 - 11, 2003 ; revised selected papers (S. 1–26). Springer.
[3] Iida, F. (Hrsg.). (2004). Embodied artificial intelligence: International seminar, Dagstuhl Castle, Germany, July 7 - 11, 2003 ; revised selected papers. Springer.
[4] Pfeifer, R., & Bongard, J. (2007). How the body shapes the way we think: A new view of intelligence. A Bradford Book. MIT Press.
[5] Sitti, M. (2021). Physical intelligence as a new paradigm. Extreme Mechanics Letters 46, 101340.
[6] Howard, D., Eiben, A. E., Kennedy, D. F., Mouret, J.-B., Valencia, P. & Winkler, D. (2019). Evolving embodied intelligence from materials to machines. Nature Machine Intelligence, 1(1), 12–19.
[7] Kramer-Bottiglio, R. (2021). Is it or isn't it: Embodied Intelligence? (this volume).
[8] Lewis, C. I. (1990). Mind and the world-order. Outline of a theory of knowledge. New York, NY, Dover Publ.
[9] Plessner, H. (1975). Die Stufen des Organischen und der Mensch. De Gruyter.
[10] Lindemann, G. (2014). Weltzugänge: Die mehrdimensionale Ordnung des Sozialen (1. Aufl.). Velbrück Wiss.
[11] Breyer, T. (2015). Verkörperte Intersubjektivität und Empathie: Philosophisch-anthropologische

Untersuchungen. Philosophische Abhandlungen: Band 110. Vittorio Klostermann.
[12] Merleau-Ponty, M. (1974). Phänomenologie der Wahrnehmung: Übersetzt von Rudolf Boehm (1966. Aufl.). Perspektiven der Humanwissenschaften. De Gruyter.
[13] Meyer, C. (2021). Co-sensoriality, con-sensoriality, and common-sensoriality. Social Interaction. Video-Based Studies of Human Sociality 4 (3).
[14] Bühler, K. (1982). Sprachtheorie. Die Darstellungsfunktion der Sprache. 1934. Aufl. Stuttgart, G. Fischer.
[15] Piaget, J. (1974). Der Aufbau der Wirklichkeit beim Kinde. Stuttgart, Klett.
[16] Hall, J. A. & Knapp, M. L. (2013). Nonverbal Communication. De Gruyter.
[17] Bourdieu, P. (1987). Die feinen Unterschiede. Kritik der gesellschaftlichen Urteilskraft. Frankfurt am Main: Suhrkamp Taschenbuch Wissenschaft.
[18] Gugutzer, R. (2002). Leib, Körper und Identität. Eine phänomenologisch-soziologische Untersuchung zur personalen Identität. Wiesbaden, Westdt. Verl.
[19] Hall, E. T. (1982). The hidden dimension. New York, Doubleday.
[20] Berger, P. L. & Luckmann, T. (1980). Die gesellschaftliche Konstruktion der Wirklichkeit: Eine Theorie der Wissenssoziologie. Fischer-Taschenbücher: Bd. 6623. Fischer-Taschenbuch-Verl.
[21] Straub, I. (2020). Zur Sozialität und Entität eines androiden Roboters: Empirische Zugänge zum Objekt- und Subjektstatus. Springer VS, Springer Fachmedien Wiesbaden.
[22] Schmitz, H. (2005). Situationen und Konstellationen. Wider die Ideologie totaler Vernetzung. Freiburg/München, Verlag Karl Alber.
[23] Gallagher, S. & Cole, J. (1995). Body Image and Body Schema in a Deafferented Subject. The Journal of Mind and Behavior, 16(4), 369–390.
[24] Straub, I. (2022): Leibliche Exzentrierung über technisch einverleibte Selbste. In: Barth, J. & Henkel, A. (Eds.): Leib. Grenze. Kritik. Weilerswist: Velbrück Verlag, 206-215.
[25] Meyer, C. (2020). Intersubjektive Entfremdung: Zur Kultur- und Technikgeschichte der Zwischenleiblichkeit. In: Kurilla, R., Kolb-Albers, K., Krämer,H., Pitsch, K. (Eds.): Sine ira et studio. Disziplinenübergreifende Annäherungen an die zwischenmenschliche Kommunikation. Wiesbaden: Springer VS, 187-211.
[26] Preester, H. de (2011). Technology and the Body: The (Im)Possibilities of Re-embodiment. Foundations of Science, 16(2-3), 119–137.
[27] Meyer, C., Streeck, J. & Jordan, J. S. (Eds.). (2017). Foundations of Human Interaction. Intercorporeality: Emerging socialities in interaction. Oxford University Press.
[28] Botvinick, M. & Cohen, J. (1998). Rubber hands 'feel' touch that eyes see. Nature, 391 (6669),
[29] Braun, N., Debener, S., Spychala, N., Bongartz, E., Sörös, P., Müller, H. H. O. & Philipsen, A. (2018). The Senses of Agency and Ownership: A Review. Frontiers in psychology, 9, 535.
[30] Ehrsson, H. H. (2020). Multisensory processes in body ownership. In Sathian, K., Ramachandran, V. S. (Hrsg.), Multisensory Perception: From Laboratory to Clinic (p. 179–200). Elsevier.
[31] Nishio, S., Taura, K., Sumioka, H. & Ishiguro, H. (2018). Body Ownership Transfer by Social Interaction. In H. Ishiguro & F. Dalla Libera (Hrsg.), Geminoid Studies (p. 317–328). Springer Singapore.
[32] Jazbec, M., Nishio, S., Ishiguro, H., Kuzuoka, H., Okubo, M. & Penaloza, C. (2017). Body-swapping experiment with an android robot: Investigation of the relationship between agency and a sense of ownership toward a different body. In 2017 IEEE International Conference on Systems, Man, and Cybernetics (SMC) (p. 1471–1476). IEEE.
[33] Cole J., O. Sacks, and I. Waterman (2000). On the Immunity Principle: A View from a Robot.". Trends in Cognitive Science, 4(5), 167–168.
[34] Dolezal, L. (2009). The Remote Body: The Phenomenology of Telepresence and Re-Embodiment. Human Technology: An Interdisciplinary Journal on Humanities in ICT Environments, 5(2), 208–226.
[35] Jarvis, L. (2019). Immersive Embodiment: Theatres of Mislocalized Sensation. Springer eBooks

Literature, Cultural and Media Studies. Palgrave Macmillan.

[36] Straub, I., Nishio, S. & Ishiguro, H. (2018). At the Café—from an Object to a Subject. In H. Ishiguro & F. Dalla Libera (Hrsg.), Geminoid Studies (p. 399–411). Springer Singapore.

[37] Nishio, S., Taura, K., Sumioka, H. & Ishiguro, H. (2013). Effect of social interaction on Body Ownership Transfer to teleoperated android. In 2013 IEEE RO-MAN (p. 565–570). IEEE.

[38] Straub, I. (n. d.). Über die Verkörperung einer eigenen Roboter-Identität: Zur kommunikativen Emergenz einer Roboter-Identität über Teleoperierende und Interaktionspartner von sozialen androiden Robotern. Unpublished manuscript.

[39] Ihde, Don (2010). A phenomenology of technics. In Craig Hanks (ed.), Technology and Values: Essential Readings. Wiley-Blackwell.

[40] Masumori, A., Maruyama, N., & Ikegami, T. (2020). Personogenesis Through Imitating Human Behavior in a Humanoid Robot "Alter3". Frontiers in Robotics and AI, 7, Article 532375.

[41] Schmitz, H. (2011). Der Leib. Grundthemen Philosophie. De Gruyter.

[42] Yee, N., Bailenson, J. N. & Ducheneaut, N. (2009). The Proteus Effect. Communication Research, 36(2), 285–312.

[43] Lewis, E. & Lloyd, D. M. (2010). Embodied experience: A first-person investigation of the rubber hand illusion. Phenomenology and the Cognitive Sciences, 9(3), 317–339.

[44] Maister, L. & Tsakiris, M. (2016). The plasticity of self-other boundaries: From body ownership to social cognition. In M. H. Fischer & Y. Coello (Eds.), Foundations of embodied cognition (S. 182–199). Routledge.

[45] Rammert, W., Schulz-Schaeffer, I. (2002). Technik und Handeln - Wenn soziales Handeln sich auf menschliches Verhalten und technische Artefakte verteilt. In: Rammert, W. & Schulz-Schaeffer, I. (Eds.), Können Maschinen handeln? Soziologische Beiträge zum Verhältnis von Mensch und Technik. Frankfurt am Main: Campus-Verlag, 11–64.

[46] Peirce, Charles S. (1985): Über Zeichen. In: Charles S. Peirce (Ed.): Die Festigung der Überzeugung und andere Schriften. Ungekürzte Ausg. Frankfurt am Main, Berlin, Wien: Ullstein, 143–168.

[47] Gennaro, R. J. (Ed.). (2004). Advances in consciousness research: v. 56. Higher-order theories of consciousness: An anthology. Benjamins.

IOP Conf. Series: Materials Science and Engineering 1261 (2022) 012025 doi:10.1088/1757-899X/1261/1/012025

On the Importance of Embodiment for Artificial Design Intelligence in the Context of Sociotechnical Systems

Thomas E. F. Witte[1], Jessica Schwarz[1], Verena Nitsch[1, 2]

[1]Fraunhofer Institute FKIE, Fraunhoferstraße 20, 53343 Wachtberg, Germany
[2]Institut for Industrial Engineering and Ergonomics, RWTH Aachen University, Eilfschornsteinstr. 18, 52062 Aachen

E-mail: thomas.witte@fkie.fraunhofer.de

Abstract. Artificial intelligence methods are increasingly used in the design of sociotechnical systems. The design of such systems can be described as an ill-defined problem. The problem space often changes while working on a design solution. Further, an anticipation of future system states is challenging, because of possible changes of organisational, technological or other environmental aspects during the lifecycle of a sociotechnical system. Embodied artificial intelligence can help to structure the problem space. It can also act as a feedback mechanism of the inner workings of a system and its surroundings for a designer. Therefore, this short article discusses the emphasis of embodied artificial intelligence for the theory of design in the context of sociotechnical systems and its implications.

1. Introduction

The process of design is essential to many domains such as engineering, education, business, and medicine, among others [1]. According to Simon "everyone designs who devises courses of action aimed at changing existing situations into preferred ones" [1, p. 111]. The statement indicates that there is an initial state, a goal state, and a procedure to find a design solution (see Figure 1).

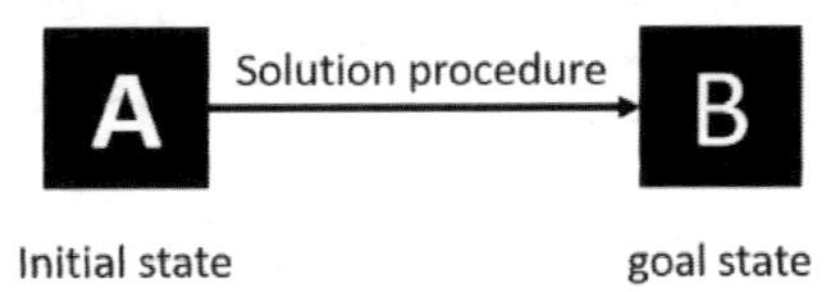

Figure 1. A basic representation of a problem [2, reproduced].

This view, on design as a problem, constitutes an information-processing-based approach to the design of cognitive systems and can be applied to the context of designing socio-technical systems [3]. The term *socio-technical system* was established to emphasize the importance of inner connections between humans (often referred to as operators), a technical system, and the environment it operates in (e.g. organizational structures) [4]. Examples are pilots interacting with the cockpit or operators working on consoles in a control room. The interdependencies suggest that the system border (the area of focus for the design process) should not focus on the human side, nor the organizational or technical aspects, but on the whole system (*holistic design*).

At first glance, designing a system with Figure 1 in mind seems to be straight forward. First, analyze and describe the initial state, then define the goal state, and use an established solution procedure that has been proven to end in the goal state. However, there are numerous fallacies and challenges that need to be considered while designing a complex system (e.g. a socio-technical system). Examples are the *ironies of automation* [5], or the *bounded rationality* [1], [6]. An information processing approach also suggests, that the design process and the cognitive functions behind the problem solving process could be modeled artificially [7], [8]. Through that Artificial Intelligence (AI) could assist during the design process and enhance human problem solving (see [9]). However, the introduction of AI to the design process implies new challenges that can make the design of systems even more complex. Embodied artificial intelligence (EAI) could mitigate those major challenges that designers of sociotechnical systems face. It could also mitigate challenges introduced by integrating AI into the design process. The following sections will focus on why EAI could do so.

First, design will be categorized in terms of problem solving. Second, the implications of the introduction of AI to the design process will be discussed. Finally, it will be laid out how embodied artificial intelligence could mitigate the challenges of artificial design intelligence.

2. Categorising Design as a Problem and its Implications

The challenge to designing a socio-technical system is that it is an ill-defined problem. This at first trivial notice has major implications for the design process and the theory of design. According to Robertson, a problem can be categorized due to its properties to get a better understanding of the structure of a problem and its elements [2]:

1. Knowledge: A problem can be knowledge rich, or knowledge lean. To design a socio-technical system, the use-context, the stakeholder, the goals and design constrains, for example organizational structures or available resources need to be known. In short a designer needs to have and understand the requirements of such a system. So, designing a system is a *knowledge rich* problem. Most of the time, it is also a *domain specific* problem. Often, a solution is not easily transferable from one domain to another. Designing a firefighting system for a frigate for example, has the same goal as a system on a tanker. However, the threat of being under attack while trying to extinguish the fire needs other safety precautions and thereby different solutions.

2. Goals: Problems can have one solution and can be evaluated by checking if the goal state is correct or not, and problems can have multiple, hard to evaluate solutions up to even insolvable problems, where the goal state cannot be reached. The design of a system usually can have *multiple solutions* and the evaluation of the solutions is often *time consuming* and not precise in the manner, that there is a definitive correct or incorrect answer to the problem.

3. Definition of the problem: A well-defined problem offers all needed information. The "Initial state, the Goal state, Operators, and Restrictions" [2, p. 9] (IGOR) need to be fully available to the solver. Designing a system is an *ill-defined* problem. Often socio-technical systems have a long lifecycle of several decades and long design periods. The initial state (which can change during the time of the design period), the goal state (how the system looks like and what it is capable to do), the operators (the methods to design a system), and the restrictions (for example organizational restrictions or a constantly changing knowledge base over a long period of time) are not fixed in time and change even during the documentation of these elements in form of requirements and a usage context analysis.

4. Semantic: Besides the knowledge needed to solve a problem at hand, a problem can be *semantically rich*, or *lean*. A semantically rich problem refers to the prior knowledge that is needed to solve a problem. For an experienced chess player, making a move for example, means following a specific pattern to reach a specific board state to come closer to the goal of winning. A less experienced player can only think one step or two ahead and is not capable to solve the problem of winning the game in an efficient manner as the experienced player could. Designing

a system is a *semantically rich* problem. The designer must know methods and techniques in how to design efficient, effective, and safe systems and how to elicit needed expert knowledge to reach the goal.

In summary, designing a socio-technical system is a knowledge rich, ill-defined, semantically rich problem with a goal state that is difficult to evaluate. One of the biggest challenges is the constant changing of the problem and its IGOR components. In other words, the solver (the designer) and the solution (the system) need to be flexible enough for change, yet stiff enough to be able to reach the goal state. The interaction design concept *Iteration by Design* suggests to introduce a design assistance system to support the designer of socio-technical systems and mitigate the described challenges (see for more information [8]). Support by an assistance system for the design process, the needed knowledge, and support for the solution evaluation can help to tackle the described challenges and will be presented in the next part.

2.1. Fog of Design

The IGOR components of the problem, for example the goal state, are in constant flux in the context of designing systems with long lifecycles [8]. Because of the ill-defined characteristic of the problem there is a knowledge gap from the beginning of the problem solving process. The knowledge gap widens over time because of the constant change of the IGOR components during their description. Figure 2 elucidates the knowledge gap.

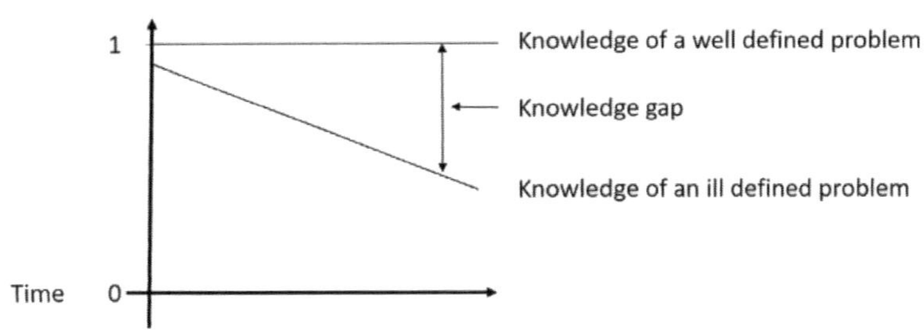

Figure 2 Knowledge about the problem structure

One solution to mitigate the dwindling knowledge is to use an iterative process as a solution procedure like suggested by the human centered design approach [10]. With each iteration, requirements can be updated, new solutions can be developed, and the solutions can be evaluated until a satisfying fit between the predicted goal state and the solution is reached. The described knowledge gap can be called the *fog of design*. Designers must make decisions with uncertainty in mind. A major challenge is to mitigate the problem of the fog of design and to be prepared for it, for example with the deliberate use of adaptation mechanisms (see [11], [12]). Another approach is to clear the fog as well as possible. Embodied artificial intelligence can help to do both, which will be discussed in the following section.

2.2. Implications of Embodied Artificial Intelligence

According to Marshall and Hornecker “embodiment typically refers to our being living, feeling, bodily entities situated in a physical world” [9, p 145]. It is an alternative to more abstract views of information processing, which are not grounded in the physical world. There are several research strands that use the term embodied cognition. It can refer to the offloading of human cognition “onto the environment, that the environment is part of the cognitive system or that the cognitive process is grounded in the bodily experience” [9, p.145], (see for example [14], [15], [16]).

Methods like thinking out loud (see for example [17]), interviews, workshops or documentation reviews of older systems for the acquisition of information often lack the direct experience of the environment, the problems of the system that need to be solved, and a continuous feedback from an embodied cognitive process. Thereby information for the design of a system is getting lost in translation from the embodied experience to the more abstract information elicitation processes.
The interaction design concept *Iteration by Design* suggests the use of an artificial design assistance system to help lift the fog of design [8]. By integrating an artificial embodied intelligence system in the design process, the designer could benefit from more accurate information. In reference to the problem-solving terms, such an assistance system for the design process could gather valid information about the knowledge rich problem. Thereby the design problem would become a better-defined problem. It could also analyze and communicate the semantic relationships between elements of the design solution, like for example the interaction between human and machine or between artificial agents. It could also help to evaluate if a satisfying goal state is reached and could track change for adaptation needs. So, to be able to support the design process, the design assistance system needs to be:

- grounded in the physical world,
- able to sense relevant information about the problem solution procedure,
- able to sense relevant information about the solution state (system state),
- able to communicate with the human problem solver efficiently and effectively, and
- trustworthy for the human problem solver.

A design assistance system that is based on artificial intelligence should be mainly embodied because of the symbol grounding problem (see [18]). For an AI system usually symbols or information are meaningless on their own, and only patterns and connections between symbols can be modeled. "The problem is analogous to trying to learn Chinese from a Chinese/Chinese dictionary alone" [16, p. 335]. There are numerous examples to simulate socio-technical systems to support the decision process during the solution process (see for example [19], [20]). The next step is to use embodied artificial intelligence principles for an artificial design intelligence system to further support the design process and to lift the fog of design by integrating it into the theory of design and simulation.

3. Conclusion

There are two major challenges of designing socio-technical systems. First, the uncertain decision-making process, which can be called the fog of design problem. Second, the need for a balance between adaptive flexibility and stiffness of a system. The changing nature of the problem space causes a *system design – system requirements gap* so that a system will not have a perfect fit to pursue its goals in a changing environment. An embodied intelligent design system, which supports the design process and tracks the fit of the system with its environment could mitigate the two major challenges. Embodied intelligence should be investigated further to be integrated into the standard repertoire of methods and techniques for the design of socio-technical systems.

References

[1] H. A. Simon, *The Sciences of the Artificial*, Third edit. London, England: The MIT Press, 1996.
[2] S. I. Robertson, *Problem solving*. New York, NY, USA: Psychology Press, 2001.
[3] A. Newell and H. A. Simon, *Human problem solving*. Englewood Cliffs, NJ: Prentice-Hall, 1972.
[4] G. Baxter and I. Sommerville, "Socio-technical systems: From design methods to systems engineering," *Interact. Comput.*, vol. 23, no. 1, pp. 4–17, Jan. 2011, doi: 10.1016/j.intcom.2010.07.003.
[5] L. Bainbridge, "Ironies of automation," *Automatica*, vol. 19, no. 6, pp. 775–779, Nov. 1983, doi: 10.1016/0005-1098(83)90046-8.
[6] R. R. Hoffman and D. D. Woods, "Beyond Simon's Slice: Five Fundamental Trade-Offs that

Bound the Performance of Macrocognitive Work Systems," *IEEE Intell. Syst.*, vol. 26, no. 6, pp. 67–71, Nov. 2011, doi: 10.1109/MIS.2011.97.
[7] Allen Newell, *Unified theories of cognition*. Cambridge, MA, USA: Harvard University, 1990.
[8] T. E. F. Witte, J. Hasbach, J. Schwarz, and V. Nitsch, "Towards iteration by design: An interaction design concept for safety critical systems," *Lect. Notes Comput. Sci.*, vol. 12214 LNCS, no. April 2021, pp. 228–241, 2020, doi: 10.1007/978-3-030-50788-6_17.
[9] S. Robert, S. Büttner, C. Röcker, and A. Holzinger, "Reasoning Under Uncertainty: Towards Collaborative Interactive Machine Learning," in *Lecture Notes in Computer Science (including subseries Lecture Notes in Artificial Intelligence and Lecture Notes in Bioinformatics)*, vol. 9605 LNCS, 2016, pp. 357–376.
[10] ISO, "Ergonomics of human-system interaction- Part 210: Human-centred design forinteractive systems (ISO 9241-210)." 2010, doi: 10.1039/c0dt90114h.
[11] J. Schwarz and S. Fuchs, "Multidimensional Real-Time Assessment of User State and Performance to Trigger Dynamic System Adaptation," *Lect. Notes Comput. Sci.*, vol. 10284, pp. 383–398, 2017, doi: 10.1007/978-3-319-58628-1_30.
[12] J. D. Hasbach and T. E. F. Witte, "Human-Machine Intelligence: Frigates are Intelligent Organisms," in *2021 IEEE International Conference on Systems, Man, and Cybernetics (SMC)*, Oct. 2021, pp. 1495–1500, doi: 10.1109/SMC52423.2021.9658640.
[13] P. Marshall and E. Hornecker, "Theories of embodiment in HCI," in *The SAGE handbook of digital technology research*, S. Price, J. Carey, and B. Brown, Eds. London, UK: SAGE, 2013, pp. 144–158.
[14] M. Scaife and Y. Rogers, "External cognition: how do graphical representations work?," *Int. J. Hum. Comput. Stud.*, vol. 45, no. 2, pp. 185–213, Aug. 1996, doi: 10.1006/ijhc.1996.0048.
[15] E. Hutchins, *Cognition in the Wild*. MIT press, 1995.
[16] G. Lackhoff and M. Johnson, *Philosophy in the flesh: the embodied mind and its challenge to western thought*. New York, NY, US: Basic Books, 1999.
[17] M. T. Boren and J. Ramey, "Thinking aloud: Reconciling theory and practice," *IEEE Trans. Prof. Commun.*, vol. 43, no. 3, pp. 261–278, 2000, doi: 10.1109/47.867942.
[18] S. Harnad, "The symbol grounding problem," *Phys. D Nonlinear Phenom.*, vol. 42, no. 1–3, pp. 335–346, Jun. 1990, doi: 10.1016/0167-2789(90)90087-6.
[19] T. Witte, *Simulationstheorie und ihre Anwendung auf betriebliche Systeme [Simulation theory and its application to business systems]*. Wiesbaden, Germany: Gabler Verlag, 1973.
[20] L. J. Hettinger, A. Kirlik, Y. M. Goh, and P. Buckle, "Modelling and simulation of complex sociotechnical systems: envisioning and analysing work environments," *Ergonomics*, vol. 58, no. 4, pp. 600–614, Apr. 2015, doi: 10.1080/00140139.2015.1008586.

IOP Conf. Series: Materials Science and Engineering 1261 (2022) 012026 doi:10.1088/1757-899X/1261/1/012026

Thinking like a State : Embodied intelligence in the deep history of our collective mind

Avel Guénin—Carlut

avel@kairos-research.org

Abstract. This article aims to show how the deep history of early State societies entails the development of a collective form of cognitive agency. It relates classical works in the anthropology of States (in particular Scott's *Seeing like a State*) with the enactive account of biological and cognitive organisation, thanks to the unified ontology for self-organisation dynamics across scales offered by the Active Inference framework.

Active Inference conceives of cognition as synchronisation across individuated sensorimotor states. It entails that biological or sociocultural constraints display a minimal form of cognition by shaping the behaviour of faster dynamics in a certain way. When such constraints collectively define a basic life form (an integrated, operationally closed system), they can therefore be said to embody adaptive knowledge properly speaking.

The (en)Active Inference account I articulate here strongly motivates and methodologically grounds a holist approach in the social sciences. Indeed, it grounds the study of human societies in the role of structural constraints, whose "meaning" depends both on the broader system's activity and in the historical context of their emergence. The present account of the dynamics of early urbanisation and State genesis aims to illustrate this approach.

IOP Conf. Series: Materials Science and Engineering 1261 (2022) 012026 doi:10.1088/1757-899X/1261/1/012026

1. Introduction

The core argument of this article is that States can be (and should be) understood as hierarchical control systems, essentially similar in their core physical architecture to brains (1,2). Such control systems are constituted by a nested ecology of (Markov-blanketed) Active Inference agents, which synchronise with their environment in a way that allows them to maximise the predictibility of their sensorimotor flow, effectively enacting adaptive control (3–5). Importantly, this drive toward optimal grip is related (although somewhat informally) to the maintenance and self-creation of the structural identity of the agent (6–8). It therefore constitutes the grounding of an emerging multi-scale theory of adaptation in biological systems (9–11).

In his outstanding books, *Against the Grain* (12) and *Seeing like a State* (13), the social anthropologist James C. Scott has provided important insight both on how States emerge and on how they understand the world. I argue these books collectively provide a solid ground for a naturalist and multi-scale understanding of the nature of States as enactive agents. Indeed, the red line permeating Scott's work is the deep relation between how States maintain their structural identity and how they exert control over their (sociocultural and material) niche. This question can clearly be reframed in terms of embodied intelligence: how is the understanding of reality enacted by States grounded in their material structure? The discussion hereby will revisit the anthropology of States in these terms, by drawing heavily from the enactivist and Active Inference (ActInf) paradigm to understand the dynamics underlying State constitution and cognition.

If this account is correct, then human societies display the same kind of complexity that biological entities display – they are, in fact, biological entities as characterised by the central criterion of operational autonomy (14). This means that all attempts to understand and control societies from the scale of individual agents are just as absurd as the attempt to predict human behaviour from the activity of single neurons. Such a realisation has a deep resonance with the politics of the early Anthropocene, as characterised by the urgency to build a post-carbon society before we lock our planet in an unlivable stable state (15). If the collective intelligence is driven by the cultural niche it embodies rather than individual behaviour, then it is the cultural niche itself we must alter to understand how to build a livable future (16).

The argument will proceed in three parts. First, I will expose the precise nature and commitments of the conceptual background I am recruiting from. Most of this discussion will bear on the (enactivised) Active Inference paradigm, and how it relates the material structure of organisms to capability for skilled agency – ie, how it conceptualises embodied intelligence. Second, I will draft the dynamics underlying the constitution of urban systems, understood as large scale settlements with high level of economic interdependancy and political hierarchy, in human history. Third and last, I will discuss how (enactivised) ActInf helps us reconceptualise key aspects of this transition, and ground the study of City and States as enactive agents. This discussion will overall demonstrate that an integrated understanding of life, minds and culture is in our reach, and expose its core insights for the social sciences.

2. Embodied intelligence in nested minds

The purpose of this section is to show how and why the Active Inference framework allows embodied intelligence research to extend beyond the scope of what we would intuitively think of as biological systems – single organisms with defined biochemical boundaries. "Intelligence" refers to the general ability for a system to understand the world, either in a reflexive / propositional way or as enacted in their ecological activity. Therefore, embodied intelligence refers to the ability for understanding / adaptation that is imprinted in an agent's physical structure rather than in some general computational ability. Although this idea has motivated an extremely productive line of research in cognitive robotics

(17,18), the most integrative research program relating mind to body in the life science has emerged from the proposition to understand "living" and "thinking" as two complementary dimensions of a single process of self-creation (*autopoïesis*) (19).

Research on autopoïesis has opened the way for a new understanding of mind and life, which was called the enactive approach due to their accent on how cognitive agents "enact" (rather than represent) the world they live in (20). Although they conserved the core commitment of studying life and cognition through their common grounding in the self-creative dynamics enacted by an organism, enactivists and associated systems biologists have refined the somewhat esoterical notion of autopoïesis into the much more physically grounded notion of autonomy (14,20–23). Autonomy refers to the property of operational closure under precarious circumstances, ie to the capability of biological systems to continuously maintain and recreate their own structural identity as a result of the collective activity of their constitutive process [1].

Interestingly, nothing in the concept of autonomy suggests that this structural identity should be imprinted in the specific carbon-based substance which constitutes woodlice and humans. If the network of constraints (understood as a slow physical process orienting the flow of faster processes) constituting a system is able to regenerate and recreate itself under normal metabolic activity, then the system is autonomous (22) and should be understood as an enactive agent. In principle, this concept could apply to patterns of fluid dynamics (as speculated in Damasio's fantasy novel *La Horde du Contrevent*) or to the material and cultural constraints embedded in human sociality. Yet, enactivists have mainly focused on the sensorimotor grounding of animal life, and their rare incursions into the realm of sociology have lacked the domain-specific knowledge to meaningfully engage with existant litterature on sociocultural organisation beyond interpersonal synchronisation (see eg (24,25)).

I hereby speculate that, beyond the historical grounding of enactivism in the life sciences, this fact can be explained by its lack of formalisation of the dynamics underlying intelligence. Indeed, while formal criterion for basic life have been devolopped by enactivists and systems biologists since the 2000's (26,27), cognition is essentially understood as an activity grounded in adaptive behaviour and sense-making (see (25)). In the absence of a more explicit criterion for cognitivity, enactivism relies on intuitive judgements about teleonomy to decide whether a given system "thinks". Such judgements should be expected to heavily favorise human-like entities, and conversely overlook the potential cognitive ability of systems whose structure is not imprinted in flesh. To fully assess the ability of enactivism to integrate the study of social systems, we should therefore introduce a formally integrated model of what "cognition" means.

Such a model has recently emerged from physicalist approaches to the cognitive science, which we will hereby call "Active Inference". Active Inference constitutes a formal ontology assimilating cognition to the process by which an agent minimises its (expected) variational free energy (VFE), an information-theoretical construct constraining the unexpectedness of its future states under ergodic assumption (28,29). In other words, ActInf agents strive to maximise their grip on their environment by continuously anticipating the sensorimotor flow they experience (30), mutually constructing their niche and the nature of their expectation regarding said niche (31,32). In virtue of this self-evidential

[1] A basic definition of the concept goes as following:

"An autonomous system is defined as a system composed of several processes that actively generate and sustain an identity under precarious circumstances. In this context, to generate an identity is to possess the property of operational closure. This is the property that among the conditions affecting the operation of any constituent process in the system there will always be one or more processes that also belong to the system. And, in addition, every process in the system is a condition for at least one other constituent process, thus forming a network." (21)

IOP Conf. Series: Materials Science and Engineering 1261 (2022) 012026 doi:10.1088/1757-899X/1261/1/012026

dynamic, any system that has individuated sensorimotor states (a Markov Blanket, formally speaking) should be expected to revisit the same core set of expected states, therefore resisting thermodynamic dissipation and enabling their survival and reproduction (6,33).

The concept of Active Inference derives from a formal result known as the Free Energy Principle, which states that the synchronisation dynamics between a Markov-blanketed system and its environment entails the minimisation of VFE (34). Since VFE is a classical measure of Bayesian model fitness, this enables an epistemic / semantical interpetation of self-organisation in cognitive systems (35,36). The attraction of the agent-niche system toward its most probable regions can therefore be interpreted as "self-evidencing" (37), in the sense that it constitutes an epistemical activity by which the agent proves that their world is indeed what they expect it to be. Because this synchronisation drive the agent's activity and structure to incorporate statistical features of their niche through behavioural and evolutionary time, they come to embody models of the world they enact (5,36,38).

The relation between Active Inference and the enactive approach remains controversed, mainly because it is grounded in predictive processing and similar neurocentric approaches - as was widely noted in the discussions between enactivists and ActInf theorists (see eg (37,39)). The notion of prediction appears to suggest that an ActInf agent would mentally represent the outside world, and to think in a way that is detached from its body (37). However, there are deep differences between the predictive motors of an ActInf mind and the kind of semantically ladden, purpose-neutral representations hypothesized by historical cognitivists. Indeed, an ActInf agent functions by adaptively anticipating future course of action and enacting them (3,4), which recruit their own body and material niche as core drivers of the very way they understand reality (40–42).

While this conceptual similarity between Active Inference and the enactive approach was widely noted, the deep complementarity between their respective formalisms remains a somewhat underground topic. Formal models grounding enactive autonomy are purely syntactical (ie they're interested in the structural properties of the system *per se*, see eg (26,43)), while the formalism of Active Inference shows how semantical content (ie meaning) can arise from basic synchronisation between an organism and their niche (36,38). This simple feature provides a clear grip on how agents can come to embody knowledge about their ecological niche, and leverage this knowledge by enacting a meaningfully integrated reality. In other words, Active Inference both grounds the enactive concept of meaning and explains its emergence by showing how basic life can integrate a structural identity, which they can consequently maintain and recreate under precarious conditions (7,8,44).

Importantly, the dynamical landscape in which Active Inference occurs is emphatically not a statespace in the usual physical sense, ie a complete representation of all variable relevant to the system's evolution[2]. The dynamical flow defining Active Inference indeed stems from the agent's Markov blanket (34), and therefore corresponds to statistical properties of its eco-cognitive niche - physical reality as perceived by the agent (38). This level of descripton is not causally closed, as events outside the scope of the statespace description (eg a novel mutation, cultural recombination, or temporary exposure to a new niche) can affect the actively inferred dynamical landscape by creating novel affordances. Conversely, biological agents may turn their subjective expectations into a physical reality by enforcing expectations entailed by their structural identity into their material niche (32) - a process known by evolutionary scientists as niche construction (46).

Therefore, even though Active Inference is a formally conservative framework conceiving of agency as a perpetual return toward a characteristic non-equilibrium steady state (34), it can ground our understanding of the self-creative processes central to the enactive approach (39). Indeed, a basic mind is operationaly conceptualised as a structurally stable hierarchy of Markov Blankets (7), where

[2]"A phase space is the space of the pertinent observables and parameters in which the theoretical determination of the system takes place. As a result, to one point of the phase space corresponds a complete determination of the intended object and properties that are relevant for the analysis." (45)

the non-ergodic (although conservative) dynamics at any scale can alter the dynamical landscape both upstream and downstream. This process is a possible explanation for the phenomenology of creative evolution, where multiscale fluctuations enable the endogenous creation (or unfolding) of the physical symmetries defining biological systems (45). Active Inference therefore draws the picture of the biosphere as a nested ecology of basic minds working continuously to enact the world entailed by their structural identity (9,11), and thereby integrating meaning into the physical fabric of reality.

As stated previously, the network of constraints defining the structural identity of basic life needs not be printed in flesh, it only needs to influence faster processes so as to continuously recreate itself (26). Such constraints can be embedded in the informationally rich sociocultural niche which enable the human ecology by affording adaptive coordination and the development of complex skills (47,48). For exemple, expectations can be imprinted as well as retrieved into this niche through immersive participation of individual humans in culturally patterned activites, a process formalised in Active Inference under the name of "thinking through other minds" (49). This echoes existing lines of research about the collective nature of cognition, and the way humans societies enact through their sociocultural activity something akin to a "collective brain" (50,51).

However, the possibility that collective cognition enacts anything beyond the activity of individual humans has never been seriously investigated, including from an Active Inference perspective. If there is such a thing as a collective brain, we do not know how it is implemented, we do not know what kind of niche it entails, and we certainly do not know what sort of biological individuality it is embeded into. We will now leverage the broad framework we articulated above, which grounds the study of intelligence in the way an organism maintains their structural autonomy, to formally assess the collective brain hypothesis. More specifically, we will discuss how the coevolution of top-down control and cultural niche construction in the deep history of the human species enabled the emergence of the City-State complex, how such collective organisations work to maintain their structural autonomy and what is the kind of world they enact.

3. The deep history of cities and States

The human species is currently dominated by a very specific ecology of dense urban settlement dominated by anonymous (often market-based or otherwise institutionaly mediated) interactions, and centrally controlled by hierarchical administrative structures known as "States". This situation is extremely recent, as it can broadly be traced to the global takeover of Western states circa 1800 and the ongoing rise of the capitalist economy. Although we lack archeological evidence to conclude on the precise timing of the emergence of Statedom (52), we do not have clear trace of their very existence before the early Holocene – approximately when climate warming enabled the rise of large-scale agriculture (53). We will hereby articulate a broad picture of how and why urban systems and hierarchical administrations came to emerge and dominate the human species, ie draft a deep history of cities and States[3].

As a preliminary remark, I should mention that the goal of this discussion is specifically not to explain the full history of how our societies came to be. Such a task would be impossible given the complexity and ambiguity of the historical and archeological evidence. In addition, there is no reason to believe that a more complex, nuanced discussion would result in a better understanding of the target phenomenon (54). The goal of this discussion is to expose what kind of dynamics were necessary and sufficient to get us where we're at, and to highlight key operational concepts that will serve as a basis

[3]The term of *deep history* typically refers to the history of the distant past of the human specie, or in other word to history before history – as the discipline of history is classically understood to refer to the study of events posterior to the invention of writing. It is here used as a shorter, more intuitive term for what is typically referred to as *histoire de temps long*, ie the study of deep structural determinants of human history typically acting at very long time scales. The term of *cliodynamics* could also be used, although it suggests a stronger accent on dynamical systems methodology.

IOP Conf. Series: Materials Science and Engineering 1261 (2022) 012026 doi:10.1088/1757-899X/1261/1/012026

to articulate an enactivised understanding of human societies. It will heavily draw from from Scott's work on the influence of States on historical patterns of cultural niche construction (12,13), as mentioned in introduction, and will also call onto an earlier, richer synthesis of the relevant literature in cultural evolution & historical anthropology (55).

The core dynamic resulting in the evolution of cities can be accounted for with no references to agriculture, States, or even to cities themselves. Indeed, denser settlements allow for both more economic codependency, and the development of nested social communities (enabling the evolution of complex cultural traits (56,57)). Consequently, denser settlements enable the development of deeper economic/functional specialisation, which consequently allow these settlements to develop an efficient export-oriented industry and capture central positions in long distance trade and migration flows (58) – priming autocatalytic growth in both size and power. The growth of cities was however constrained by the agricultural capability of the backland they relied on for food production. As grain can be stored much longer and transported affordably over much longer distances than other staple crops (12), the development of cereal monoculture essentially lifted this constraint for the early cities of the Fertile Crescent, Indus Valley and Yellow river basin.

Scott's *Against the Grain* (12) discusses in much more depth how agriculture interacted with the emergence of early States. His central remark is that while cereal cultivation allows for an unparallel productivity per unit of land, it shows poor performance in productivity per hour of human labour relative to earlier forms of horticulture (see also (59)). Cereal farmers were additionally locked into their dense settlement due to the extreme labour intensity of field work, which (compounded by their reliance on a main crop for subsistance) exposed them to extreme material insecurity through epidemics and famine driven by crop failure. Most importantly, sedentarity of farmers, predictibility of crop harvest, and transportability/storability of grain made cereal monoculture an ideal target for captation by armed outsiders. All these properties made farming an extremely fragile niche for a tribal society to adopt, which certainly explains why non-urban societies typically prefer to rely on hunting or mixed horticulture.

However, those very same properties also explain why agriculture would later succeed and even become the globally dominant mode of food production. First of all, the sedentarity of farmers indeed make them vulnerable to famine and illness, but it also allows much faster reproductive growth than nomads could afford. Second, the viability of agriculture effectively depends on social institutions which could reward long term investment on resources that would be easy to appropriate. This entails the development of efficient communal organisation for niche construction (or at the very least for defense against outsiders), which grain farming could piggy back on over the course of human history. Third, urban centers critically depended on cereal imports to survive, and had ample opportunity to support the development of agriculture by trade and - most importantly - by coercion. Scott indeed documents that early States would routinely coerce population into providing agricultural labor, either through taxation, corvée, deportation, or slavery.

The capability of those States to enforce anything relied on an intricate array of material, economic, normative, and symbolic means. The concentration of resources and population, and economic codependancy characteristic of urban systems indeed affords a lot of ways for aspirant elites (be they rich merchants, appointed officials, or a warrior caste) to build grip over society. They could leverage communal institutions and control over the inflow and distribution of grain to organise rituals legitimising their rule, build monuments to their own glory, and generally alter the urban landscape in any way they see fit. However, these very same features meant that a general strike or insurrection could bring the regime down in a matter of days if it tried to enforce unpopular decisions or appeared to lose control. A key prerequisite to enforcing State control over urban systems is consequently to build acceptation in their population through perceived competency, wealth redistribution, or symbolic legitimacy.

IOP Conf. Series: Materials Science and Engineering 1261 (2022) 012026 doi:10.1088/1757-899X/1261/1/012026

It is therefore entirely unsurprising that the earliest documented form of dominant institutions in the history of cities, the Levant temples, have seemingly emerged around a dual economic and religious mandate. The historical role of Levant temples in helping organise agriculture (58), as well as the contemporary analogue of Balinese water temples (60), suggests more specifically that those institutions may have emerged as a tool for building a socio-ecological niche adapted to agriculture. Communal redistribution of the crop, added to the inscription of social norms in a collective identity and sacred values built through religious ritual (61–63), indeed grounds the kind of efficient large-scale coordination necessary for landscaping cereal fields. Temples are therefore well equipped to catalyse the growth of human settlements into urban centers, help coordinate their more complex organisation, and to assume *de facto* rule over their population without any form of coercion.

However, those same institutions that helped organise collective action could also serve to institute coercive rule and labour extraction, and so they did. Indeed, temples were pivotal in legitimising the secular rule of the warrior caste, although they tended to regain domination in periods of peace (58). Most importantly, they have introduced a form of debt decoupled from interpersonal relations: lending with interest (64). Lenders could extract labour from borrowers through high interest rates and debt slavery, and become immensely powerful due to the multiplicative effect of lending. More importantly, the control they collectively exerted over monetary flow / creation canalysed economic development toward activities which could ensure a sufficient return on investment to meet interest rates. This mechanism catalyses the emergence of institutions priorising wealth extraction over human well-being or sustainability, and (when financial elites successfully infiltrate the State) drives the development of European-style imperialism.

Debt provides a key exemple of how extractive institutions can develop in relatively democratic societies such as early Levant City-States. As settlements grow beyond the size where each resident roughly knows each other, their stability come to depend on coordination at greater scale than is practical by the means of face-to-face interactions only. This offers an opportunity for the evolution of mediating institutions capable to generate adaptive decision-making from aggregate anonymous interactions. Therefore, the very process of urbanisation entails the shaping of the urban niche - be it the material infrastructure of cities, or the social norms they enforce – by supra-human agency. Such agency can exert tyrannical rule over population as Scott suggests early States did, or can result from recognisably democratic institutions (65). In any case, its constitutive mechanisms (or what we could call the logic of their activity) start shaping urban development and other political decisions, feeding back onto every level of human activity.

However, there is no prior warranty that those mechanisms allow an accurate perception of the consequences of their own activity on human life, or on society at large. Effectively integrating information beyond the scale afforded by interpersonal relations entails the development of new forms of communications, which implies new constraints on what information can effectively be transmitted. The necessity of perceiving and enacting decisions at scale has repeatedly driven the development of writing (or close analogues such as quipus) in urban systems, and of the administrative, accounting, and legal technologies it afforded (66). But any information that does$ not fit into their forms is essentially useless to any instutions that work on the basis of adminstrative systems. The same technologies that enabled Levant temples to keep track of debts and manage economic activites did not allow them to see concrete living conditions in the backcountry, and understand how the threat of bankrupty could drive the erosion of living standards, political stability, and soil productivity.

Consequently, States have historically devoted huge amounts of energy to attempt to build a social and material environment they could manage. This process of course includes landscaping for cereal agriculture, and the construction of material infrastructure (such as roads and aqueducts) underlying urban metabolism. But most importantly, it includes attempts to engineer human societies itself from the top down, as documented in Scott's *Seeing like a State* (13). Indeed, States have routinely attacked communal institutions so as to atomise society and/or bring it under the control of State-sponsored

IOP Conf. Series: Materials Science and Engineering 1261 (2022) 012026 doi:10.1088/1757-899X/1261/1/012026

institutions. The Bolshevik imposition of collective farming, as well as the process of enclosure that created the worker class fueling early English capitalism, provide two examples of how the law effectively imposed direct control of the elite over farmers. However, this process could also retroactively shape the material landscape of the urban system itself by physically preventing any social life outside work and family, or displaying State power through monumental architecture.

This discussion did not highlight a typical trajectory for the development of urban systems, because we have no reason to believe there exists one. It did however gave a coherent (although complex) picture of what drove the development of cities and States in human history, and what kind of transformations accompanied this process. We saw how cultural niche construction allowed humans to alter their socio-cognitive ecology, in a way that allowed extensive functional specialisation and the evolution of supra-individual agency through mediating institutions. We saw how specific technologies, mainly cereal agriculture and writing, allowed States to build their niche by enabling the development and control of urban systems. We will now turn back to the (enactivised) Active Inference paradigm we articulated in the first part of this paper, and try to understand whether human societies actually are enactive systems, and how exactly they understand their world.

4. From the urban metabolism to the collective mind

In this part, we'll finally turn to the fundamental question of this article: what kind of intelligence is displayed by City-State systems, and how is it embodied within their structure? Answering this question in terms of (en)active inference entails revisiting our account of the urbanisation dynamics as the results of City-States systems actively managing to get a grip on their world - ie minimise their (expected) Variational Free Energy. As the collective mind is effectively constituted of whatever process modulates the activity of human sociocultural organisation so as to sustain its integrity, developping a principled understanding of its purpose, mechanisms and material grounding entails the study of how City-State systems actively produce their own structural identity under precarious conditions. We will therefore turn to the basic metabolic activity of urban systems and work our way up the nested hierarchy of adaptive constraints to finally explain what it means exactly to be *thinking like a State*.

As we saw earlier, the process of urbanisation follows the polarisation of human activites around specific settlements which manage to grab a central position in the flows of products, cultural traits and political power. This core dynamic is what simultaneously enables the developpement of mediating institutions, and the wide cooptation of whichever forms of organisation manage to control those flows efficiently. These flows of informations and ressources, collectively underlying the activity and maintenance of urban organisation, are strongly reminiscent of metabolic activity in carbon-based life. Basic autonomy is indeed formally defined as operational closure, or more precisely the capability of a network of constraints to canalise faster processes (ie metabolic flows) into maintaining and (re)creating itself (22,67). Since the material and sociocultural landscape of urban systems canalises human activities so as to maintain itself, the process of urbanisation clearly entails the constitution of cities as basic life.

In fact, the deep history of urban systems display much stronger signs of the emergence of cities as biological individuals. The emergence of new forms of individuality, commonly studied in evolutionary transition theory (68), occurs when a group of biological individuals cooperate so closely that they progressively become dependant on each other. The collective progressively becomes an adaptive unit, where each individual fullfils a specific function as imposed by emerging regulation mechanisms enacting supra-individual agency within the collective organisation (69). A full fledged evolutionary transition typically entails the emergence of new forms of information systems mediating regulation at the collective level and/or the heritability of collective traits (70). The emergence of City-States, with their professional classes and their mediating institutions based on writing technologies, constitute a clear case of such a transition – especially in the case of the late agricultural civilisations (71).

To be fair, the entire history of the human species is best characterised as a continuous process of evolutionary transition. The human ecology is indeed chararacterised, from their early evolutionary history, by eusocial cooperation, reliance on social learning, and language. All these capabilities have probably emerged from a system of collective regulation through embodied coordination, evolved so as to extend our cooperative abilities beyond what is possible (72,73). They rapidly allowed the emergence of an inheritance system orthogonal to genes, capable of evolving adaptive knowledge (47) and group-level norms (74), with drastic feedback on the ecology of human individuals (48). This process enabled an explosive scaling of human cooperation (74,75), and most importantly the emergence of group-level organisational traits selected at the scale of human collectives (76,77).

In this sense, urbanisation only constitutes a prolongation of preexistant human tendancies, which agriculture-driven scaling was accidentally enabled by the Pleistocene-Holocene climate change (78). This quantitative change in scale however entailed a qualitative change regarding the status of human agency. Cultural evolution in humans typically entails inter-individual synchronisation through active inference in the collective sociocultural niche (49). This allows the definition of the "collective brain" (ie collectively enacted cognitive activities) as the emerging structure of individual-level innovation and diffusion patterns of socially acquired cognitive traits such as language, crafting skills, or even social norms (51). However, the material constraints embedded in urban systems canalyse human activities at a structural scale, and direct collectively enacted decisions so as to maintain and (re)create their own collective structure - for exemple by weighting for a road network fitted to a city's interests (79,80).

It is worth noting here that each structural constraints developped throughout human history effectively act as an ActInf agent. Indeed, constraints are causally effective in virtue of shaping faster processes (26), and therefore enforcing specific expectations over the activity of system components. Additionally, they are "structural" in the sense that the expectations they enact happen to directly map onto the system structure, by helping produce its constitutive constraints. The system's architecture itself is therefore constituted of "nested minds", although in a somewhat figurative sense since simple ActInf agency does not entail cognitive abilities (7). However, if the nested hierarchy of regulation systems enables adaptive behaviour at the system's scale, it effectively enacts a collective mind – regardless of whether system components retain their cognitive ability (81). In other words, the process of evolutionary transition entails the constitution of a collective mind embodied in the concrete mechanisms of its adaptive regulation.

The closest analogue of our intuitive picture of a mind in urban systems is their most disembodied mechanism for adaptive regulation: administration. Indeed, just like neuronal cognition, it entails the top-down regulation by a specialised population with the help of a specific-purpose information system. Most importantly, Scott's work has shown that administrative systems restrict their interests to specific observables which are informative for their specific purposes (generally resource extraction, taxation, or drafting) while being tractable by their cognitive means (ie standardised written documents) (13). In other words, they distinctively implement a form of predictive processing, an cognitive architecture which corresponds most directly to the mechanical implementation of the abstract principle of active inference (82). As predictive cognition is an intrisically embodied and contextual activity (4,42), the question of embodied intelligence in urban systems reduces to the documentation of how constitutive traits prolong or support the activity of administrative systems.

The deep history of urban systems we articulated above incidentally provides us with numerous instances of administrative systems externalising their expectations, which directly corresponds to the core theme of Scott's *Seeing like a State* (13). An obvious example is of course the way cities and States have worked to impose an ecological niche they knew how to exploit, both in the deep history of cereal agriculture in Neolithic societies and in modern attempts to manage forest at the advantage of

IOP Conf. Series: Materials Science and Engineering 1261 (2022) 012026 doi:10.1088/1757-899X/1261/1/012026

military industry. But more importantly, cities themselves have been shaped to enforce State-sponsored social organisation. Scott provides the extreme exemple of Brazilia's urbanism, explicitly meant to segregate family, leisure, and work, and therefore promote productivity for dominant institutions at the expense of serendipity. The same logic operates far back in time with the role of monumental architecture in the legitimation of power and the promotion of dominant norms.

Importantly, the externalisation of admistrative expectations does not limit to the alteration of the material niche by States. Indeed, the institution of State-sponsored legal codes affords a way to directly transcribe administrative expectations into individual minds. Because it consists of a publically available set of social norms expected to apply to all, and to regulate decisions of the State itself, the law modulates behaviour at a slower time scale (and independantly from) any individual executive or judicial decision (83). Consequently, it affords top-down social engineering by the State [4] through the enforcement of novel sociocultural constraints *ex nihilo*. Most importantly, those constraints define the affordances the State has in its interaction with the wider society – such as taxation, drafting, or economic management. Rule of law is therefore a core tool by which States create their world through building the basis of their own intelligent behaviour into their sociocultural niche.

This discussion should have established that the normal cognitive activity of States is prolonged in the structure of the landscapes and sociocultural constraints they interact with. However, talking of landscapes or sociocultural constraints as bearers of "embodied intelligence" implicitly entails that they are not passive recipients of an external cognitive agent's expectations, but constituant part of its organism. It is worth noting here that our notion of intelligence is based on Active Inference, a scale-free framework applying to any system with individuated boundary states. That States define implicit expectations through their activities and offload them onto their environment is therefore insufficient to make them the proper locus of urban system's cognitive activity. As urban systems build structural constraints into their world (like an agricultural landscape or a civil code) and become dependant on their maintenance, they integrate those constraints themselves into the boundaries of its operationally closed socio-ecologic structure, and their life goes on.

We can finally turn back to our central question, what it means to think like a State. By our account, the cognitive activity of States is defined mechanically by the standardised administrative systems they use to understand and manipulate their world, and functionally by their drive to build and enact grip over their material and sociocultural niche. As it ultimately relies on a form of predictive processing, their phenomonology must consist of a "controlled hallucination" made of the expectations they build in the course of their ecological activity (84). This activity is of course determined by their administrative mind, but it is also deeply coloured and oriented by the broader cognitive and metabolic activity of the broader self-producing structure of the urban societies they emerge from. Therefore, a naturalistic account of a State's mental activity cannot circumvent a motivated analysis of the structural identity it participates to enact.

We have now exposed how the historical development of urban systems can be described as a process of (en)active inference, and how this process affords the integration of adaptive knowledge in their material and sociocultural structure. This discussion should establish that the collective mind does not directly emerge from interindividual interactions, but from slower evolving dynamical constraints which supervene on (but do not reduce to) human activity. Therefore, the proper scale of analysis of collective cognition is the integrated activity of entire urban systems, and should more specifically be grounded in the broader processes by which they maintain and (re)produce their structural identity – understood as an operationally closed set of constraints underlying their metabolic, cognitive and ecological activity. This perspective enables a unified understanding of the historical emergence of

[4]This best describes civil law systems, where State legislative institutions have the monopoly over the definition of the law, and judges are only allowed to define it. In contrast, common law systems recognise a legally binding role to judicial precedents, therefore allowing individual judges to make law. As judges remain State officials, I don't consider this distinction to be critical here.

urban societies, where the progressive evolution of a basic form of life entailed the development of increasingly complex forms of regulation.

5. Conclusion

We have hereby articulated an account of embodied intelligence as adaptive knowledge integrated into an organism's structure, and explained how this account could apply to human societies. By recruiting a formal notion of inference as synchronisation across a Markov Blanket, therefore applying to any scale that displays individuated boundary states, we were able to motivate a unified understanding of the nested scales of sociocultural evolution as a single process of active inference. Accordingly, we exposed how urban systems infer adaptive knowledge through behavioural, developmental and evolutionary time, and integrate this knowledge into their material and sociocultural structure so as to effectively constrain their activity to metabolically and ecologically viable domains. Our (en)active inference account relates in a principled manner the subjective understanding of their world States enact to the historical development of the objective structure that enables their activity, by showing how both processes participate in the maintenance and (re)production of the structural identity of urban systems.

This discussion is meant to introduce and illustrate a proposal to articulate active inference and the enactive approach in a single framework drawing from their conceptual and formal synergy. Both understand cognition as a fundamentally multiscale and relational process, permeating all forms of biological organisation. While the enactive approach grounds its study of cognition in the way an organism enacts its structural identity, Active Inference formalises the relation between the structural architecture of a cognitive system (ie what constraints define it) (9) and its functional architecture (ie what world it enacts) (85). This feature allows us to understand extended intelligence as the externalisation of an ActInf agent's expectations as constraints in its environment. As this ActInf agent can consequently integrate these external constraints in its structural identity, or even be a mind acting upon its own body (rather than an individuated organism), this also provides us with an explanatory mechanism for the emergence of embodied intelligence.

The ability of (en)ActInf agents to constantly redefine their boundaries and recreate their structure makes it difficult to identify the proper scale of analysis for the study of human societies. Indeed, we have showed that they display an autonomous structure, which could either be understood in enactive terms as an operationaly closed set of constitutive constraints, or in ActInf terms as an implicit world model they embody and adaptively work to evidence through their activity. Both entail a multiscale approach encompassing the whole autopoietic structure rather than a single arbitrarily chosen trait (such as cereal agriculture) or scale (such as the activity of individual humans). Importantly, nothing entails that a single structural identity could account for the activity of any human society, as symbiotic multiscale integration (eg between cities and States, or individuals and organisations) remains a distinct possibility. This provides a critical argument for the necessity of importing in the social sciences the integrative multiscale approach characteristic of the life sciences.

In particular, I argue that it is critical to adopt what is traditionally understood as a holistic approach by replacing the study of structural traits in the context of the broader system's activity and of its historical emergence, as is standard in the evolutionary science (86). While I have focused my argument on early urbanisation dynamics, when the most determinant features of later urban societies arguably emerged, a lot of structural traits have since disappeared from or emerged in human societies. In particular, the modern period has been the pivotal moment when European institutions of early capitalism and State control emerged, and consequently took over the world. The study of its dynamics should provide key insights into the prospective of our socio-political trajectories in the early Anthropocene era. The (en)active inference framework I have hereby introduced, motivated, and illustrated could facilitate such an integrative historical demarch by providing a core formal ontology for describing in a unified manner the metabolic, behavioural, developmental, and evolutionary timescales of the system's dynamics.

Acknowledgements
I am grateful to Kairos Research and the Active Inference Lab for affording a research demarch oriented toward reflexive understanding of our sensible reality. I would like to thank Virginia Bleu Knight, Daniel Friedman, Iona Brenac, Manon Job, and Sergi Lozano Perez for their valuable comments.

References

1. Badcock PB, Friston KJ, Ramstead MJD. The hierarchically mechanistic mind: A free-energy formulation of the human psyche. Physics of Life Reviews. 2019 Dec 1;31:104–21.
2. Hipólito I, Ramstead MJD, Convertino L, Bhat A, Friston K, Parr T. Markov blankets in the brain. Neuroscience & Biobehavioral Reviews [Internet]. 2021 Feb 16 [cited 2021 Feb 18]; Available from: https://www.sciencedirect.com/science/article/pii/S0149763421000579
3. Bruineberg J, Kiverstein J, Rietveld E. The anticipating brain is not a scientist: the free-energy principle from an ecological-enactive perspective. Synthese. 2018 Jun 1;195(6):2417–44.
4. Hipólito I, Baltieri M, Friston K, Ramstead MJD. Embodied skillful performance: where the action is. Synthese [Internet]. 2021 Jan 28 [cited 2021 Feb 5]; Available from: https://doi.org/10.1007/s11229-020-02986-5
5. Ramstead MJD, Kirchhoff MD, Friston KJ. A tale of two densities: active inference is enactive inference. Adaptive Behavior. 2020 Aug 1;28(4):225–39.
6. Friston K. Life as we know it. Journal of The Royal Society Interface. 2013 Sep 6;10(86):20130475.
7. Kirchhoff M, Parr T, Palacios E, Friston K, Kiverstein J. The Markov blankets of life: autonomy, active inference and the free energy principle. Journal of The Royal Society Interface. 2018 Jan 31;15(138):20170792.
8. Kiverstein J. Free Energy and the Self: An Ecological–Enactive Interpretation. Topoi. 2020 Jul 1;39.
9. Hesp C, Ramstead M, Constant A, Badcock P, Kirchhoff M, Friston K. A Multi-scale View of the Emergent Complexity of Life: A Free-Energy Proposal. In: Georgiev GY, Smart JM, Flores Martinez CL, Price ME, editors. Evolution, Development and Complexity. Cham: Springer International Publishing; 2019. p. 195–227. (Springer Proceedings in Complexity).
10. Ramstead MJD, Badcock PB, Friston KJ. Answering Schrödinger's question: A free-energy formulation. Physics of Life Reviews. 2018 Mar 1;24:1–16.
11. Ramstead MJD, Constant A, Badcock PB, Friston KJ. Variational ecology and the physics of sentient systems. Physics of Life Reviews. 2019 Dec 1;31:188–205.
12. Scott JC. Against the Grain: A Deep History of the Earliest States. Yale University Press; 2017. 335 p.
13. Scott JC. Seeing Like a State: How Certain Schemes to Improve the Human Condition Have Failed. Yale University Press; 2020. 462 p.
14. Moreno A, Mossio M. Biological Autonomy. A Philosophical and Theoretical Enquiry. 2015 May [cited 2020 Nov 25]; Available from: https://www.academia.edu/12363315/Biological_Autonomy_A_Philosophical_and_Theoretical_Enquiry
15. Steffen W, Rockström J, Richardson K, Lenton TM, Folke C, Liverman D, et al. Trajectories of the Earth System in the Anthropocene. PNAS. 2018 Aug 14;115(33):8252–9.

16. Gowdy J, Krall L. The ultrasocial origin of the Anthropocene. Ecological Economics. 2013 Nov 1;95:137–47.
17. Braitenberg V. Vehicles: Experiments in Synthetic Psychology. MIT Press; 1986. 174 p.
18. Brooks RA. Intelligence without representation. Artificial Intelligence. 1991 Jan 1;47(1):139–59.
19. Varela FJ, Thompson E, Rosch E. The Embodied Mind: Cognitive Science and Human Experience. MIT Press; 2016. 390 p.
20. Di Paolo E, Thompson E. The enactive approach. In: The Routledge Handbook of Embodied Cognition. Routledge; 2014.
21. Di Paolo E. Extended Life. Topoi. 2008 Dec 18;28(1):9.
22. Montévil M, Mossio M. Biological organisation as closure of constraints. Journal of Theoretical Biology. 2015 May 7;372:179–91.
23. Thompson E. Mind in Life: Biology, Phenomenology, and the Sciences of Mind. Harvard University Press; 2010. 561 p.
24. Dumas G, Kelso S, Nadel J. Tackling the social cognition paradox through multi-scale approaches. Frontiers in Psychology. 2014 Jul 29;5:882.
25. Froese T, Di Paolo E. The enactive approach: Theoretical sketches from cell to society. Pragmatics & Cognition. 2011 Jan 1;19(1):1–36.
26. Montévil M, Mossio M. Biological organisation as closure of constraints. Journal of Theoretical Biology. 2015 May 7;372:179–91.
27. Ruiz-Mirazo K, Moreno A. Basic Autonomy as a Fundamental Step in the Synthesis of Life. Artificial Life. 2004 Jun 1;10(3):235–59.
28. Friston K. The free-energy principle: a unified brain theory? Nature Reviews Neuroscience. 2010 Feb;11(2):127–38.
29. Friston K. A free energy principle for a particular physics. arXiv:190610184 [q-bio] [Internet]. 2019 Jun 24 [cited 2021 Apr 22]; Available from: http://arxiv.org/abs/1906.10184
30. Bruineberg J, Rietveld E. Self-organization, free energy minimization, and optimal grip on a field of affordances. Front Hum Neurosci [Internet]. 2014 [cited 2019 May 16];8. Available from: https://www.frontiersin.org/articles/10.3389/fnhum.2014.00599/full
31. Bruineberg J, Rietveld E, Parr T, van Maanen L, Friston KJ. Free-energy minimization in joint agent-environment systems: A niche construction perspective. Journal of Theoretical Biology. 2018 Oct 14;455:161–78.
32. Constant A, Ramstead MJD, Veissière SPL, Campbell JO, Friston KJ. A variational approach to niche construction. Journal of The Royal Society Interface. 2018 Apr 30;15(141):20170685.
33. Constant A. The Free-energy principle: It's not about what it takes, it's about what took you there. 2021.
34. Da Costa L, Friston K, Heins C, Pavliotis GA. Bayesian Mechanics for Stationary Processes. arXiv:210613830 [math-ph, physics:nlin, q-bio] [Internet]. 2021 Jun 25 [cited 2021 Jul 7]; Available from: http://arxiv.org/abs/2106.13830
35. Constant A, Clark A, Friston KJ. Representation Wars: Enacting an Armistice Through Active Inference. Frontiers in Psychology. 2021;11:3798.
36. Ramstead MJD, Friston KJ, Hipólito I. Is the Free-Energy Principle a Formal Theory of Semantics? From Variational Density Dynamics to Neural and Phenotypic Representations. Entropy. 2020 août;22(8):889.
37. Hohwy J. The Self-Evidencing Brain. Noûs. 2016;50(2):259–85.
38. Ramstead MJD, Hesp C, Tschantz A, Smith R, Constant A, Friston KJ. Neural and phenotypic representation under the free-energy principle. Neuroscience & Biobehavioral Reviews. 2021;
39. Di Paolo E, Thompson E, Beer RD. Laying down a forking path: Incompatibilities between

IOP Conf. Series: Materials Science and Engineering 1261 (2022) 012026 doi:10.1088/1757-899X/1261/1/012026

enaction and the free energy principle [Internet]. PsyArXiv; 2021 [cited 2021 Apr 19]. Available from: https://psyarxiv.com/d9v8f/

40. Clark A. Whatever next? Predictive brains, situated agents, and the future of cognitive science. Behavioral and Brain Sciences. 2013 Jun;36(3):181–204.
41. Clark A. How to Knit Your Own Markov Blanket. In: Philosophy and Predictive Processing. 2017.
42. Nave K, Deane G, Miller M, Clark A. Wilding the predictive brain. WIREs Cognitive Science. 2020;11(6):e1542.
43. Aguilera M, Bedia MG. Adaptation to criticality through organizational invariance in embodied agents. Sci Rep. 2018 May 16;8(1):7723.
44. Kirchhoff MD, Froese T. Where There is Life There is Mind: In Support of a Strong Life-Mind Continuity Thesis. Entropy. 2017 Apr 14;19(4):169.
45. Longo G, Montévil M. Extended criticality, phase spaces and enablement in biology. Chaos, Solitons & Fractals. 2013 Oct 1;55:64–79.
46. Laland K, Matthews B, Feldman MW. An introduction to niche construction theory. Evol Ecol. 2016 Apr 1;30(2):191–202.
47. Boyd R, Richerson PJ, Henrich J. The cultural niche: Why social learning is essential for human adaptation. PNAS. 2011 Jun 28;108(Supplement 2):10918–25.
48. Richerson PJ, Boyd R. The human life history is adapted to exploit the adaptive advantages of culture. Philosophical Transactions of the Royal Society B: Biological Sciences. 2020 Jul 20;375(1803):20190498.
49. Veissière SPL, Constant A, Ramstead MJD, Friston KJ, Kirmayer LJ. Thinking through other minds: A variational approach to cognition and culture. Behavioral and Brain Sciences [Internet]. 2020 ed [cited 2020 Oct 7];43. Available from: https://www.cambridge.org/core/journals/behavioral-and-brain-sciences/article/thinking-through-other-minds-a-variational-approach-to-cognition-and-culture/9A10399BA85F428D5943DD847092C14A
50. Falandays JB, Smaldino PE. The Emergence of Cultural Attractors: How Dynamic Populations of Learners Achieve Collective Cognitive Alignment [Internet]. PsyArXiv; 2021 [cited 2021 Jul 29]. Available from: https://psyarxiv.com/v5fsr/
51. Muthukrishna M, Henrich J. Innovation in the collective brain. 2017;
52. Singh M, Glowacki L. Human social organization during the Late Pleistocene: Beyond the nomadic-egalitarian model [Internet]. EcoEvoRxiv; 2021 [cited 2021 Mar 15]. Available from: https://ecoevorxiv.org/vusye/
53. Richerson PJ, Boyd R, Bettinger RL. Was Agriculture Impossible during the Pleistocene but Mandatory during the Holocene? A Climate Change Hypothesis. American Antiquity. 2001 Jul;66(3):387–411.
54. Healy K. Fuck Nuance. Sociological Theory. 2017 Jun 1;35(2):118–27.
55. Guénin--Carlut A. Beyond State - drafting a prospective anthropology. 2020 Mar 9 [cited 2020 Mar 9]; Available from: https://osf.io/ehqjs/
56. Migliano AB, Battiston F, Viguier S, Page AE, Dyble M, Schlaepfer R, et al. Hunter-gatherer multilevel sociality accelerates cumulative cultural evolution. Science Advances. 2020 Feb 1;6(9):eaax5913.
57. Romano V, Lozano S, Pablo JF-L de. A multilevel analytical framework for studying cultural evolution in prehistoric hunter–gatherer societies. Biological Reviews [Internet]. 2020 [cited 2020 Apr 18];n/a(n/a). Available from: https://onlinelibrary.wiley.com/doi/abs/10.1111/brv.12599
58. Thomas A. Urbanization before Cities: Lessons for Social Theory from the Evolution of Cities. Journal of World-Systems Research. 2012 Aug 26;211–35.

59. Bowles S. Cultivation of cereals by the first farmers was not more productive than foraging. PNAS. 2011 Mar 22;108(12):4760–5.
60. Lansing JS. Balinese "Water Temples" and the Management of Irrigation. American Anthropologist. 1987;89(2):326–41.
61. Atran S, Norenzayan A. Religion's evolutionary landscape: Counterintuition, commitment, compassion, communion. Behavioral and Brain Sciences. 2004 Dec;27(6):713–30.
62. Purzycki BG, Sosis R. The extended religious phenotype and the adaptive coupling of ritual and belief. Israel Journal of Ecology & Evolution. 2013 Jun 1;59(2):99–108.
63. Sterelny K. Religion: costs, signals, and the Neolithic transition. Religion, Brain & Behavior. 2020 Jul 2;10(3):303–20.
64. Graeber D. Debt : the first 5000 years [Internet]. Melville House; 2011. Available from: https://en.wikipedia.org/wiki/Debt:_The_First_5000_Years
65. Graeber D, Wengrow D. The Dawn of Everything: A New History of Humanity. Farrar, Straus and Giroux; 2021. 704 p.
66. Goody J. The Logic of Writing and the Organization of Society. Cambridge University Press; 1986. 236 p.
67. Mossio M, Moreno A. Organisational Closure in Biological Organisms. History and philosophy of the life sciences [Internet]. 2010 [cited 2021 Sep 11]; Available from: https://www.academia.edu/661020/Organisational_Closure_in_Biological_Organisms
68. Szathmáry E. Toward major evolutionary transitions theory 2.0. 2015;
69. Stewart JE. Towards a general theory of the major cooperative evolutionary transitions. Biosystems. 2020 Dec 1;198:104237.
70. Jablonka E, Lamb MJ. The evolution of information in the major transitions. Journal of Theoretical Biology. 2006 Mar 21;239(2):236–46.
71. Gowdy J, Krall L. Agriculture as a major evolutionary transition to human ultrasociality. J Bioecon. 2014 Jul 1;16(2):179–202.
72. Dunbar R. Grooming, Gossip, and the Evolution of Language. Harvard University Press; 1998. 244 p.
73. Shilton D, Breski M, Dor D, Jablonka E. Human Social Evolution: Self-Domestication or Self-Control? Front Psychol [Internet]. 2020 [cited 2020 Jul 23];11. Available from: https://www.frontiersin.org/articles/10.3389/fpsyg.2020.00134/full?utm_source=F-AAE&utm_medium=EMLF&utm_campaign=MRK_1254058_69_Psycho_20200225_arts_A
74. Boyd R, Richerson PJ, Henrich J. Rapid cultural adaptation can facilitate the evolution of large-scale cooperation. Behav Ecol Sociobiol. 2011 Mar 1;65(3):431–44.
75. Chudek M, Henrich J. Culture–gene coevolution, norm-psychology and the emergence of human prosociality. Trends in Cognitive Sciences. 2011 May 1;15(5):218–26.
76. Richerson P, Baldini R, Bell AV, Demps K, Frost K, Hillis V, et al. Cultural group selection plays an essential role in explaining human cooperation: A sketch of the evidence. Behavioral and Brain Sciences [Internet]. 2016 ed [cited 2020 Jan 14];39. Available from: https://www.cambridge.org/core/journals/behavioral-and-brain-sciences/article/cultural-group-selection-plays-an-essential-role-in-explaining-human-cooperation-a-sketch-of-the-evidence/638ED0187A9727D9D327661A91DE0759
77. Smaldino PE. The cultural evolution of emergent group-level traits. Behavioral and Brain Sciences. 2014 Jun;37(3):243–54.
78. Richerson PJ. Institutional Evolution in the Holocene: The Rise of Complex Societies. British Academy. 2001;38.
79. Fulminante F, Prignano L, Lozano S. The urbanization process in Central Italy. A network approach. In: Social Simulation Conference. 2014.

80. Fulminante F, Prignano L, Morer I, Lozano S. Coordinated Decisions and Unbalanced Power. How Latin Cities Shaped Their Terrestrial Transportation Network. Front Digit Humanit [Internet]. 2017 [cited 2020 Apr 18];4. Available from: https://www.frontiersin.org/articles/10.3389/fdigh.2017.00004/full
81. Sims M. How to count biological minds: symbiosis, the free energy principle, and reciprocal multiscale integration. Synthese [Internet]. 2020 Sep 24 [cited 2021 Aug 11]; Available from: https://doi.org/10.1007/s11229-020-02876-w
82. Friston K, FitzGerald T, Rigoli F, Schwartenbeck P, Pezzulo G. Active Inference: A Process Theory. Neural Computation. 2016 Nov 21;29(1):1–49.
83. van Rooij B. Homo Juridicus: Questioning Legal Assumptions About Behavior [Internet]. Rochester, NY: Social Science Research Network; 2020 Jan [cited 2021 Sep 16]. Report No.: ID 3572602. Available from: https://papers.ssrn.com/abstract=3572602
84. Ramstead MJD, Hesp C, Sandved-Smith L, Mago J, Lifshitz M, Pagnoni G, et al. From generative models to generative passages: A computational approach to (neuro)phenomenology [Internet]. PsyArXiv; 2021 [cited 2021 Feb 23]. Available from: https://psyarxiv.com/k9pbn/
85. Friston KJ, Parr T, Vries B de. The graphical brain: Belief propagation and active inference. Network Neuroscience. 2017;
86. Bateson P, Laland KN. Tinbergen's four questions: an appreciation and an update. Ecology & Evolution. 2013 Dec 1;28(12):712–8.

AUTHOR INDEX

Institute of Physics
Dirac House, Temple Back
Bristol BS1 6BE UK

ISSN: 1757-8981
ISBN 978-1-7138-6157-7
9 781713 861577